Management for Professionals

More information about this series at http://www.springer.com/series/10101

Stefano Tonchia

Industrial Project Management

International Standards and Best Practices for Engineering and Construction Contracting

Second Edition

Foreword by Russell D. Archibald

Springer

Stefano Tonchia
Department DPIA
University of Udine
Udine, Italy

ISSN 2192-8096 ISSN 2192-810X (electronic)
Management for Professionals
ISBN 978-3-662-56327-4 ISBN 978-3-662-56328-1 (eBook)
https://doi.org/10.1007/978-3-662-56328-1

Library of Congress Control Number: 2018931190

Printed on acid-free paper

This Springer imprint is published by the registered company Springer-Verlag GmbH, DE part of Springer Nature.
The registered company address is: Heidelberger Platz 3, 14197 Berlin, Germany

Foreword[1]

Projects and multiproject programs are vital to all business, industrial, and governmental organizations. They are the vehicles for achieving the strategic goals of every complex organization. How to best manage these projects and programs has been the subject of hundreds of books, articles, conferences, and seminars over the past several decades. This book presents a unique view of project management from the total company viewpoint based on successful experience within a large Italian corporation with global operations.

It is useful to view the project management discipline from two perspectives within any large organization: from the strategic perspective and from an operational one. The strategic project management perspective deals with selection, prioritization, and allocation of money, people, and other scarce resources to the organization's projects at the project portfolio level. This is usually accomplished through the strategic management processes in place within the organization. The operational perspective deals with the planning, financing, scheduling, monitoring, executing, controlling, and closing out of each project individually, and within multiproject programs, always keeping in view the strategic purposes to be achieved by each and every project. Operational project management principles and practices are concerned with developing and improving the best: (1) project life cycle models and processes for each major category of projects within the organization; (2) methods for planning, scheduling, budgeting, authorizing, monitoring, controlling, and closing out projects, including IT tools and effective use of the

[1]Russell D. Archibald, PhD (Hon), MSc, PMP, Fellow PMI and APM/IPMA, is a globally recognized author, consultant, and educator on project management. With a career spanning more than 50 years, he has held executive, project management, and engineering positions in defense/aerospace (USA), refinery construction and operation (USA, Venezuela), automotive manufacturing (USA, France, India), and telecommunications industries (USA, Mexico). Since 1982, he has consulted for companies, agencies, and development banks in 16 countries on 4 continents and has taught project management principles and practices to thousands of managers and specialists around the world. He is an original founding trustee and member No. 6 of the Project Management Institute. He is the author of *Managing High Technology Programs and Projects* (1st edition 1976, 3rd edition 2003) and the coauthor of *Network Based Management Information Systems (PERT/CPM)* (1967). He received the Jim O'Brien Lifetime Achievement Award from the PMI College of Scheduling in 2006. His Web site is www.russarchibald.com.

Internet; (3) effective project and program risk identification and mitigation methods; and (4) development, selection and assignment of project managers and skilled project planning, scheduling, and control specialists.

Change and innovation are always required to achieve the strategic goals of a company or other organization. Small changes may be achieved in a simple manner, but significant changes must be viewed and managed effectively as projects. In many cases, a group of related projects are best managed when organized and managed as multiproject programs. The approach presented in this book—if properly applied—will assure that all projects and programs within the company are planned, budgeted, scheduled, authorized, executed, controlled, and completed while meeting the schedule, cost, quality, and technical project objectives to the satisfaction of all key stakeholders in the projects. That is the overall objective of effective project management at the operational level.

There are very few, if any, books in the project management field that describe in detail a company-wide approach to project management within a large, multinational corporation, with the amount of detail and giving the useful, practical understanding of the underlying principles and specific processes and practices that are presented here.

Professor Stefano Tonchia has drawn together a concise description of the fundamental principles and practices of modern project management and shows how these are effectively applied on a company-wide basis in Danieli & C., Spa, for a variety of project categories. Danieli is one of the three largest worldwide suppliers of plants and equipment to the metals industry with ten worldwide subsidiaries in as many countries on three continents.

Professor Tonchia's extensive experience in both process and project management gives him a unique perspective regarding how the discipline of project management relates to the overall field of business and corporate management. His book, *Process Management for the Extended Enterprise*, coauthored with Andrea Tramontano (2004), describes (p. 15) three basic roles of a business executive today: as director of an organizational unit, as process manager, and as project manager. There is a need today for ownership of the project management processes, and we see that need being filled by the emergence of the project management office (PMO) in many organizations under the direction of a director or vice president of project management. Project managers *own* the specific processes being used for their current project, but the person in charge of the PMO usually holds responsibility for continually improving the processes to be used by all project managers in the organization. The most important of these processes, in my view, is the project life cycle management process, which will be unique for each major category of project being executed within the organization. Professor Tonchia focuses on three of these major categories in this book: product design and development projects; service design projects; and engineering-to-order production, construction, and services projects. This book is unique not only because of its focus on the company-wide project management methodology but also because of the integration of process management concepts and practices with those of project management.

Guanajuato, Mexico Russell D. Archibald

Contents

About the Author

Stefano Tonchia, Ph.D., is full professor of Innovation & Project Management at the University of Udine, Italy, and former dean of the Project Management School of Alenia Aeronautica, the Italian Aeronautics Company. He has written many articles on international journals and published several international books on Lean/Agile PM (including *Process Management for the Extended Enterprise* in 2004 and *Performance Measurement* in 2008, both for Springer). He works jointly with major international institutions and leading companies.
www.tonchia.it

Part I

Project Management: Fundamentals and Perspectives

Introduction

1

The organization and management of an enterprise is increasingly considered from the viewpoint of competitive advantage: the technical equipment of firms may be the same and technology can often be acquired. Organisation and management are more difficult to imitate and so determine the characteristic of a winning strategy: uniqueness. If several firms possess a Ferrari, success will depend on their ability to drive it.

The continuous evolution of technology and market dynamics push managers to devote more attention to the management of *change and innovation*. It should be deemed competitive and strategic rather than confined to the mere role of *supplier of specifics* for the routine activities.

The risks increase but also the opportunities. It is important to be prepared to play a proactive role; if a *project* could be considered a *wager on the future*, this should be the best that could be done and its pursuit the propellant of all the activities.

The power to innovate and a firm's success are becoming closely linked to its ability to manage product development projects, service planning, and improvements in the firm's internal and strategic processes—in short—management by objectives. The change should be managed not just to avoid being swept away but to use as a competitive weapon.

If we define a *project* as a group of integrated activities aimed at carrying out one or several objectives of quality within a certain time and with a limited budget and availability of resources, it is clear how important *project management* is and its close correlation with the firm's performances.

This "modus operandi," besides being the basis of industrial production, is essential in civil and industrial construction sectors, engineering services, and the increasingly popular engineer-to-order and turnkey contracts. Its use can be extended to the reorganization of public offices and various research projects, whether commissioned by local, national, or European Community authorities, for which a strict organizational–managerial approach is required.

It should be remembered that though product design and development (integrated with manufacturing and marketing techniques), design and service management (including public administration), and engineer-to-order production are very

S. Tonchia, *Industrial Project Management*, Management for Professionals,
https://doi.org/10.1007/978-3-662-56328-1_1

different activities, they have a common denominator that is they are strongly oriented toward *customer satisfaction*.

This book describes—in a precise but practical way—the most recent principles and techniques of project management, at the highest international standards. The book deals with the project management applied to the industrial sectors, in particular the engineer-to-order and plant construction according to international contracting (Engineering & Construction Contracting—ECC). In fact, it is unique because of the integration of project management fundamentals with the practices of *international contracting*, which characterize planning, design, and construction of large works (such as plants and machinery) in the industrial sectors.

The challenge faced, having analyzed numerous books on project management, was of making a contribution to a subject that is, on one hand, going through a phase of rapid growth—at times almost explosive—and on the other quite broad and varied. Up-to-now published books on this topic highlight how difficult it is to link the exactitude of *scientific management* with the practical needs and variety of the projects themselves. The objective is then to present a text that is methodologically exact but at the same time manageable!

The methodology, tested in numerous qualified firms, describes a *company-wide* and *multiproject* approach to management, involving all areas of the firm in an integrated system of projects.

In particular, the *company-wide project management* (*CWPM*) approach aims at implicating the innovative processes, adopting a philosophy resembling the one total quality management (TQM) applies chiefly to operational activities (remember the importance of procedures in the Manual of Quality), in that case extended upstream as far as the design function, thanks to the concepts of *upward quality* and reduction of the sources of variability. Similarly, that notion (*in all and with all the areas and skills of the firm*) can lead now to a greater diffusion of project management even in the opposite direction, that is *from upstream to downstream*, stretching from the traditional sites (technical and planning offices) to the whole enterprise (from top management to the factory and sales units), with everyone making some contribution.

In such a situation, first the principles and instruments of project management are developed internally and then exported, having been reclassified as *systemic* and made available to all. They have now become an occasion of integration and a method of coordination. In addition, the selfsame creation of projects is modified. They can arise and/or be required anywhere in the firm because they are linked to change, improvement, and the firm's need to be a dynamic organism in a dynamic environment. If TQM identifies and involves all the subjects that can contribute to quality and service, CWPM has the onerous but determinant task of offering tools to these subjects so that they can improve *quality system* and make it dynamic and flexible: in other words *to manage change and innovation*.

The greater complexity of the environment now necessitates managing several projects at the same time, i.e., as a *project portfolio*, using other principles and techniques determined by multiproject management. These concentrate on the organization of human resources shared among the projects and pervade the strategies to the point where the so-called multiproject/multiproduct strategies (such as the platform projects, design transfer, etc.) are side by side with the traditional ones of product and project strategies.

The book aims at integrating a fully company-wide, *process-based*, multiproject management with the specific practices adopted by international leaders winning complex contracts all around the world.

With this new edition, the book has been completely revised and expanded, and it is up to date with the very recent (September 2017) 6th edition of PMBOK, the current standard ISO 21500, and worldwide PM best practices, including innovative approaches such as Agile Project Management (SCRUM et al.). A specific annex is provided on the professional use of the reference software *Microsoft Project*.

The book is now divided into six Parts.

Part I presents the international standards of Project Management (PMI, IPMA, Prince 2), starting from a strategic and competitive perspective. Some aspects are also depicted, such as the professional certifications and the maturity models. New trends in PM are then illustrated. Project management is considered as an organizational process—although different from operations management—aimed at managing change and innovation.

Part II focuses and describes the characteristics of Engineer-To-Order projects and the management of contract works. It includes legal aspects, organization of ETO, life cycle of work orders, and product and service design.

Part III concerns project scope and structuring (WBS) and includes strategy and quality issues. Particular attention is dedicated to the measurement of project performances.

Part IV illustrates the management of ETO/ECC projects, based on the managerial variables of time, costs, and resources. The organisational and HR aspects of the projects are examined in detail. Furthermore, the financial, investment implications are strictly considered. A great importance is reserved to project communication.

Part V is about project and program management—the most current challenges—together with the *hot* topic of risk management, dealt both at the individual project and portfolio level.

In Part VI, finally the rigorous academic approach is mixed with the managerial contribution of Danieli, one of the largest worldwide suppliers of equipment and plants to the metals industry. The part describes project management in Danieli, and how it can be effectively applied to win and manage large international contracts.

The book is aimed at project managers, CEOs, general and technical managers, directors, and personnel responsible for activities in medium-large enterprises operating worldwide through international contracts, consultants, and researchers in the areas of innovation business, process, and project management.

A book is always the result of activities and experience developed in collaboration with many people, but it would be impossible to remember all and fairly in over 25 years of teaching and consulting. One person I want to thank in particular, for the Preface to this book: Russ Archibald—one of the fathers of modern project management, PMI founding member No. 6, with a career now spanning more than 70 years.

It is an honor for me to collaborate with him to promote the worldwide diffusion of the discipline of PM.

I furthermore wish to thank Ruth Milewski and Prashanth Mahagaonkarfor at Springer International Publishing.

I dedicate this book to my lovely Simonetta and our *best projects*, Nausica and Federico.

2 The International Standards of Project Management

2.1 Project Management for Business Results

Project Management (*PM*) is a discipline that has long established itself internationally within companies operating in highly dynamic competitive environments, and in markets that require continuous product and process innovation, quickly and with high quality standards.

The matter is strictly topical, due to the felt need, on the part of companies and institutions, for a greater and better managed planning, also seen in strategic terms, as a competitive weapon. More precisely, the importance of PM today is due to:

- The now strategic role of "design" (no longer only supporting)—performance is not in operations but upstream, in project's decisions
- The need to have a "methodology" to develop innovation (PM is not the content of innovation, but the indispensable tool to reach the market based on innovative ideas)
- The presence of a portfolio of company's projects and consequent predominance of the management aspect (time and costs)
- The extension of the "design mentality" to other company's areas/functions (compared to the traditional technical area) and to other areas of application (not only technical projects but also projects of re-organization, lean programs, etc.).

Today the areas of application of PM are in fact very broad:

1. New Product Development—N.P.D.
2. Service Design
3. Contract or Engineering-To-Order
4. Company's Re-Organization and Performance Improvement
5. Reforms of Public Administrations (Public Sector)
6. Public and Private Research, and International Cooperation.

S. Tonchia, *Industrial Project Management*, Management for Professionals,
https://doi.org/10.1007/978-3-662-56328-1_2

The doctrine of the PM is based on some principles and techniques from a long history of managing large, successful and unsuccessful, projects. The success of a project is reflected in the achievement of performance targets of cost, time and quality that cannot be managed through routine operational tools. These objectives can only be achieved through effective and efficient management of the entire project's lifecycle.

What distinguishes a project—which can be more or less innovative—from all other company activities is that: (1) it is *complex* and (2) there is (or should be) a *start* and an *end*: it is temporary, sooner or later it ends (for the achievement of the objective, or for the elapse of time, or for the exhaustion of financial resources, or the dissolution of the organization/persons involved).

In this sense, projects can be considered not only the design of an Airbus, the construction of the Channel Tunnel, the re-layout of a factory, the introduction of a new information system, the revision of the sales network, the renovation of a building, but also scientific research, space missions, an advertising campaign, the production of a television programme or a movie, writing a book, the design of a training course; and in the personal sphere: the organization of a wedding, a move, a holiday...

There is in the concept of project its projection towards the future rather than the reference to past standards and experiences; sometimes the same word "project" is used as a synonym of the will or desire to do something, until in some cases it takes on contours of little concreteness with respect to everyday life (as in the famous "Life is what happens to you while you're busy making other plans").

Working for projects, i.e. pursuing a clear objective while respecting the constraints of time, cost and quality, is complex and requires the implementation of organisational and management principles and the use of specific techniques.

The competitive value of PM is linked to:

- *Organisation and management*, which are becoming more and more order-winning (technical know-how, or product and production process technologies, are not enough)
- *Proactive* behaviour, which shifts from risks (more and more) to opportunities (multiple and differentiated from competitors)
- *Change* as a lever for performance sustainability
- The "design" of *performance improvement*, which cannot be random or disorderly (in the face of increasingly complex scenarios)
- The greatest *need for "planning"* (inside the firm and by HR skills).

PM has deep historical roots. Three thousand years ago, the most civilized people were already managing the projects of major architectural works. Just think of the seven wonders of the ancient Mediterranean area, from the Egyptian pyramids to the giant of Rhodes, and the Chinese wall built in the Far East. However, PM as we know it today was born in the twentieth century in the United States.

In the early 1900s Henry Gantt developed a bar chart which, in his honour, will be called *Gantt diagram*. This instrument, designed for shipbuilding projects, greatly increased efficiency in the construction of transport vessels during the First World War.

Beginning in the 1950s and throughout the 1960s, the first techniques were devised, developed and formalized to support the management of large complex projects in the industrial field, but above all within military and aerospace projects. For example, chemical manufacturer Du Pont designed the Critical Path Method (CPM) to plan shutdowns in the maintenance of its production plant and Rand Corporation developed the Program Evaluation and Review Technique (PERT) to plan missile design. Work Breakdown Structure (WBS) and Earned Value (EV) approach were developed in these same years.

In the 1970s, the project manager's profession developed. The *International Project Management Association (IPMA)* was founded in 1964, and in 1969 the *Project Management Institute (PMI)*, for the certification and development of the profession, through codes of ethics, standards and accreditation.

At the end of the 1980s, the PMI produced the first edition of the "Project Management Body of Knowledge"—PMBOK®, a text which has reached its sixth edition (published on 6 September 2017) and which contains tools and techniques for PM through an integrated and structured approach.

In September 2012, the first ISO standard dedicated exclusively to PM was published: *ISO 21500—Guidance on Project Management*, a substantial "rapprochement" between PMI and ISO (see the following section).

2.2 PM International References

Among the international institutions dealing with Project Management, the *Project Management Institute* (*PMI*) certainly deserves the forefront, both in terms of history and scope (number of certified project managers), not least the "contribution" to the recent ISO 21500, published in 2012, which will be discussed in the next section and which has represented, after decades, the reunification between the international community of PM and the ISO (International Organization for Standardization).

The PMI (www.pmi.org) is a non-profit organization whose strength lies primarily in the number of members, project managers or interested in various ways to PM, all over the world; born in the United States in 1969, it has almost 500,000 members worldwide.

The objectives and activities of the PMI are essentially as follows:

- Establish the world's leading network of PM professionals
- As a result of this, hold and update the professional knowledge of PM, which is written down in the volume *A Guide to the Project Management Body of Knowledge* (*PMBOK®*)
- Develop research, education and dissemination of the culture, principles and techniques of PM as an international standard
- To certify the knowledge and skills of PM.

The PMBOK is probably the greatest result of the PMI's many years of activity; it is not a text to "learn" the PM, it does not have this aim: PMBOK—the title says—is the collection of the body of knowledge developed and consolidated over the years, such as "best practices", by project managers all over the world (without references to specific sectors). More than 30 years have passed since the first edition (1983) and now there is the sixth edition (September 2017), translated into many languages. PMBOK's contents have been transposed by the American National Standards Institute (ANSI), and by the Institute of Electrical and Electronics Engineers (IEEE), until they are summarised in the already mentioned international standard ISO 21500.

The *individual* certification of PM is certainly one of the most important and impactful activities. Such certification is an increasingly important opportunity for professional development and growth for project managers at all levels and in every private and public sector. The value that the market attributes to "certified" subjects is constantly growing and is becoming a distinctive element during staff selection, an important proof to clients of the company's professional skills, not least a prerequisite for participation in tenders.

By far the most widespread and well-known certification of the PMI, issued since 1984, is the "Project Management Professional"—PMP®, which has over 700,000 certified professionals worldwide. The other certifications, always issued by the PMI, are far less widespread (in the order: CAPM®, which certifies a lower level than the PMP, if only because it does not require experience in the role; PgMP®, which certifies the program managers; other certifications, such as Risk Management Professional, Agile Certified Practitioner etc., have even smaller numbers). In other words, PMP® is the *true* certification of PM:

- Applies and is the same worldwide
- Has no sectoral or context references
- Check your knowledge through an online closed-question test (multiple answer), 200 questions in up to 4 h
- Requires 3 years (if graduated) or 5 years (if under-graduated) of demonstrable experience in PM processes, and to have attended a PM course of at least 35 h beforehand
- Unlike CAPM®, it has unlimited validity provided that at least 60 "PDU" training credits are acquired every 3 years (through participation in conferences, events, courses or public activities).

The other major global PM organisation is the *International Project Management Association*—IPMA (www.ipma.ch), founded in Vienna in 1964 and with over 120,000 members worldwide. As in the case of PMI, its two main activities are on one hand the definition and maintenance of PM standards [named "Competence Baseline"—ICB®, arrived at version 3.0, consisting of 46 "competence elements" covering: (1) the techniques of project management (20), (2) the professional behaviour of project management personnel (15), (3) the relations with the project's context (11)] and, on the other hand, the certification of PM skills (with over 250,000 certificates worldwide).

IPMA's certifications are more differentiated and "scalable" than PMI's certifications. Apart from an "Introductory Certificate" (which is a simple certificate of participation in training courses on PM, including university courses), the official levels are 4: from "D" (without experience, *project manager associated*) to "C" (with 3 years, *project manager*), to "B" (with at least 5 years of experience, of which 3 years in managing "highly complex" projects, *senior project manager*), to "A" ("able of directing a project portfolio or programme", *project director*). Levels A–B–C also require a project report and an assessment interview, while level D only requires a written test (both closed and open questions). IPMA certifications also have a duration of validity (5 years) and an assessment of the permanence of the requirements.

Even if it is not really a "standard", it should be mentioned also Prince2® (acronym of "Projects in controlled environments, version 2"): is not a certification of the knowledge/competences possessed by a person to be able to carry out the profession of project manager (as in PMI and IPMA cases), but certifies that the person knows the Prince methodology2. In fact Prince2 is a PM methodology, which also explains why more than a few months of study is needed to certify PMP, and only a few days for basic certification Prince2 ("Foundation").

Prince2 was initially developed as a PM method for the UK Government's Information Systems (IT), but immediately applied regularly outside the purely IT environment. Prince2 was released in 1996 as a generic method for PM in the UK and more than 50 countries worldwide. The latest version was released in 2009 by the Office of Government Commerce—UK.

The contents of the methodology are set out in two textbooks: "*Managing Successful Projects*" and "*Directing Successful Projects*". The certification of knowledge of the methodology—on two levels ("Foundation" and "Practitioner")—is based only on the first textbook. Starting from 2012, a third level—called Prince2 "Professional"—was introduced, and assessed the competence in applying PM (Prince2) and not only the knowledge of methodology (Prince2), so much that its examination is far more demanding and includes 2 and a half days of residentiality at an Assessment Centre. For further information: www.prince-officialsite.com.

2.3 PM as an International Standard: PMBOK and ISO 21500

On 3/9/2012 it was issued the international standard "ISO 21500:2012—*Guidance on Project Management*".

In reality, an ISO standard—specifically dealing with PM quality—was already in effect (ISO 10006), but it had never achieved the same popularity as PMBOK, which has always been considered the international "reference", even if not officially an "ISO standard".

ISO 21500 can be used by any type of organisation, public or private, and for any type of project, regardless of complexity, size or duration. It describes, at a high level, concepts and processes considered good practices of Project Management. Projects are considered in the context of the Program and Portfolio Management, but the standard is limited to the context of Project Management.

From a content point of view, ISO 21500 can be considered a kind of synthesis of PMBOK, with the difference that one takes up more than 450 pages and the other only 47 pages (The PMBOK describes in detail the inputs, outputs, tools and techniques of the various PM's processes).

The "knowledge body"—both PMBOK and ISO 21500—includes 10 "*knowledge areas*" (including time, cost, quality and resource management variables) and 5 *macro-processes or* "*groups of processes*" (these latter divided into "*processes*", for a total of 49 processes).

The 10 "*knowledge areas*" are:

1. *Project Integration Management* includes the activities necessary to ensure that the many elements of a project are effectively and efficiently coordinated and integrated.
2. *Project Scope Management* includes the processes required to ensure that the project includes all the work required, and only the work required, to complete the project successfully.
3. *Project Resource Management* includes the processes that organize, manage, and lead the project team. The project team is comprised of the people with assigned roles and responsibilities for completing the project. Latest version also includes the management of non-human resources, i.e. materials and technologies.
4. *Project Quality Management* includes policies, procedures and all the activities so that the project will satisfy the needs for which it was undertaken.
5. *Project Schedule Management* includes the processes required to manage the timely completion of the project.
6. *Project Cost Management* includes the processes involved in planning, estimating, budgeting, financing, funding, managing, and controlling costs so that the project can be completed within the approved budget.
7. *Project Risk Management* includes the processes of conducting risk management planning, identification, analysis, response planning, and controlling risk on a project.
8. *Project Communications Management* includes the processes that are required to ensure timely and appropriate planning, creation, collection, distribution, storage, retrieval, control etc. of project information.
9. *Project Stakeholder Management* includes the processes required to identify the people, groups, or organizations that could impact or be impacted by the project, to analyze stakeholder expectations and their impact on the project, and to develop appropriate management strategies for effectively engaging stakeholders in project decisions and execution.

10. *Project Procurement Management* includes the processes necessary to purchase or acquire products, services, or results needed from outside the project team.

The five *macro-processes* ("*groups of processes*"), which correspond to precise moments in the project's life cycle, are:

1. *Initiating*: starting-up, define and authorise the project or a phase of the project.
2. *Planning*: defines and develops project's objectives and plans the carrying out of the actions necessary to achieve them.
3. *Executing*: integrates people and other resources for the implementation of the project plan (including the acquisition of products and services from outside).
4. *Monitoring and Controlling*: measure, check and regularly monitor progress, to identify deviations from the project plan or to accept formal changes from the client, so as to allow corrective actions or re-planning.
5. *Closing*: formalise the acceptance of the product, service or result, and allow for the correct closure of the project (or phase of the project).

ISO 21500 is a substantial convergence between ISO and PMI, to the benefit of PM professionals, companies and their customers. They will make international standards increasingly essential, making even those countries and companies that are not yet aligned with international best practices "grow". Standards, but also professional certifications, will increasingly be a differentiating element and a competitive plus.

2.4 New Perspectives on Project Management

The constantly evolving technologies and competitive market dynamics mean that company management increasingly focuses its attention on the management of *change and innovation*, seen from a strategic point of view rather than confined to a mere role of "supply of specifications" for *operations* activities.

Risks are growing but also the opportunities. It is therefore important to be prepared and play a proactive role, making sure that if a project can be considered as a "bet on the future", this is the best that can be done, and its pursuit is the driving force behind all activities. The innovative capacity and success of a company is therefore increasingly linked to its ability to "manage projects". Managing change not only to avoid to be overwhelmed, but as a competitive weapon.

The formalisation of the PM from *reference* (for the international community of project managers) to ISO *standards* confirms and strengthens:

- The now defined *convergence of practices* by the worldwide community of professionals and operators within the organizations in which they operate.

- A *standard now uniform at international level*, which means that PM, if only because it developed a few decades before the more recently understood globalization, is probably the most standardized discipline of Management (in the positive sense of the term).
- The existence of an *ISO standard* to which you can (or should) report.

It will then become important to be able to measure the capacity of an organisation to manage projects, also using the so-called Project Management Maturity Models (such as CMMI by the Software Engineering Institute, P3M3 by the UK Office of Government Commerce, and P2M in Japan), the best known being the *Organizational Project Management Maturity Model—OPM3®* by PMI (third edition, 2013), which considers five "Levels of Project Management Maturity":

- Level 1: Initial Process
- Level 2: Structured Process and Standards
- Level 3: Organizational Standards and Institutionalized Process
- Level 4: Managed Process
- Level 5: Optimizing Process.

But PM is not only method, techniques, tools, it is first and foremost a *culture*:

- *Project culture*, i.e. a focus on activities to achieve an objective, in time and at predetermined costs, with resource constraints
- *Management culture*, i.e. planning and control
- *Culture of the customer*, as PM is not aimed at achieving pre-established standards but rather is the methodological instrument for innovation (a product in a market segment, or the response to specific one customer's needs).

That's why PM has today and increasingly:

- A strategic corporate value
- And is a managerial skill of individuals.

More precisely:

- "Design" is not only technical, but also of the way of operating (organisational-managerial and strategic projects, pervasive throughout the organisation)
- The ability to conduct projects then becomes a competitive weapon and source of performances
- From the management OF projects to the management BY projects (company as a breeding ground for projects; corporate management as a portfolio of projects or initiatives)
- Investment decisions and their implementation become projects
- With these meanings, there is an increasing need for companies to adopt, extend and increase PM

- Beyond flexibility and customization, there is a need for greater PM
- Needs which concern services, the private service sector, government and public administrations too.

This ability to innovate and in general to face technological and market dynamics constitute a guarantee of success over time, which is self-fed ("learning organization") and which is difficult to imitate (as opposed to products and production), because it is related to the organization and integration of its primary resources, the human ones.

2.5 Agile Project Management

But projects are changing, becoming more and more.... *extreme*. Through the attribute "Extreme" (sometimes replaced by "Agile" or "Liquid"), in international practice many of the current projects are now characterized, those that have:

- Objectives that are not entirely clear a priori, with a focus on benefit and added value for the customer, constantly evolving
- Under unstable and turbulent (internal and external) environmental factors
- With time targets often at the limit, and frequently re-planning needed
- With human resources that are difficult to find and manage (sometimes even virtual teams, collaborative at a distance through new web technologies), also because of the stress caused by high uncertainty and frequent changes.

The differences between extreme and traditional projects are therefore *dimensional* (the extreme projects are decidedly more complex), *temporal* (the time-to-market is largely shorter than traditional projects), *environmental* (the context is considerably subject to change) and *human* (it is necessary to manage carefully the relations with all those involved in the project). Table 2.1 shows some dimensions that characterize the main differences between the two types of projects:

However, an extreme project is often also of great interest!

The risk and uncertainty in such projects is greater than in traditional projects, but classical Risk Management is not sufficient, even though it belongs to the baggage of Project Management, for extreme projects because:

- Most of the risks are not identifiable in time
- It is not possible to plan a risk response in good time
- Often there is no recognisable separation between planning and control (which is an assumption of the PM base), but continuous adaptation, even if it is governed.

If, then, the innovation required is becoming more and more pronounced, how can we manage increasingly extreme projects?

Table 2.1 Differences between traditional and extreme projects

	Traditional Projects	Extreme Projects
Predictability	High	Low
Objective + Content	Defined, stable and clear from the beginning	Clarified later
Changes	Few	Many
Focus	On product & service	On benefit and value
Environment	Stable	Turbolent
Duration	Long	Few weeks/months
Final success	Plan complied	Satisfied stakeholders
Implementation strategy	Sequential	Quick

- Using "lighter" techniques (this is the Lean concept of simplification, now in vogue), but to be applied rigorously
- Renouncing total planning and control to concentrate on the essential things, in other words adopting the philosophy of "little but without compromise".
- With methods that from *adaptive* become *predictive*, moving from a logic of pursuit of a *plan* to a logic of prediction of *scenarios*
- Pursuing flexibility and not necessarily the plan, whereby the output of the project (deliverable) is customer-driven in progress, until the end of the project
- With a *Rapid Planning* (focusing only on what is really predictable at the moment, which can be little and therefore the planning is quick) and frequent *Quick Meetings* of advancement (also short, but frequent, given that this is the pace of the market)
- Taking a Project Tracking rather than a Time Tracking, i.e. counting mainly the results achieved so far rather than the time spent as a measure of progress
- Considering the project's Life-Cycle, i.e. also the period of the product on the market or post-contract assistance, as a justification for the costs and other project's variables.

The "Agile Manifesto" (2001), then published by the Agile Alliance (www.agilealliance.com), considers four basic principles of agile methodologies, which are constructed in order to consider important:

- Individuals and interactions more than processes and instruments
- The product working more than the complete documentation
- Collaboration with the customer rather than contract negotiation
- Responding to change rather than following plans.

Due to present importance of Agile PM, the very recent release of PMBOK—6th edition (September 2017) has been accompanied by the "*Agile Practice Guide*" (new edition, October 2017), partnered with the same Agile Alliance® and provided as a complimentary copy together with the PMBOK.

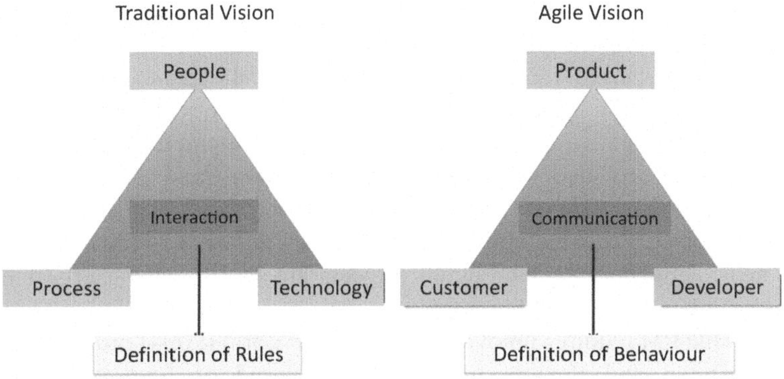

Fig. 2.1 Traditional vision versus agile vision

The basic premise of agile practices is the need to be able to limit the costs of planning, execution, monitoring and control of the project while increasing predictive reliability, an approach that increases process flexibility and reduces the cost of change.

Agile approaches propose a set of guidelines and principles which, in some contexts (e.g. software development), provide for a reduced effort during the initial definition phase of requirements and their continuous updating. In this way, some strategic decisions relating to the development process are postponed in order to reduce the cost of change and make choices under conditions of maximum information availability.

People's skills and abilities are essential to adopt such a strategy: the reduction of project's documentation (explicit knowledge) increases the value of a correct sharing of implicit knowledge (the one in people's heads).

As illustrated in Fig. 2.1, the central role of people therefore contributes to shifting emphasis from "rules" to "behaviours".

The personal qualities and human abilities of those who are part of the project team assume a fundamental importance: if you want to adopt an agile methodology, you need to think and design the agile process in line with the level of preparation and experience of the professionals involved and trying to always emphasize the skills of the team members rather than the need to respect the procedures.

Traditional ("heavy") PM methodologies, suitable for traditional projects, are therefore contrasted with "agile" methodologies.

PM *heavy* methodologies have as their main objective the description of reality and all its variables in a single project plan, which must then be implemented in order to achieve the objectives planned on the basis of the specifications defined during the start-up phase. In other words, through this plan, the main objective of the project managers is to plan in detail, over a rather broad time horizon, a large part of the project activities, including in their planning all the technical details and the various milestones identified. It is therefore easy to understand how this type of methodology is effective only for those projects characterized by stable, linear and predictable contexts, where all the variables at stake are fixed, predictable and immutable (therefore they are suitable for "traditional" projects).

The *iterative/adaptive* ("*agile*") approaches envisage the management of time and costs with a logic that tends to adapt the project's scope to changes, placing flexibility as a prerogative and encouraging rapid development and the maximum value that can be provided with respect to agreed constraints. Agile models are aimed at a continuous customer's involvement in all the stages characterising the life cycle of the project. These methodologies also use an informal communication style; unlike "heavy" methodologies, "light" projects are in fact characterised by few documents, few practices and few rules.

The Extreme PM requires innovative models. De Carlo (2004) proposed a highly customer-oriented and flexible framework composed of four iterative cycles (*Visionate*, *Speculate*, *Innovate* e *Re-evaluate*) and a final element called *Disseminate*. Iteration within a cycle continues until the desired result is achieved. In particular:

- The *Visionate* cycle includes the activities of identifying and agreeing on the final project objective, identifying the economic benefits and risks associated with the project as well as identifying possible future scenarios that could influence it, both positively and negatively.
- The *Speculate* cycle consists of structuring the project plan including deliverable and temporary milestones; during this cycle, the activities and documents produced are updated in accordance with the latest information available to the team.
- The *Innovate* cycle deals with the rapid development and timely generation of feedback with project's stakeholders (the customer/end user), trying to obtain their approval on the work carried out. At the end of this phase, the results are formalised and prepared for the revision ("Re-evaluate") phase.
- The objective of the *Re-evaluate* cycle is to decide on the future of the project; questions like this are therefore answered: is it still right to continue with the project? Can it be improved?
- The last element, *Disseminate*, aims to release the final output to the customer and start the process of realizing the economic benefits.

Among the agile methodologies, SCRUM is the closest to achieve an agile extension of the traditional practices of PM: it uses the PM methods prescribed by the PMBOK (at macro level) while maintaining an iterative character of agile management (at micro level). In other words, it is possible to overlap the SCRUM methodology on the life cycle proposed by PMBOK, with two different operating levels. The first, "coarse-grained", is the *level of release*, where the correspondence is almost the same: the start-up, planning, execution, monitoring and control, and closure phases are interpreted from a managerial point of view, expressly directing areas such as the management of the scope, cost, time, quality, human resources, communications and all the managerial processes described in the PMBOK. The second level, "fine-grained", is the *level of iteration(s)*.

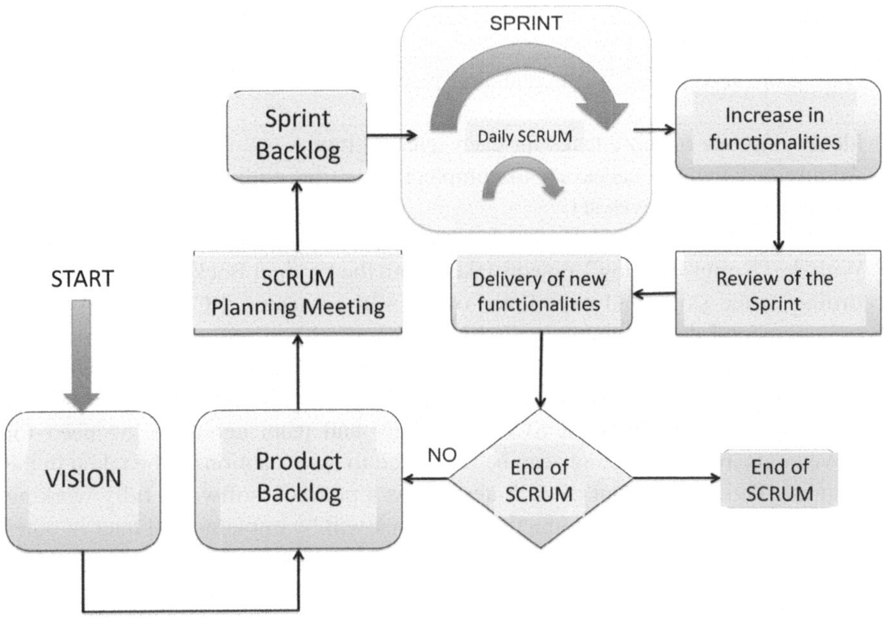

Fig. 2.2 SCRUM process flow

SCRUM (like the scrum in the rugby sport) is perhaps the agile methodology actually most successful. Although born in the context of software development (applied by companies such as IBM, Microsoft, SAP, Google), SCRUM is a methodology that can be extended to other contexts where *agility* is required.

As illustrated in Fig. 2.2, a SCRUM project begins with the definition of a Vision, which includes forecasts on return on investment, and the definition of releases and main milestones. Activity planning is based on the so-called "Product Backlog", an initial list of tasks to be performed that includes all the features and technological improvements specifically designed for the project.

The main activity is "Sprint". The SCRUM process is, in fact, iterative and divided into a series of successive *sprints*, each of which can last from one week to one month. Each sprint starts with a "planning meeting" organised by the SCRUM master. The meeting is divided into two parts. In the first part, SCRUM master and SCRUM team meet in order to:

- Identify a sprint target and provide a general description of the work to be done during the sprint, in order to concentrate everyone's efforts in one direction
- Prioritise the remaining objects to be done ("items") or introduce additions to the product backlog
- Obtaining answers from the product owner on any questions that the team may have about the content, purpose and meaning of the objects in the product backlog

- Select the highest number of objects that can be created by the end of the sprint.

During the second part of the planning meeting, activities are organized:

- Identifying one or more tasks for each selected item
- Adding tasks where necessary to complete the sprint within the time limits and achieving the objectives set.

With the "Sprint Backlog", objects taken from the Product Backlog are developed according to the skills and experience of the various teams. SCRUM uses rather short incremental development cycles—*sprints* indeed—in order to implement the items in the Product Backlog. At the end of each sprint, the new function developed is delivered. A review meeting of the sprint will be held in which the customer must also be present. In addition, if SCRUM master and team agree on the need for improvements, review meetings can be organised to assess actions to be taken in the next sprint. The main output of each iteration is a product (software) fully working and of quality, but without the functionalities that will be implemented later in other sprints. The process is iterative and stops when all required features have been implemented or if the customer considers the project completed.

The Project Management Process 3

3.1 Project Management: A Process-Based Approach

A project can be defined as a set of complex, coordinated activities with a clearly defined objective that can be achieved through synergetic, coordinated efforts within a given time, and with a predetermined amount of human and financial resources.

What distinguishes a project (no matter its degree of innovation) from all other activities carried out in a firm is that it always has a *beginning* and an *end*. It is temporary and sooner or later will finish (either because the objective has been achieved, the scheduled time has passed, the financial resources have been used up, or the organization involved has closed down).

But what does *managing a project* mean? Managing means dealing with variables that can be more or less influenced: it is impossible to manage when no variation is allowed or, likewise, when the factors are beyond control, since managing means making decisions and acting accordingly (i.e. planning and executing interventions).

What can be managed are those *variables* that are typical of a project, since they are linked to its definition. The main variables are as follows:

1. Quality
2. Time
3. Costs
4. Resources

The first three variables are project performances, whereas the last represents the (human and technological) restraints limiting the activities needed to execute a project (e.g. the availability or saturation of the resources within 24 h, etc.) (Fig. 3.1).

A project is an organizational process, and hence has all its features.

According to the Oxford English Dictionary, a process can be defined as 'a continuous and regular action or succession of actions, taking place or carried on in a definite manner and aimed at achieving some result'. A business process consists

S. Tonchia, *Industrial Project Management*, Management for Professionals,
https://doi.org/10.1007/978-3-662-56328-1_3

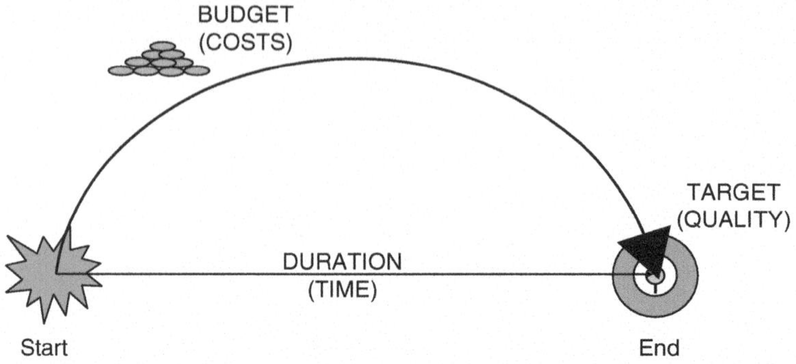

Fig. 3.1 Characteristics and performances of a project

of a set of activities; each activity in turn is formed by elementary operations, requiring specific resources, and is aimed at achieving an objective that concurs, with those of all the other activities, to the overall goal of the process, a goal that integrates the objectives of the various activities.

It must be stressed that all these activities have a synergic effect on the process (i.e. the overall result is greater than that sum of the single, partial results) and require the involvement of a variety of departments/units (which can be considered as reservoirs/incubators of the expertise needed to carry out the process) within the organization.

To produce outputs destined for downstream, the activities forming a process require upstream inputs, as well as the competency of the resources carrying out the activities (resources *borrowed* from the departments/units of the company): upstream and downstream can coincide—as described in more detail in the next few pages—with other processes carried out inside the company or by external suppliers and external customers, respectively (Fig. 3.2).

A process thus structured helps distinguish between output (or *result*), internal, and input (or *received*) *performances*, making it easier to determine where to carry out measurements and therefore estimate the performances with a greater degree of precision.

Moreover, by assessing the output and taking into account the cause–effect relationships between received and internal performances, it is possible to establish where to intervene in the most effective manner, selecting and managing the portfolio of improvement projects.

A process can also be viewed as the place where value addition is created, or, in other words, every process generates value addition. Therefore, process logics combine the typical input/output approach of the system theory with an economic approach, taking into account that 'a process is a combination of activities requiring one or more inputs and creating an output with a value for the customer' (Hammer and Champy, in their *Reengineering the Corporation: A Manifesto for Business Revolution* published in 1993) and that 'processes constitute a network where the

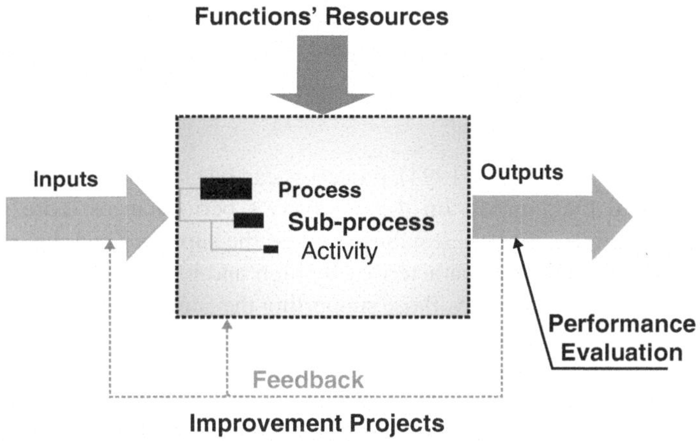

Fig. 3.2 A process consists of activities transforming inputs into outputs, thanks to resources loaned by the functions/units of the company

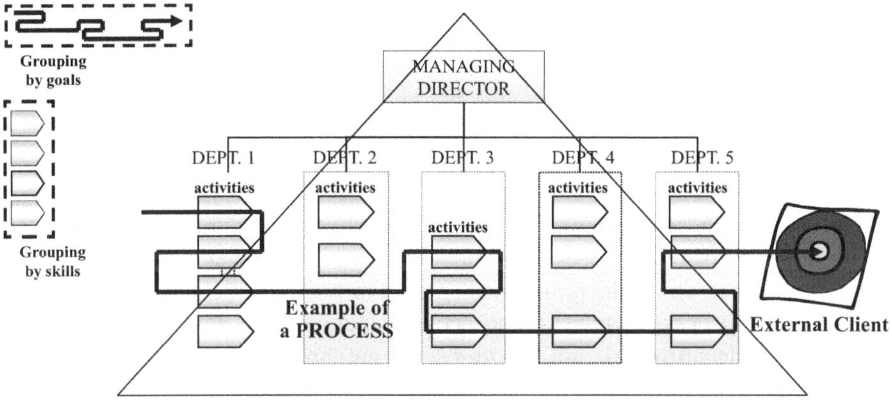

Fig. 3.3 Integration between functional organization (by departments) and process management: a process is a set of activities requiring resources that belong to departments

activities of a certain process serve to add value to the inputs deriving from the previous process' (Armistead and Rowland 1996).

Figure 3.3 shows the integration between processes and departments (each department—the rectangles—covers part of the organizational chart). An example of a process is shown, spanning the organization to reach its targeted external customer: in order to do so, a variety of activities (depicted as oriented polygons) must be coordinated, exploiting the resources of the different departments. Human resources can be grouped *according to skills* (with respect to the departments) *and goals* (with respect to the processes). Consequently, a single human resource necessarily belongs to only one organizational unit (and therefore holds a unique position on the company's organizational chart), but may contribute to more than one process.

Rummler and Brache (1990) identify three types of processes: (1) primary (production, logistics, order management, product development, marketing and sales, etc.), (2) managerial (such as strategic planning or total quality programmes), (3) support processes (planning the use of resources, managing information, staff, technology, etc.).

According to Earl and Khan (1994), processes can be classified according to their structurability and their impact on the company's performances: core processes (internal) and network processes (established with the suppliers and the customers) have a direct impact, and are characterised by high and low structurability, respectively; support processes (namely, those supporting the core processes) and management processes (the direction) have an indirect impact, and a high and low structurability, respectively.

Gilbreath (1986), on the other hand, distinguishes between *flow* processes (such as the manufacturing ones, characterized by repetitiveness in their execution and standardized output) and *impulse* processes (such as product innovation).

The *process owner* operates transversally with respect to the company's departments (although he may also be in charge of a specific unit). The process owner defines the goals of the process, coordinates the activities, manages the resources exploited for the process, establishes procedures and interventions, identifies the characteristics of the process and the performance indicators, and chairs all activities aimed at improving the performances of the process. In other words, the process owner acts as an *entrepreneur* within the clearly marked boundaries of the process.

The process owners differ, in expertise and tasks, from the managers of the organizational units (function/board/department/office), as illustrated in Table 3.1.

But there are also differences between a *generic* process owner and that *particular* process owner who is the project manager. In fact, if it is true that a *project* is an organizational *process* (having inputs/activities/outputs and involving resources from various departments), not all *processes* are *projects*, because not all the processes have a precise beginning and end (for example, a manufacturing process is not a project because manufacturing continues for many months/years, until the manufacturing specifications change through an innovation project).

As a consequence, a project manager has all the responsibilities of a process owner, plus those deriving from starting and ending a job.

3.2 Operations Management and Project Management

Company management can be differentiated into *specific* management (with activities that are specific for the sector/compartment) and *extraspecific* management (with activities—usually of a financial nature—that are independent of the field/sector where the company operates).

Many departments require specific management: Production (including the delivery of services, if the company operates in this field), Procurement, Trade, Marketing, Distribution/Supply, Design, Process Engineering (Technology/Machines or similar), Research & Development (R&D). For the sake of simplicity, without

Table 3.1 The three managerial *hats*

Authority	Skills	Activity/ performances
Director of an Organizational Unit	• Ability to maximise function results in relation to objective-aimed parameters • Technical specialisation	• Preside over technical quality • Comply with function budget • Manage resources efficiently
Process Owner (process management)	• Ability to act as entrepreneur of one's process • Responsibility for results • Relational skills with upstream and downstream • Team leadership skills	• Satisfy downstream customers • Involve/motivate human resources • Manage resources effectively
Project Manager (project management)	• Ability to manage change • Forecasting talent • Wide knowledge (even if not in depth) • Team leadership skills	• Achieve project objectives • Comply with project budget • Comply with project time schedule

considering the supporting Functions/Boards (Administration and Control, Personnel—Human Resources, Information Systems, etc.), the functions listed previously can carry out two types of *processes* (Fig. 3.4):

- *Order execution* (or production on forecast), characterised by more or less repetitive activities complying with standards
- *Innovation*, a process characterised by change and the absence of reference standards

Because of the repetitiveness of operations, the type of management required for the former process is also known as *Operations Management*, and consists in managing the (logistic) flow of materials and information that starts from the upstream supplier/company interface (the Procurement Office) and flows through the various Production Departments, finally reaching the downstream interface with the customers (i.e. Marketing, Trade and Distribution).

All the activities of this process refer, in a more or less formalised manner, to the standards defined during the process of innovation/change. The management of the latter process is known either as *Innovation Management*, with reference to the *result* of the activity characterising it, or *Project Management*, with reference to *how* these activities—the projects—are carried out.

Innovation/Project Management must therefore produce *new* standards for *routine* activities or, every so often, define new and better standards for the products/

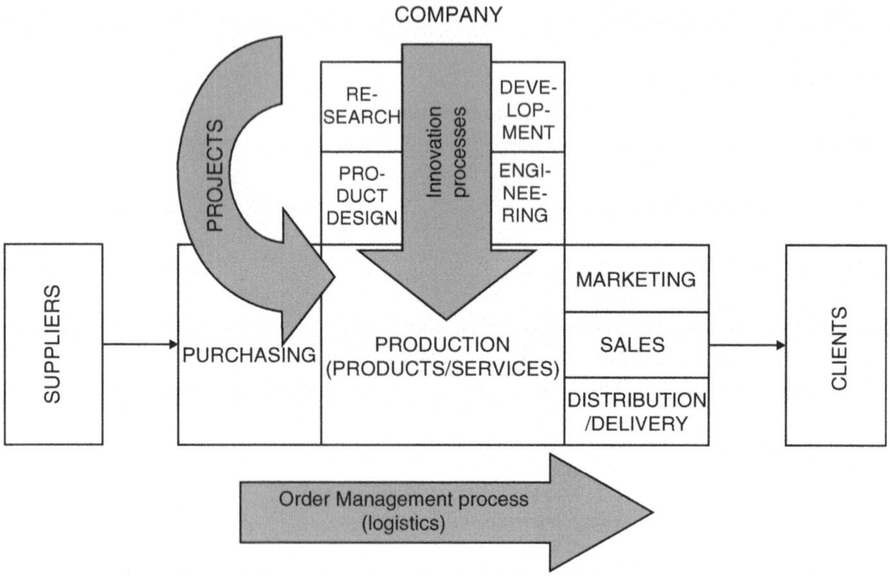

Fig. 3.4 Functional units and key processes in a firm

services delivered (Design department) and the production processes involved (Engineering office), on the grounds of various factors, including the results obtained by the upstream R&D unit.

The Project Management Institute—PMI (in *PMBOK—A Guide to the Project Management Body of Knowledge*, 6th ed., 2017) explains how *operations* share various characteristics with *projects* (both are carried out by persons, exploit limited, available resources, and are planned, executed and controlled); however, 'operations are ongoing and repetitive while projects are temporary and unique' ... 'A project can be defined in terms of its distinctive characteristics: it is a temporary endeavour undertaken to create a unique product or service. Temporary because each project has a definite start and finish. Unique because the product or service somehow differs from all other similar products and services.'

Table 3.2 illustrates some distinctive traits of the two types of management: Operations Management on one hand and Innovation/Project Management on the other.

The main difference between the two processes emerges when analysing the meaning of the performance triad (cost/time/quality) characterising both processes (Table 3.3).

In the case of Operations Management, *quality* is first of all synonymous with conformity. With reference to one of the key principles of Total Quality Management, quality is produced *upstream*, during the stages of design/engineering, and an effort must be made to reduce, and possibly eliminate variance along the production chain, so as to conform to the standards established for the product/process. It is the task of

Table 3.2 Distinctive characteristics of operations management and innovation/project management

	Operations management	Innovation/project management
Activities	Continuative	Intermittent
Focus	The periods	The single project
Reference	Stable	Uncertain
Scope	Productive	Creative
Control	Feedback (actual)	Feed-forward (forecast)
Centres	Cost centres	Investment centres
Departments	Purchasing–Production–Sales	R&D–Design–Engineering

Table 3.3 Differences in the performances between operations management and innovation/project management

	Operations management	Innovation/project management
Quality	Conformance	Level (objective)
Time	Standard time	Duration
	Delivery time	Time-to-market
Costs	Standard costs	Budget of expenditure

the Innovation/Project Management to *produce quality* (objective quality), while quality certifications require a standardization of the procedures—purchasing, production, and so forth, but also when delivering services, establishing trade relationships, etc.—and compliance with the given specifications.

As regards *time*, Operations Management establishes reference standard times for every operation, and defines delivery times that the customer can rely on: longer times mean bad performance; shorter times mean that the specifications made by the Design and Engineering departments can be improved. Project Management, on the other hand, refers to duration of the activities: there are no specific references, because by definition these activities are new, liable to change and their duration can only be estimated.

There are also considerable differences as regards *cost* performances: in operational activities, the evaluation of this parameter usually depends on whether standard costs have been respected, whereas it is impossible to do so in the case of innovative activities, since there are no standards, and the only restraint is the budget.

In practice, however there is rarely a clear distinction between the performances of the two types of processes: even in the case of operational activities it is often necessary to make some slight changes in response to unpredicted situations, whereas in innovative ones—in order to limit the uncertainty characterising them—it is advisable to have some degree of standardisation, by relying for example on past experiences to establish the duration of the activities, exploiting part of pre-existing projects to limit the magnitude of the new project or estimating costs on the grounds of reference standards.

3.3 Types of Projects

Company-Wide Project Management (*CWPM*) considers the firm as a source of integrated and coordinated projects.

Quality is not enough: products and services must be continuously improved, hence the need for *innovation*. The latter can be regarded as a competitive weapon, a strategy that requires careful planning. Innovation concerns both the design and development of the product/service, and the definition/updating of the manufacturing processes (and/or service delivery).

In all these activities, the customer must be the focal point, the target at which companies must aim with all their might and main. In many cases, customer satisfaction takes the form of customisation, and projects are carried out in compliance with the customer's specifications; this is known as *Engineer-To-Order* (*ETO*) or *Engineering and Construction Contracting (ECC)*, and is particularly common in the fields of civil constructions, industrial engineering and engineering services.

But nowadays this is considered insufficient, because flexibility has also become an essential feature. *Time-Base Competition* is already a reality in many sectors. *Planning flexibility* means *managing change*, making plans to improve performances (by reducing delivery times, increasing productivity, etc.), revising the existing organizational structure (internal units, the pool of suppliers and the sales network), modifying procedures (relative to the production planning and control, procurement, plant maintenance, technical assistance services, etc.—Business Process Reengineering—BPR) and even rethinking/revising strategies (*strategic turnaround*).

In brief, there are at least three spheres of business where Project Management methodology can be successfully applied, determining different types of projects:

- Product/process innovation (design and development of new products and services)
- ETO/ECC contracts (for production, construction, services)
- Strategic, organizational and managerial change (aimed at improving the company's performances).

3.4 Managing Innovation

The relationship between market demand and innovation can occur in five different ways:

- The development of technology incorporated into a product triggers a demand (a past example of technology push was the demand for microprocessors).
- The demand already exists, and product innovation responds to it (as in the case of many pharmaceutical products).
- Product innovation stimulates and develops a pre-existing demand (as in the past, in the case of photocopying machines using normal paper).

- Product innovation makes a latent demand emerge (as in the case of mobile phones).
- Innovation can give new life to a pre-existing demand (even in mature sectors, as in the case of Swatch watches).

Clark and Fujimoto (1991), although mainly referring to the automobile sector, observe that sometimes the evolution of pre-existing technology is preferable to the definition of a new product, whereas in other cases, it is the other way round.

In the case of technology push, the *development of technology* should precede that of the product, because of the following:

– Because of its increasing level of sophistication, technology requires much longer development cycles than products do.
– The dynamic, volatile nature of the market requires the companies to have a vast amount of technology at their disposal, from which to choose in order to start production.

On the other hand, in the case of demand pull, *product development* should precede that of technology, because of the following:

– Customers are increasingly more demanding; hence, the companies must focus more on the product, its concept and the harmonious, functional integration of its various parts.
– The great technical and functional interdependency of the various components, as well as the need to economize, using parts and sub-assemblies from other projects, past and/or present, make the management of a product portfolio a key priority, and privilege the development of new products instead of technology.

The development of (product and process) technology follows the traditional path of *R&D*: from *pure or basic research* (usually public and always pre-competitive), which is aimed at advancements in the various fields, and so shed light on phenomena and define the principles governing them, to *applied research* (incorporating a wealth of knowledge and expertise for a specific purpose), which will finally lead to the development of innovative solutions by means of prototypes and pilot runs.

It can be stated that while the processes of Operations management and Product/Project management refer to the customers of the present and the near future, respectively, R&D processes are aimed at satisfying the customers of the distant future.

R&D is what links research to the market: it traces a path along which many projects will perish (R&D projects have in fact a high death rate) and in its last stage includes the area typical of *technological innovation*, regarding both product and process. It is in this area that the more specific processes of *product design and development* are carried out (see Chap. 5).

Except for the case of process innovations that are not aimed at a specific product (but are made for instance to increase productivity—the introduction of automated

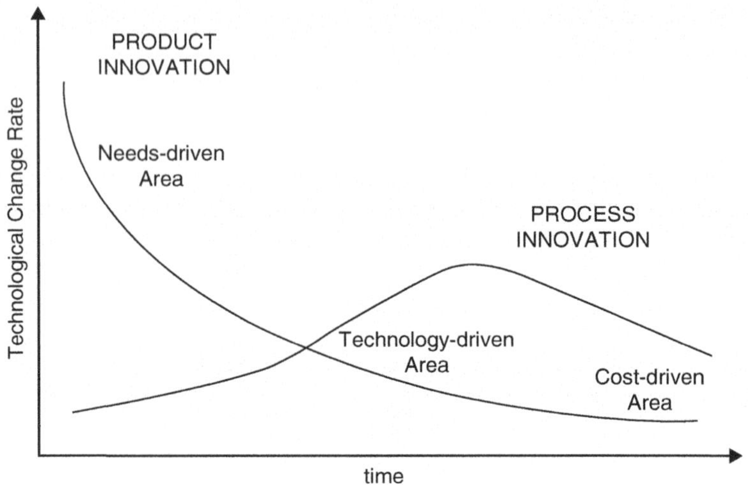

Fig. 3.5 Development stages for products and production processes (Source: Abernathy and Utterback 1978)

plants is an example—or to improve management practices), the processes of product design and development include both product and process innovation, usually with a prevalence of the former during the first stages of development and of process innovation in the latter stages (Fig. 3.5).

Also, the source of technological innovation must be taken into account: this may be *endogenous* to the company (thanks to its R&D department) or *exogenous* (*not invented here*—NIH).

In the latter case, innovation reaches the company through various channels, such as those established with the suppliers of plants and materials, at trade fairs, or observing the competitors (including small but innovative niche companies), collaborating with universities, public or private research institutes, joint ventures and so forth. On one hand, in joint ventures, the actual innovative content may be inferior and more easily imitated, but on the other hand, also the risks are more limited.

When innovation derives from an exogenous source, there can be problems concerning the *transfer of technology*, as well as the difficulty of properly spreading and exploiting it throughout the company. An important figure that should be present in the firm is the so-called *gatekeeper*, a person possessing great relational and technical skills who is in charge of keeping in contact with the potential sources of innovation.

Since human resources play a key role in creating innovation, it is also important to keep in mind those factors that can hinder *creativeness*. Choffrey and Dorey (1983), in a detailed analysis, describe various types of obstacles:

- Of a perceptive nature (difficulties in defining the problem or a tendency to define it in too rigid a manner, inability to perceive all the various aspects of a problem or the tendency to consider only some)
- Of a cultural nature (inclination to resist change, tendency to consider the problems from a scientific and economic viewpoint only, neglecting social and cultural aspects)
- Of a psychological nature (the fear of making mistakes, the inability to accept the initial ambiguity of a project, a propensity to curb one's creativeness, a tendency to evaluate, rather than generate, ideas)
- Of an intellectual nature (no flexibility, insufficient ability to learn from previous experiences, lack of relational skills)
- Of an environmental nature (inability to trust colleagues and collaborate with them, the presence of an autarchic chief, lack of material and financial means)

As well as these hurdles on the path to innovation, there can also be problems in the way technical competencies develop and spread inside a company; *technological innovation* must therefore go hand in hand with *organizational innovation*: the organization/innovation relationship is bidirectional, in the sense that innovation modifies the organization producing it. It is also important to keep in mind that nowadays, innovations are mostly organized in *portfolios* and consequently, each innovative project must be considered from a *multi-project management* point of view.

References

Abernathy, W. J., & Utterback, J. M. (1978, June/July). Patterns of industrial innovation. *Technology Review*, 40–47.

Armistead, C., & Rowland, P. (1996). *Managing business processes: BPR and beyond*. Chichester, UK: Wiley.

Choffrey, J. M., & Dorey, F. (1983). *Développement et gestion des produits nouveaux*. Paris: McGraw-Hill.

Clark, K. B., & Fujimoto, T. (1991). *Product development performance: Strategy, organization and management in the world auto industry*. Boston, MA: Harvard Business School Press.

Earl, M., & Khan, B. (1994). How new is business process redesign. *European Management Journal, 12*(1), 20–30.

Gilbreath, R. D. (1986). *Winning at project management*. New York: Wiley.

Hammer, M., & Champy, J. (1993). *Reengineering the corporation: A manifesto for business revolution*. London: Nicholas Brealey.

Rummler, G., & Brache, A. (1990). *Improving performance: How to manage the white space on the organizational chart*. San Francisco: Jossey-Bass.

Part II

Engineering and Construction Projects

Management of Contract Work

4

4.1 Legal Aspects and Contract Administration

A contract is a legally binding agreement stipulated by two or more parties and characterised by specific obligations. Contracts usually consist of a list of conditions, which may be general (conditions that are standard to all, similar contracts), special (disciplining that particular contract) and technical (regulating how the contract is materially executed). Contractual clauses are listed both in the general and special conditions.

It is possible to distinguish between the following:

- *Sales contracts*, i.e. contracts having as object the transferral of the property of standard goods, services or rights at a certain price
- *Contract work*, where one party, by organizing and managing the necessary means at his own risk, becomes responsible for delivering a good or a service—*whose characteristics are specified by the client*—in exchange for a certain sum of money

A sales contract provides for the mere transferral (*obligation to give*) of goods or services *that already exist* in the firm's catalogue (*standard* goods or services); in the case of goods, it is unimportant whether these are made before or after the customer's purchase order (*make-to-stock* or *make-to-order*, respectively).

Contract work, on the other hand, provides for the delivery of a good or a service that *does not exist in a catalogue*, and whose characteristics are *specified by the client* (*obligation to do*).

In practice, contracts are often a combination of both, so—given the legal implications—it is necessary to define the prevailing component. This assessment must be based more on the object of the contract than on economic considerations (i.e. whether the value of the work is greater than that of the materials): for this reason, it is appropriate to consider *contract works*—also known as *work orders* or *job orders*—as those that are necessary to deliver works that are *not* ordinarily mass

S. Tonchia, *Industrial Project Management*, Management for Professionals,
https://doi.org/10.1007/978-3-662-56328-1_4

produced, but are custom-built in response to a *specific* order (because in this case, work prevails on materials).

According to this type of contract, the output must be delivered by an enterprise; for this reason, the contractors may decide to temporarily join together in order to execute the work. In this case, all the contracting companies must answer to the customer, although to different degrees:

- There may be a clear distribution of tasks (as in the case of a *consortium*).
- The organization and management are common to all, and there is a sort of *shareholding* of the contract work, which is not however officially divided among the various parties (this is the case of a *joint venture*, which can therefore be considered by all means a real new company, although of limited duration, whose scope is to fulfil the terms of the contract).

4.2 Types of Contract Work

The contracting companies can do either of the following:

1. Receive the specifications directly from the client (or, if they are subcontractors, from the contracting company).
2. Design and/or engineer the product in-house (namely, *Engineer-To-Order*—ETO).

In the former case, a given company is required to deliver an output that complies with the specifications *defined by the client and listed in the contract:* this is often the case of civil construction companies, subcontractors, or firms offering specific services. These firms receive a job order, legally defined by a contract, requiring little or *no design/engineering*.

In the latter case, the customer *also* requests design and engineering, or engineering alone. The *design and engineering is done* ad hoc *for the customer,* hence ETO companies only carry out these activities when receiving an order, in the form of a contracted job (*contract work* or *work order*).

For this reason, this industry is also known as *Engineering and Construction Contracting (ECC)*, sometimes *Engineering, Procurement and Construction (EPC)*.

ETO companies usually manufacture/deliver/supply, as well as design/engineer, the following (Fig. 4.1):

- *Buildings/infrastructures* and *industrial equipment/plants*
- *Products* of remarkable size and value, manufactured and/or assembled in an *island* mode (i.e. not along a manufacturing/assembly line) such as ships, aeroplanes, machinery, etc., and *special products*, which, although smaller, are engineered to order: their specifications are only in part provided by the customer, hence the contractor must design them ad hoc

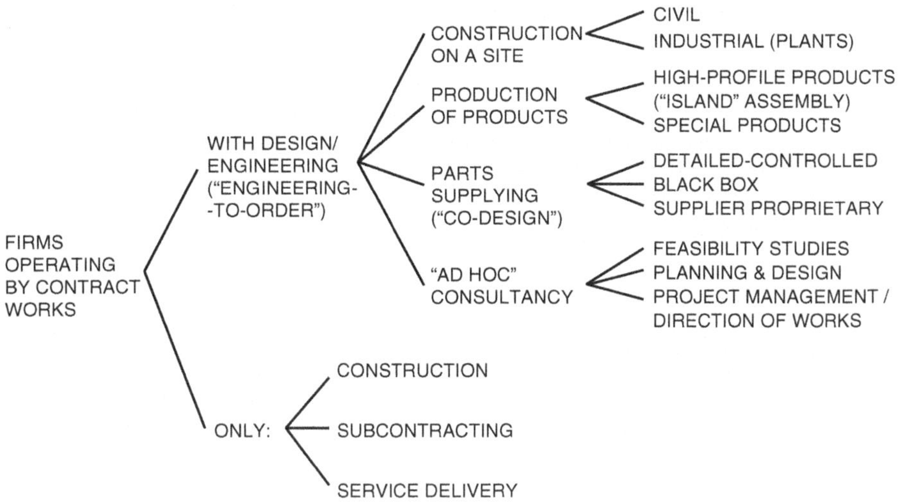

Fig. 4.1 Firms operating by contract works

- *Parts and sub-assemblies*, which are made-to-order by the suppliers, who must *also* design and/or engineer them (*Co-design*)
- *Services* alone, of an engineering type, which must respond in detail to the customer's requirements

Construction companies receive an order to carry out a job (usually, buildings and public infrastructures, such as roads, railways, airports, etc., or equipment and plants for the industrial sector) on a certain site.

Usually the customer/owner only defines one contract, namely that with the *General Contractor*. This person is responsible for the entire construction job, is paid in proportion to the value of the building/plant/work, may carry out part of the work (usually the structural part) and find subcontractors for the remaining jobs (fittings and finishes). The owner can commission an architect/engineer to design the building/plant: the project will determine the technical specifications, which will be listed in the contract put out for competitive bid to find the General Contractor. Alternatively, the General Contractor may directly design the building/plant: this is the case of Design–Build Contractors (as specified, for example, by the Committee on Construction Management of the American Society of Civil Engineers—ASCE).

Over the last dozen years, a new figure has become popular in English-speaking countries—that of the *Construction Manager*. This person/company, although director of the works, does not act as an intermediary: the various enterprises involved stipulate a contract directly with the owner, hence the Construction Manager is no longer a contractor but a *consultant*, and is paid in proportion to the value of the building, without earning the markups received by the General Contractor from the subcontractors.

This type of contract has opened the road to a range of new legal agreements:

(a) The Management Contract (MC), according to which the client delegates the contractual side to the Manager (as in the case of the General Contractor), who in turn receives a fixed percentage (as in the case of a Construction Manager): the Managing Contractor has a greater margin of risk, and only delivers a service, without carrying out any tasks
(b) The Design and Management Contract (DMC), where, as well as providing the competency needed to manage the project, the contractor must also coordinate the designers
(c) The Design, Management and Construction Contract (DMCC), providing for a figure similar to that of the General Contractor; only that in the case of DMCC the path to completing the offer starts from managerial services to which design and construction are added, and not the other way round.

Manufacturing companies can respond to the market demands with different lead times (Fig. 4.2): the response can be more or less immediate (when operating in a *Make-to-Stock* mode—MTS, namely when production is carried out on prevision, and the goods are stocked whilst the company awaits orders), take longer times (including assembly time in the case of *Assemble-to-Order*—ATO, the entire production time in the case of *Make-to-Order*—MTO, and even procurement times when operating in a *Purchase-to-Order* mode—PTO), and reach maximum lead time when the customer also specifically requests product design and engineering (firms operating in an ETO mode).

ETO companies do not possess a *catalogue of products*, but can offer a vast range of custom-made products starting from a basic *catalogue of competencies*; it is therefore possible to distinguish between the following:

1. Single *custom* orders (items from a basic, unofficial catalogue—usually unknown to the customer—are adapted to the specific needs of the client)
2. Repetitive *custom* orders (like the previous ones, but repeated in time—a typical example is subcontractor orders)
3. *Differentiated* orders (always single, where design and engineering play a more important role than in the former two cases—these can be considered as real engineering orders).

The *suppliers* of parts and sub-assemblies may also be requested to design and/or engineer them (*Co-design*). In this case the customer can provide detailed specifications, requiring engineering only (*detailed controlled*), more generic ones (a so-called *black box*, where the property of the drawings either passes to the customer or remains in the hands of the supplier—*consigned drawings* and *approved drawings*, respectively), and even greater margins of freedom (the typical *supplier proprietary*). When the specifications are *entirely* provided by the customer and are already engineered, the supplier can be considered by all means a *simple subcontractor*.

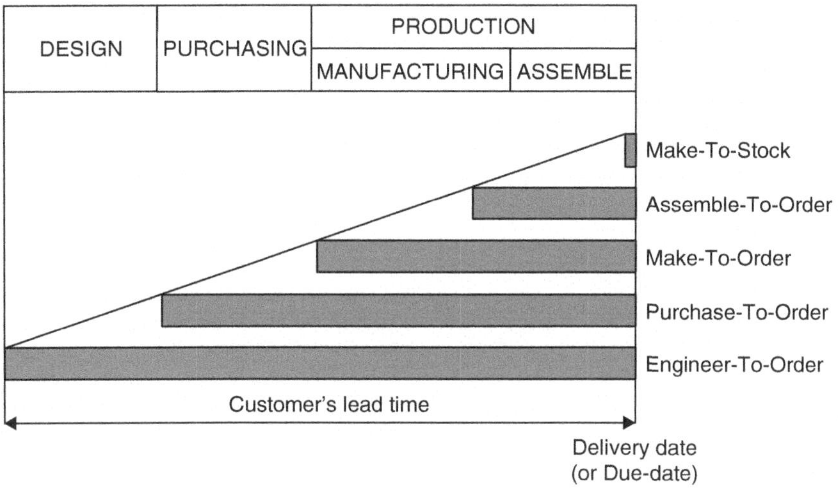

Fig. 4.2 How manufacturing companies can respond to the market demands

ETO companies can also offer engineering services alone. Typical examples are the engineering companies that offer technical and professional services in the fields of industrial plants, infrastructures and engineering work in general.

The services relate to these areas: (1) preliminary studies and investigations (feasibility, project financing, risk evaluation, geological surveys, etc.), (2) planning (in all its stages, from the preliminary ones to execution, including the preparation of documents for RFOs—Request for Offer), (3) management processes (project management in a stricter sense, providing assistance during production/construction, direction of the works, purchases, inspections, routine maintenance, etc.).

These companies can also deliver *turnkey* plants or high-profile infrastructures: in this case, they not only act as consultants, but also as General or Main Contractors, directing and coordinating the activities of all the subcontractors.

4.3 The Life Cycle of Contract Work Orders

The input for ETO companies is the opportunity to take part in a bid or the request to make a proposal, whereas the outputs can be housing estates, industrial plants or high-profile products, all designed ad hoc for the customer and built either in-house or by directing and coordinating the activities of subcontractors.

What characterises these companies is that they not only design the product, but also *manufacture* it; hence, the project is only completed when production ends, not after defining the product's specifications. ETO firms usually operate on an international level, often acting as main contractors, and can therefore work with various subcontractors or establish partnerships; they possess a know-how that ranges from understanding the basic technology of the end product to that relative to the

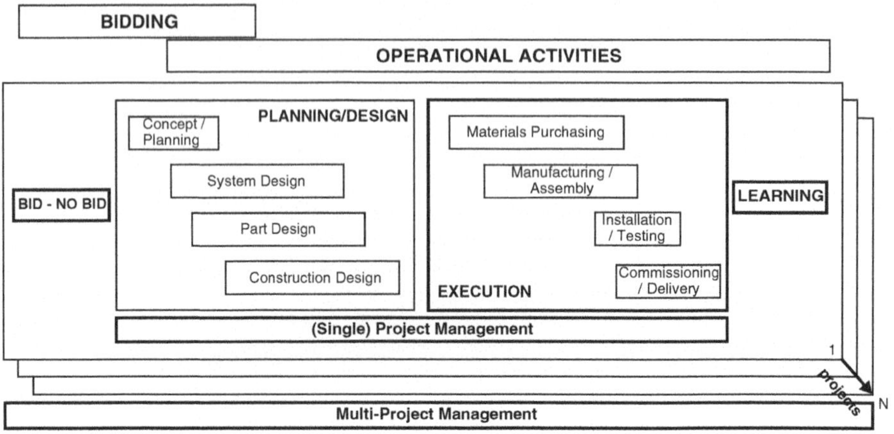

Fig. 4.3 Stages of contract works

product's architecture, as well as having the managerial competency to direct and coordinate the entire project.

Contract jobs can be divided into two stages: (a) *bidding*, which starts with the company's decision to present an offer for a certain job and can end with the stipulation of a contract; (b) *operational activities*, which include all the tasks needed to complete the job. The decision to respond or not to an RFO (*bid–no bid*) is usually supervised by the board of directors, since it is an important decision, which implies a variety of trade-offs and strategic opportunities for the company: visibility on the market, economic and financial convenience, availability of internal and external resources over a medium–long period of time.

These companies usually require Multi-Project Management, because they are often in charge of various projects at the same time, each of which develops in two macrostages: *planning/design* and *execution* (Fig. 4.3).

The *macrostage of planning/design* evolves from the initial concept to the planning of the output, which must respond to the functional characteristics specified by the customer; the overall system and the single parts are then designed, and this stage is followed by plans for its construction, defining the specifications, procedures and methods for setting up and performing operations and maintenance.

The *macrostage of execution* includes various stages: the purchase of materials/components/systems, the manufacturing/assembly activities, finding subcontractors, if needed, installation and/or testing, and delivery—in other words, a set of activities that are carried out in different places according to the type of sector where the firm operates (housing estates, industrial plants or products, etc.).

In all these cases, the project boundaries are well defined, for they are specified in the contract, the stipulation of which may be preceded by a certain degree of negotiation. The contract in fact lists the customer's needs, which are directly dictated by the client. When the contract also describes the designing and operational

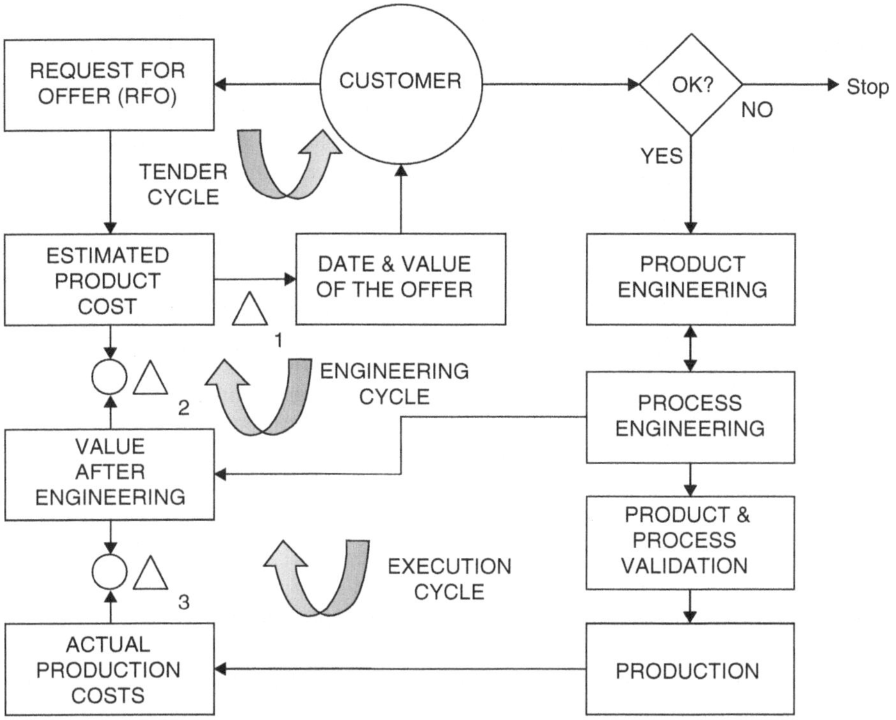

Fig. 4.4 Cost cycles and variations in contract works

activities to be carried out, these are known as *Contract Work Breakdown Structures* (CWBS).

Particular care is devoted to defining the specific objectives (*scope*) of the project, because the company's overall performance depends on the results obtained in the single projects it carries out. There is an office in charge of managing the often intense and continuous relationship with the customer; the firm on the other hand always follows certain internal procedures, based on previous experience, to carry out the commissioned projects and find the solution to the many problems encountered. The existence of standard work breakdown structures (WBS) contributes to remarkable *project economies*.

ETO companies possess a *technological capacity* that allows them to bid for a certain job; the designing activity starts once the company has received the order; in other words, the resources are only activated once a contract is signed. Drawing up a contract is a critical moment, because decisions must be made concerning prices and the date of delivery (Fig. 4.4). Estimating the price can be more or less simple, depending on whether the project can be broken down into standard structures (or *chunks*) that have already been developed and executed in the past. On the other hand, when defining the date of delivery, the company must consider the orders that have already been placed and those that are being negotiated, and whose workload is still undefined.

The importance of this stage is underlined by its other definitions, i.e. *competitive bidding* or *scope management* (the management of contractual goals as a compromise between customer satisfaction and the strategies of the company); in this context, an important figure is that of the *proposal manager* who is responsible for promoting and defining an offer, and for the following stage of contract stipulation.

Once the offer has been accepted, or when the bid has been awarded, it is necessary to engineer both the product and the processes needed to manufacture it. It is in this way that a *costed cycle* is obtained, illustrating the activities to be carried out and their estimated costs, which are examined in an analytical manner: obviously, the costed cycle and the estimate should coincide at the end.

Figure 4.4 illustrates the lifecycle of a work order, from engineering/costing to its execution; as depicted, there can be variances between estimated costs and value of the offer (Δ_1), between the value after engineering and that estimated beforehand (Δ_2), and between the actual manufacturing costs and value after engineering, a parameter indicating production efficiency (Δ_3).

4.4 The Organization of ETO/ECC Companies

To carry out the projects, ETO companies are organized in a matrix structure, resulting from the intersection of the company's permanent structure—consisting of functional units—and various temporary structures that are specifically set up for the project/job (Multi-Project Management).

If the resources of the various departments are allocated full time to an order—once this decision has been made—the traditional conflicts characterising the matrix structure tend to diminish: the resources must answer to the directors of the permanent (functional) structure as regards the achievement of the goals specified in the *internal contract* (concerning the allocation of the resources for a certain order), and the quality of the technical solutions for the project, which must conform to the company's standards and the specifications listed in the (external) contract, namely that made with the customer.

The *project structure,* set up by the project manager and the other managers and coordinators under his/her supervision, is responsible for the achievement of the global objective of a project (the schedule and the technical, qualitative, economical, and financial aspects), using the resources allocated for this purpose, unless the board of directors intervenes, requesting a different allocation of the resources amongst the open orders.

The *permanent functional structure* usually consists of the following offices or departments:

- (Product and process) engineering
- Procurement or purchasing
- Production/construction

- Quality
- Safety
- Management control
- Administration/contract office
- (Multi-)project planning

The latter unit is in charge of formalising and centralising the planning of orders as provided by a Multi-Project Management approach; its members include the company's top executives, who are in charge of defining order priorities and coordinating—for all contracts—the activities of the various offices/departments. This unit is also the source from which the *project culture* spreads throughout the company, a culture that is the key *raison d'être* of the enterprise.

The fundamental, technical expertise is instead in the hands of the engineering unit(s); the term *engineering* is usually preferred to design, because it describes better the activities carried out, which are based on the know-how possessed and calibrated to the specific demands made by the customer.

The *(temporary) project/order structure* includes the following figures (although not all may be present, depending on the type of order):

- The Project Manager, responsible for the work order
- The Contract Administrator, responsible for the contractual relationships with the customer
- The System Engineering Manager, responsible for the architecture and systemic integration of the building/plant/product
- The Planning Manager, responsible for the detailed planning of activities
- The Procurement Manager, responsible for purchases
- The Production Coordinator, responsible for in-house production
- The Expediting Coordinator, responsible for expeditions to a specific building yard
- The Subcontracting Manager, responsible for the jobs carried out by the subcontractors, who in turn are managed by a Coordinator
- The Field/Construction Manager, responsible for the building yard and the construction
- The Resident/Site Manager, in charge of verifying the availability of materials on the building site, coordinating labour and subcontractors, complying with the rules and legislation, and managing the relationships with the local institutions
- Specialist Leaders, directing the task forces working on a specific project
- The Quality Coordinator, ensuring compliance with the technical, quality related and contractual specifications of an order
- The Safety/Reliability Coordinator, in charge of ensuring safety and reliability of the works and the product/plant made to a specific order
- The Commissioning Manager, responsible for ensuring the starting of the plant
- The Project Controller, monitoring the progress of times/costs for a specific order

In companies working to order, project managers must have greater crossfunctionality than that requested for any other project (such as product development or company improvement). Now project managers must use all their organizational, administrative, commercial, financial and legal skills, as well as their technical and managerial know-how. In short, they act as managing directors, though only for a limited period of time, during which they have full ownership of the project.

4.5 From Engineering to Manufacturing and Construction

The specifications made by the customer and stated in the contract are formalised in a *design/engineering WBS*. Since orders usually include manufacturing (or production-by-parts), there is a bidirectional link between this type of WBS and the *manufacturing WBS*.

All the activities revolve around a *Master Production Schedule* (*MPS*) which states the following:

- Which items are to be produced
- In what quantities
- When they are required

The first MPS issued is usually *tentative*. To ascertain its feasibility, it is necessary to check the production capacity requirements, which are considered at an aggregate level, i.e. only the gross requirement of resources per unit of end product (Rough Cut Capacity Planning—RCCP). Once this has been verified the schedule is *approved*.

Three factors are required to formulate a production schedule: (1) materials, (2) machinery/equipment, (3) human resources. Since machinery/equipment and human resources form the productive capacity (and their mix determines the *level of automation*), what must be scheduled and managed are the following:

1. The single items of materials (Material Requirements Planning—MRP)
2. The capacity of each work centre (Capacity Requirements Planning—CRP)

From a logical viewpoint, materials and capacity can be considered on the same level, although it makes no sense to schedule the capacity in the absence of materials. Hence, material requirements are scheduled *beforehand*, and once their presence is ensured, productive capacity can be planned. In the event of overload, workloads can be levelled or material is allowed to queue.

The requirement of materials is planned through the *MRP* technique, except in the case of materials of less value or importance for which the *Re-Order Point* (ROP) technique is used: in this case, the stock level displays a typical sawtooth pattern. MRP calculates the net requirement of materials from the gross needs, subtracting the available stock and the amounts that are being processed and are not yet assigned

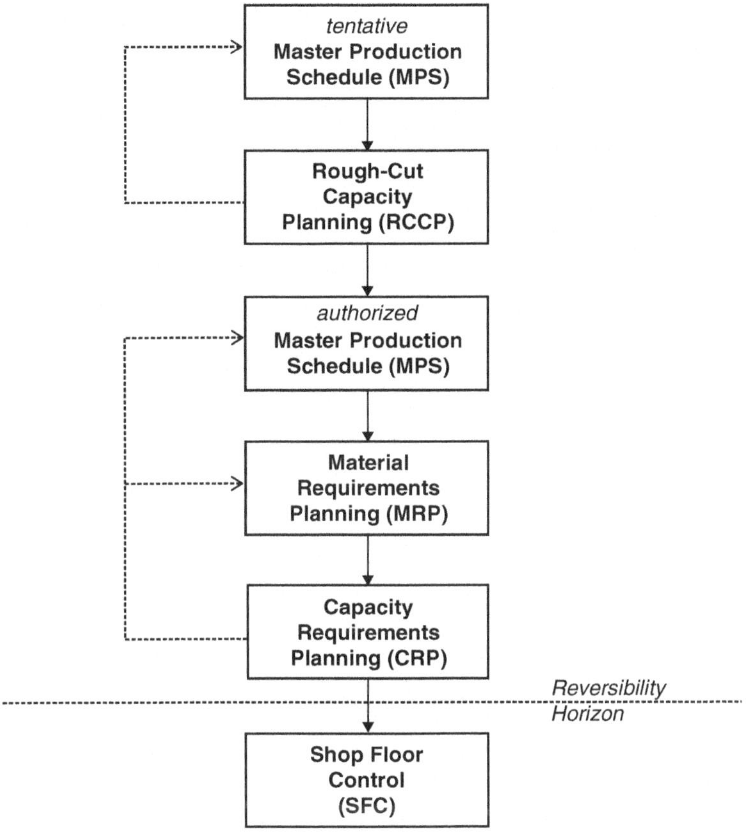

Fig. 4.5 Logical flow for Production Planning & Control

(materials exceeding previous requirements, and released for reasons of minimum lot size); production, assembly or purchase orders are released with an *as late as possible* logic, taking into account the lead times (relative to production or supply).

Figure 4.6 shows an example of MRP calculation. The typical MRP record (one for each item) consists of four rows of information, periodically updated: (1) gross requirements, (2) scheduled receipts (existing replenishment orders not yet assigned), (3) stock availability (from a certain *time zero* on), (4) planned order release (this last row is the result of a calculation). To calculate this value, certain parameters must be specified, such as lot size (if operating by lots), production lead times (the production order of the 40 units in Fig. 4.6 is released two periods before reaching the minimum stock threshold level, as the lead time is of 2 periods), safety stock, etc.

It must be noted that when receding from a future date (i.e. the time of release) towards the present, time availability may be insufficient (i.e. production should have already started): in this case, the MPS must be revised (the dotted feedback lines in Fig. 4.5). If there are parent items, their release primes a gross requirement

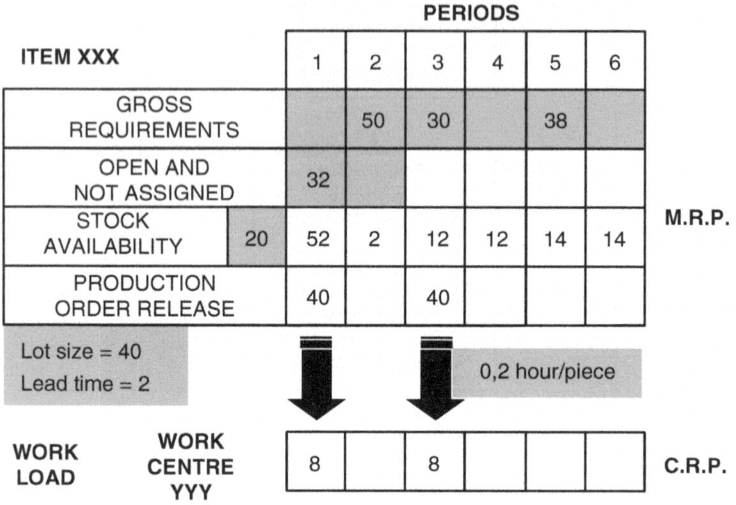

Fig. 4.6 An example of record MRP (above) and CRP (below): the data of the problem are shaded; the other data are the solutions

for child items. These parent–child relationships between the various items are described in the *Bill of Materials* (*BOM*).

In the case of *single contract orders*, it is possible to plan material requirements through an *MRP focusing on critical items*, and in that of *repetitive contract orders*, through an *MRP with planning bills*; these facilities are already present in the more updated MRP modules, which are valid for *cone* and *mushroom/sandglass*-shaped bills of materials.

In cone-shaped BOMs, a limited number of end products, if not only one, are obtained by combining a large number of items (typical examples can be found in the aeronautical industry, shipyards, industrial plants, etc.). Items may be considered *critical* for a variety of reasons: technological (difficulties in their production), managerial (long production or procurement lead times), economical (high unit value) or structural (linked to other critical items). The MRP focusing on critical items gives them priority, and tries to limit subsequent variations (depicted by the dotted, feedback lines in Fig. 4.5) as much as possible.

CRP is used to calculate the load profile for each work centre starting from planned order releases, routing files and open orders. Planned order release is indicated in the last row of the MRP record. These values are translated from physical quantities into workloads or *backlogs* (the lower section of Fig. 4.6), which are generally expressed in hours per work centre. The routing files indicate the work centres and their requirements in terms of labour and equipment to complete the order. The CRP thus identifies the load profiles for released and planned orders in the various work centres. Load *spreading* among the work centres is carried out presuming that the latter have an infinite capacity; in the event of

WORK CENTRE YYY		PERIODS 1	2	3	4	5
C.R.P. PLANNED INPUT		15	15	0	10	10
ACTUAL INPUT		14	13	5	9	17
CUMULATIVE VARIATION		–1	–3	+2	+1	+8
MNG PLANNED OUTPUT		11	11	11	11	11
ACTUAL OUTPUT		8	10	9	11	9
CUMULATIVE VARIATION		–3	–4	–6	–6	–8
ACTUAL BACKLOG	20	26	29	25	23	31

Fig. 4.7 Input/output control for a work centre: data are expressed in man-hours (the darkened cells contain the data; the white ones the solution)

capacity overload, it is essential to level the loads (*finite loading*) and thus modify the MRP and, if necessary, even the MPS (the dotted feedback lines in Fig. 4.5).

To avoid excessive levelling and changes of plans, it is advisable to use certain procedures to *control the input/output* loads on the work centre. Figure 4.7 illustrates an example, indicating the planned and actual inputs and outputs. The input depends on the CRP, the output on managerial decisions and/or constraints due to a constant level of work; the interaction of actual inputs and outputs, starting from an initial backlog (equal to 20 in Fig. 4.7) determines the actual load throughout time (in the example, it passes from 20 to 31 at the end of the fifth period, but the procedure makes it possible to analyse the four possible causes—the rows corresponding to planned/actual and inputs/outputs—and find a remedy).

Up to this stage, planning can still be changed: in other words, only a *what-if* simulation has been made to obtain an Available-to-Promise (APT) date of delivery. With the passage of time, it soon becomes indispensable to *freeze* the decisions that have been made; else it would be impossible to respect the delivery date for the order. This step leads into the *horizon of irreversibility* (Fig. 4.5): all the orders regarding production, material purchasing and processing by outside contractors are thus *placed*. Planning is over, and from this moment that which has been scheduled must be performed. These last stages are governed by *Shop Floor Control* (*SFC*).

SFC—often aided by local software—has the following tasks: (1) to verify and dispatch the definitive production orders (by checking documents, quantities, presence of materials), (2) allocating the resources to the production centres in compliance with the *priority rules* for *scheduling*, namely the definition of the start and end date for each activity, (3) collecting data, monitoring and tracking (i.e. constant traceability of the orders and their progress), (4) if required, interventions and corrective actions (e.g. extra work, alternative cycles, lot breakdown, subcontracting, etc.), (5) order reporting, until it is closed.

These rules of *scheduling priority* concern activities that can be managed autonomously within the given period (for instance, in a day) in accordance to global planning (in other words, respecting the scheduled WBS, MRP and CRP for each period) and leaving the necessary margins for operational flexibility. These are rules such as *minimum slack* (the difference between the deadline and the time needed to complete processing), FIFO (*First-In–First-Out*), etc.

In a multi-project environment, it is also necessary to take into account the *project* (*order*) *priorities;* the most typical are (a) the closest delivery date, (b) the importance of the customer, (c) magnitude of the penalty, (d) the importance of the order in terms of payoff or profitability. The system used to manage them is based on *project core categories*, namely filters that make it possible to classify information according to (a) department/unit/team, (b) contracts, (c) penalties, (d) the project manager.

The earlier stages are executed by single information modules, which are integrated by means of *MPCS* (*Manufacturing Planning and Control Systems*). These systems, originally devised for the management of materials (MRP Systems), are now used to manage capacity, complete the overall planning and carry out SFC. Nowadays they are commonly known as *ERP* (*Enterprise Resource Planning*) *Systems,* because they extend to all the areas of the company—not only production—including trade, general and analytical accountancy, treasury, financial management, administration, personnel, and so forth. The latest step of this evolution is the *NERP* (*Networked Enterprise Resource Planning*) *System,* which integrates the planning systems of different enterprises (suppliers, customers, subsidiaries) by means of Web technologies (Extranet, XML language, etc.).

Product Design 5

5.1 Delivering a New Product

Rather than merely referring to *design*, nowadays it is preferable to use the expression *New Product Development* (NPD—Clark and Fujimoto 1991): this must not only process design, but also develop, update and integrate the product. For this purpose, it can also use parts of pre-existing product projects, or convert them from similar ones manufactured by competitors. In other words, these are designs characterized by a remarkable degree of *carry-over* (i.e. the percentage, in value, of the parts of a previous project that have been adapted and used for the new one).

There are three fundamental reasons for this:

- The high level of technological differentiation, and its constant evolution, increases the risk of investing in completely new products, namely those with low carry-over rates.
- The customers, ever more exacting and oriented towards customisation, and the market situation in general, recommend the development of new products sharing a common platform and not ones that are radically new.
- The strong international rivalry between many competitors and the adaptation to different local markets make the systematic use of carry-over and common parts practically inevitable when designing new products.

Today, there is a strong need to innovate the management of the product design/development process, and the term *Lean Design*, derived from the concept of Lean Production (Womack et al. 1990), has become popular to describe a global objective of design improvement.

An international research project carried out by the MIT and coordinated by Cusumano and Nobeoka (1998) documents the need to *slim down* designs that are too *bulky* in terms of both unnecessary parts and features, but especially in terms of efforts and resources invested in the single, often uncoordinated, projects.

S. Tonchia, *Industrial Project Management*, Management for Professionals,
https://doi.org/10.1007/978-3-662-56328-1_5

Table 5.1 Comparison between traditional and lean design

Traditional design	Lean design
Slow introduction of new models	Frequent introduction of new models
Slow expansion of product lines	Rapid expansion of product lines
Long development times	Short development times
Few innovations and of a radical type	Numerous innovations and of incremental type
Long testing times	First-pass quality design
Rigidity in product/process changes	Flexibility in product/process changes
Large information stock between development stages	Small information stock between development stages
Slow rates of information transmission from upstream to downstream	Fast rates of information transmission from upstream to downstream
Feedback and problem-solving through milestones	Feed-forward and early problem-solving
Research and Development push	Market demand pull
Specialist designers	Job enrichment and enlargement for designers
Lightweight project managers (coordinators)	Heavyweight project managers
Special departments	Multi-function teams
Design is mostly internal	Co-design with the suppliers
Sequence of development stages	Shorter and simultaneous development stages

In short, Lean Design can be achieved by shifting the focus from single-project to multi-project management (by means of product platforms and modularity), in the context of a precise portfolio strategy, featuring flexible team organisation, a high degree of concurrent engineering, frequent exchange of information and co-design with the suppliers, so as to reduce expenses while preserving and improving product integrity (Bowen et al. 1994; Brown and Eisenhardt 1995). Internal and external customer satisfaction can be thus achieved by employing such procedures as the Variety Reduction Program (VRP), Quality Function Deployment (QFD), Failure Mode and Effect Analysis (FMEA) and so forth.

Table 5.1 illustrates the differences between conventional design and lean design.

5.2 The Stages of NPD

The product innovation process that starts from the *concept idea* and leads to the launch of production (by delivering product and manufacturing specifications so as to start operations and execute custom or stock orders) can be divided into various stages. These can be summarized as follows:

1. The starting point is the definition of the *concept idea* (a mix of technological inputs and market needs/receptivity). The concept of a product is contained in a small number of key elements, and only at a later stage will the product features and performances (i.e. those described in the catalogue) be defined; the concept

also describes the price range of the product, the way to market it and the image that the company wants to project. It can be stated that a product is successful because the concept idea is good, just like a good portfolio of products is one that contains a good mix of innovative concepts that respond to the demands of the market.

2. It is impossible to deduce design specifications—product size and features—directly from the concept idea: an intermediate stage is required, where product functionality is examined in relation to the concept expressed. The output of this stage, known as *product planning*, is a Product Function Structure (PFS), namely a tree or breakdown structure describing the general features and functions that the product must have.
3. Stage 2 is followed by *product design*, which details the main *deliverables* of the project, i.e. the product specifications. These include all the physical-dimensional features of the product and the Bill of Materials (BOM). The level of detail must be such as to ensure identical products, even when manufactured by different persons or groups; the specifications, in their final version, will determine the reference parameters for the measurement of product conformity, once production has started.
4. The decisions made in the previous stages are not definitive and must be verified in practice, by means of *engineering*. While design can be carried out independently of the availability of machinery and plants, engineering requires at least laboratories and/or testing units or pilot runs. Engineering is aimed at testing the effectiveness of the solutions identified for the product (type of materials, geometry of the different parts, assembly, etc.)—*product engineering*—and in a second stage, verifying work instructions and analysing manufacturing problems—*process engineering*—also by means of prototypes (*partial* products, in the sense that they only have certain functions).
5. Engineering and even prototyping are still insufficient for defining the final, definitive product and process specifications, as they are chiefly aimed at assessing the effectiveness of the product rather than the manufacturing process. Another stage is required, *industrialisation*, which leads to production on a vaster (*industrial*) scale at lower costs and with saturation of the resources (which can be translated into high productivity).
6. It is now possible to start mass production, which may be preceded by a *pre-series*.

All these stages should benefit from continuous feedback cycles, keeping in mind the golden rule of design: *the later the stage at which changes are made, the more expensive they become* (Fig. 5.1).

Figure 5.2 illustrates the levels of resources and attention required by the various stages of project development, and specifies where it is still possible to make changes.

Design should also take into account the environment (*Green Design*), because public opinion is becoming increasingly sensitive to this issue, as reflected by the current laws. Enterprises must therefore strive to design and engineer environmental-friendly products because green design can become a competitive weapon to

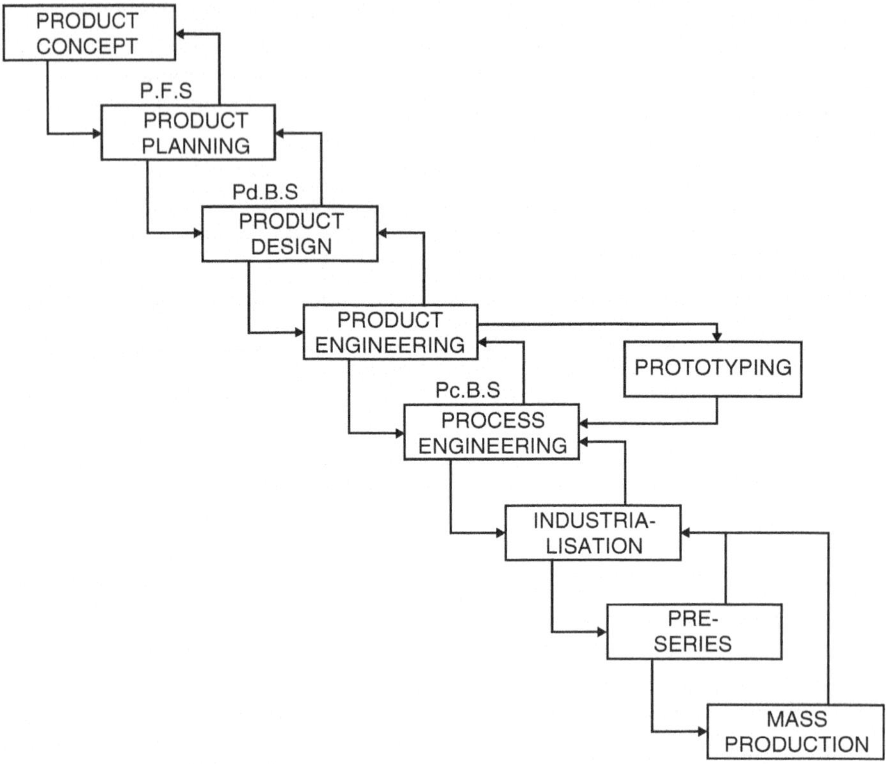

Fig. 5.1 Product development stages

improve the company's image. In practice, the firm must design products, and also their manufacturing processes, so that they are *eco-efficient* in the use of resources and respect the environment for their entire life cycle (for this reason, we must talk about *Life-Cycle Design*).

A Green/Life-Cycle Design implies lengthening the project's lead times, because they must also include disassembly and recycling:

1. *Design for Long Life* (*DFL*) is aimed at easily substituting and/or updating the components of the product, thanks to the latter's simple, modular architecture featuring standard interfaces.
2. *Design for Disassembly* (*DFD*) aims at simplifying disassembly by reducing the number of connections, making them more accessible and privileging unidirectional disassembly, hence reducing the number of operations required.
3. *Design for Recycling* (*DFR*) is aimed at recycling materials and parts, so that they can be used for other products. The choice of materials becomes fundamental: these must be of a limited number, easy to identify and group together so as to avoid contamination. It is also essential to provide clear instructions for their recycling.

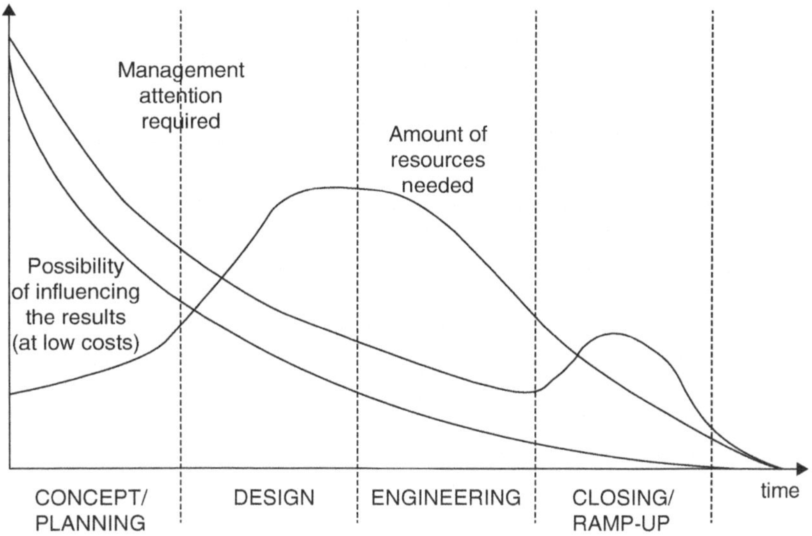

Fig. 5.2 Resources and project management

5.3 Managing NPD

Product design and development are changing dramatically—thanks to Information and Communication Technologies (ICT), which aid the engineering activities of designers and technicians. These instruments have proved to be particularly useful in four fields:

- *Design and experimentation*: they ensure greater capacity at more limited costs and times (CAX instruments such as CAD, CAE, CAM, CAPP, etc., as well as specific technologies for Rapid and Virtual Prototyping).
- *Management of technical documentation*: these systems are used to store, transfer and retrieve technical data. They ensure coherence and an efficient reuse, and avoiding duplication (PDM systems).
- *Communication*: in this case, the systems simplify teamwork, even when the team is formed by members belonging to different companies, who co-design at a distance (*Groupware*).
- *Project management*, i.e. tools helping to plan and control the project and integrate the project portfolio (Project Management software).

Following is a brief description of these four classes of tools. The most popular systems for *designing and experimenting* are as follows:

– *CAD—Computer-Aided Design*, which produces a model of the product (two- or three-dimensional—2D and 3D, respectively)

- *CAE—Computer-Aided Engineering*, which simulates how the product works (FEM—Finite Elements Method—to verify the mechanical resistance of the various parts, CFD—Computerised Fluid Dynamics—to test the behaviour of products when immersed in a fluid, etc.)
- *CAM—Computer-Aided Manufacturing*, namely computerised systems to plan, manage and control manufacturing operations (by translating CAD models into *part programs* for the machines, and defining the necessary tools, tests, etc.)
- *CAPP—Computer-Aided Process Planning*, namely software to plan the manufacturing processes (work cycles, times and procedures, etc.)

The *management of technical documentation* can be aided by specific *Product Data Management* (*PDM*) software, devised to manage, archive and retrieve all information relative to a product (BOMs, drawings, manufacturing and assembly plans, catalogues, etc.); this is done in a dynamic manner, so that it is possible to keep track of all project and product improvements.

PDM systems integrate upstream with CAD, and downstream with Enterprise Resource Planning (ERP) systems, used to manage production. Data interrogation is based on the *hypercube* and *drill-down* facilities (which can track any dimension/category of data and explode it to reveal the details) that are typical of Decision Support Systems and, more in general, of Business Intelligence technology.

Communication is aided by Groupware instruments, such as Lotus Notes, which helps promote communication and exchange among work groups.

Communication among the teams in charge of design can also be based on e-mails and networks in general, such as Intranet, Extranet or Internet. E-mails are mostly used to exchange data concerning task assignment and activity updates (using for example Microsoft Project and Outlook). The Web, on the other hand, offers a wider range of possibilities:

- The members of the team can visualise diagrams relative to their activities.
- They can access information relative to the entire project.
- They can also create new activities and send them to the project manager so that they can be included in the project file.
- The project manager can summarise the reports of the team members in a single, roundup document.
- The project manager can automatically accept updates relative to the project from certain team members.
- The members of the team can delegate the activities assigned to them by the project manager, who will therefore only interact with a limited number of *project team leaders*.

From a technical viewpoint, it is necessary to install a *groupware* on a server; the project manager has the Project Management software installed on his/her computer, whereas the team members can access the homepage of the groupware by means of a user ID.

The administrator of the groupware (usually, the project manager) controls a series of settings relative to protection (hence, deciding who is authorised to visualise and modify certain documents). Accounts can have different properties; it is also possible to choose whether to show the updates immediately or in a second moment. Usually there is also the possibility of managing delegated work, so the person who delegates a task can verify the data, before sending them off to the project manager.

Support to project management is given by special software—also known as Project Management Information System—which is defined as a *set of documents and procedures that help define, plan and execute projects in a firm*. Most software packages on the market include a database, a Word/sheet programme, an activity schedule, graphic interfaces based on Gantt charts and network diagrams and a generator of reports. The more advanced programmes may also include the following:

- Activity scheduling integrated with resource management
- Algorithms to spread the workloads among the resources
- Integrated management of various projects (multi-project environment)
- Integration with the company's accounts and budget
- Risk analysis and management
- Relational database, on-line accessibility to data on local Webs or the Internet
- Advanced graphics, with the possibility of managing breakdown structures (such as WBS) and prepare presentations such as those typical of business management
- Web-based integration with suppliers and subcontractors
- Great customization opportunities, including the preparation of documents having pre-established formats (as, for example, in the case of civil orders)

The advantages of these software packages, in addition to ease of calculation (early and late dates, slack, cost/workloads, etc.) are as follows:

- The integration of the projects in a context characterized by shared resources and simultaneous activities
- The sharing of the same formats and procedures, which enables to establish a *common language*
- Rationality and transparency of the planning and control activities
- Easy, swift communication and inter-team learning

Design management does not only exploit ICT instruments. Other *managerial* procedures are used, which can also be applied to other fields and products and—in a company-wide framework—can involve, in addition to the design office, other functional units of the company. According to who is involved, these techniques can be classified as follows (Tonchia et al. 1998) (Fig. 5.3):

1. Techniques for the involvement of suppliers and the purchase office (*Co-Design—COD/Early Supplier Involvement—ESI*)

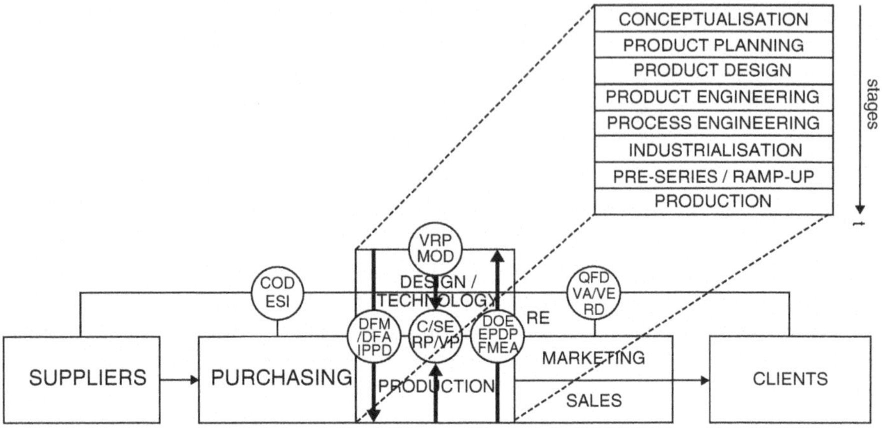

Fig. 5.3 Functional units and product development techniques

2. Techniques used exclusively by the design department (*Variety Reduction Program—VRP* and *Modularization—MOD*)
3. Techniques involving both design and production, and aimed at coordinating and integrating the efforts of both (*Design for Manufacturability and Assembly—DFM/DFA* and *Integrated Product-Process Design—IPPD*)
4. Techniques involving design and production, targeted at reducing the duration of the project by carrying out as many contemporaneous activities as possible (*Concurrent/Simultaneous Engineering—CS/SE* and *Rapid/Virtual Prototyping—RP/VP*)
5. Techniques involving design and production, aimed at revising decisions and analysing problems in advance (*Reverse Engineering—RE*: *Design of Experiments—DOE*, *Early Problem Detector Prototyping—EPDP*, *Failure Mode & Effect Analysis—FMEA*)
6. Techniques for the involvement of customers and the marketing office in product design (Quality Function Deployment—QFD, Robust Design—RD and Value Analysis/Value Engineering—VA/VE).

Since these techniques are mainly aimed at achieving the goals of the project, they will be described in greater detail in Chap. 9 devoted to Project Quality Management. It is important that they go hand in hand with other techniques, such those for Project Time Management (e.g. CPM and PERT) and Project Cost Management (e.g. the Earned Value Method).

5.4 Platform Management

Henderson and Clark (1990) classify product innovation according to both the type of improvement/changes made on the *core concept* of the product and the preservation or not of its components. They therefore distinguish between (Table 5.2) *radical innovation* (leading to a new *dominant product*, stable in time), *incremental innovation* (perfecting the dominant product, although its architecture is preserved), *modular innovation* (which modifies the concept, but not the architecture of a product) and *architectural innovation* (which reconfigures the combinations and interfaces of the modules, giving rise to a new architecture, without modifying the concept of the product).

Product innovation usually leads to four different types of projects, which should alternate harmoniously throughout time:

1. Projects for the development of platforms
2. Projects for *derivatives*
3. Projects for *shelf innovation*
4. Projects for the development of core components

A project that develops a *product platform* gives rise to a new architecture from which the single products are derived. *Product architecture* describes the various functions of the product, their attribution to the parts forming the product, and the interactions between the different parts. In other words, the architecture of a product is the relationship between PFS and PdBS. It is also possible to define it as a map of functional elements on *chunks* interacting with each other (Ulrich and Eppinger 1994). It can be stated that the performances of a product, its cost, manufacturability, and the extent of a range of products largely depend on its architecture. A substantial increase in a product's performances or significant reductions in its price can be usually ascribed to a new architecture (for instance, front-wheel drive in cars or the introduction of scooters on the motorbike market).

There is no univocal definition of platform in the literature: Wheelwright and Clark (1992) consider it as a set of architectural solutions from which all products are derived and developed; Cusumano and Nobeoka (1998) define it as a critical, expensive sub-system that defines the performances of a product; Meyer and Utterback (1993) observe it from a more structural point of view, and consider it

Table 5.2 Product innovation (Source: Henderson and Clark 1990)

		Core concept	
		Improved	Changed
Links between concept/ components	Unvaried	Incremental innovation	Modular innovation
	Modified	Architectural innovation	Radical innovation

as a set of project solutions (core concepts, procedures and rules for the integration of the various sub-assemblies) and parts common to and shared by a *family of products*.

The latter definition poses two important questions:

- The *structural* approach distinguishes between *integral* and *modular architecture*, which will be further discussed in a subsequent part of this chapter.
- The sharing of solutions and parts poses the problem of clearly defining what *families of products* are.

Whilst in the past, families were defined as such on the grounds of commonality (technical in the case of product parts, technological in that of manufacturing processes); in recent years, it is more common to consider the analogies on the market, in specific market segments and in the needs expressed by the customers. A family hence becomes a set of products that share two factors ensuring efficient and effective production, delivery and services: a *network of market applications* and the core technology, represented by the product platform (Meyer and Lehnerd 1997).

In the automobile sector for instance, according to Cusumano and Nobeoka (1998), a platform consists of (a) underbody/floor panels, (b) suspensions/brakes, (c) rocker panels, (d) firewall (for the engine); the macromodules assembled onto the *super-module* of the platform are (1) the upper body (body panels, doors, glass, etc.), (2) interiors (dashboard, seats, etc.), (3) the power train (engine, transmission, etc.), (4) electronic equipment.

The projects for the development of new platforms, entrusted to platform managers, have longer life cycles than those of the single products, because their obsolescence is not linked to changes in the tastes/demands of the customers or to technological innovations of limited significance, but rather to the establishment of new standards and innovative technologies that cannot be incorporated into the product by simply adding new parts to it (Meyer 1997).

Starting from platforms, the enterprises can produce a variety of products/models (known as *derivatives*), so as to meet the market demand, in all its variety and variability, and respond to the technological evolution that has meanwhile occurred (Tatikonda 1999). This derivation has more limited costs and requires less time than the new, single projects.

Derivatives can have different features: they can improve the performances of a previous version by adding new functionalities and extend a line of products so as to offer new options; they can be custom-made products targeted at specific market niches or allow significant cost/price reductions—thanks to a revision/simplification of the initial project and/or the exploitation of more efficient technology, etc. However, the actual potential of a platform must always be kept in mind, because on one hand there is the risk of continuing with incremental innovations that in the long run are not appreciated by the market, and on the other—in order to differentiate the products as much as possible—there is the risk of sustaining greater costs than those needed to design a new platform.

Projects regarding the *innovation of components*, on the other hand, serve to develop new parts and sub-assemblies that are not necessarily used immediately but can be shelved (hence the name *shelf innovation*) for future models or platforms. Hewlett Packard terms this type of approach *pizza bin*, because the innovative parts and technologies are developed, tested and made available for new projects, as if they were the ingredients for a pizza.

The fourth type of projects—namely, those aimed at developing *key components*—is a hybrid case, because these parts are designed for shelf purposes, but can also become the key feature of a platform, in which case they will be known as *core products*. These components are the parts or sub-assemblies of the product that have a well-defined function and are at the heart of a stable, competitive advantage. Although affected by changes in the market demands, their obsolescence mostly depends on technological evolution.

Examples are the optical equipment produced by Canon, the small Black&Decker electric motors, etc. It is also important to distinguish between the market share held by the producers of these key components—the so-called *Original Equipment Manufacturers* (OEM)—and that owned by commercial brands: for instance, the Matsushita group retains 45% of the market of key components for video recorders, a larger share than that of its brands, Panasonic and JVC, which is 20%.

Core components/products are the link between end products and *core competencies*. A core product is the tangible result of a core competency: the perceptibility of a competency is ensured by this product (or part, or sub-assembly), more than the actual end product. Examples are the miniaturization ability of Sony, the skill needed to integrate microelectronics and precision mechanics, typical of Casio, or the extra-thin films produced by 3M, etc.

In brief, a product platform:

- Is a core design, relatively stable in time, on which a company (or a joint venture) invests significant resources and from which it is possible to develop a number of different products, sharing the same concept, design, manufacturing and assembly procedures
- Has one or more key sub-systems that characterise all its products and ensure the competitive advantage of the company for a long period of time
- In combination with distribution and commercial synergies, gives rise to a family of products that can exploit one platform for various generations of products, improving and increasing their range (hence the platform can be considered as the genetic imprint common to a family of products)
- It has a remarkable impact both on the company's strategies and performances and on organization by projects

As mentioned previously, the horizontal and vertical expansion of a family—in other words, the products developed contemporaneously or introduced throughout time, respectively—poses the problem of a correct balance between product integrity (*integral architecture*) and modularity (*modular architecture*).

In *modular architecture*, each *chunk* (a physical block, which in this case is also termed *module*) implements one or few functions; there is direct mapping, and the interactions between the modules are well defined, simple and if possible, standard.

This type of architecture is also known as *open design*. Personal computers are a typical example of modular architecture: although the structure has remained more or less unchanged throughout the years (featuring a hard disk, various cards, drives, peripheral ports, etc.), the performances of the single parts evolve incessantly.

In *integral architecture*, on the other hand, each chunk has numerous functions and very complex interactions/interfaces with the other ones: this type of architecture is therefore also known as *closed design*.

The integrity of a product should be preserved in all projects relative to product innovation. It is possible to distinguish between the following:

- *Internal integrity*, referring to the coherence between the structure and the function of a product; in other words, it can be defined as the coherence of the product's architecture.
- *External integrity*, referring to the conformity of the product with the expectations and perceptions of the customers.

The integrity of a product ensures that it does not lose its personality and distinctive features. Its advantages are complementary to the risks linked to modularization: excessive modularization risks making all products too similar to each other, and may give rise to displeasing aesthetics (as in the case of certain dashboards, whose basic models have all the button lodgements but lack various buttons), makes it easier for the competitors to imitate the design, and can reduce the product's performances (Clark and Fujimoto 1990).

On the other hand, modularization offers unquestionable advantages:

- In product innovation—product management is more rapid and simple, because the design of a module can be delegated to relatively autonomous teams or even to the suppliers; the costs are lower because a new technology affecting only one module can be easily adopted without having to redesign the whole product; finally, it is possible to use the module for other products, current or future.
- In the manufacturing process—its standardization ensures economies of scale, with an increase in the production volumes of the single items; there is a more rapid response to the market, because the standard modules can be stored and the end product assembled on order; finally, the amount of stock is reduced, because the demand for the single modules is less uncertain than that for the end products.
- For the customer—there is a greater variety and adaptability to the specific needs of the customer, thanks to the possibility of combining the modules; the product can be easily improved and expanded after its purchase by updating only certain modules or adding new ones; maintenance is more simple, because only the faulty/damaged module needs to be changed.

In brief, it would be better to modularize those parts of the architecture that are less visible to the customers, whereas the integrity of the interface between product and consumer should be preserved as much as possible (external integrity). This is the reason why *a platform, considered as a core sub-system (or super-module) can represent a valid compromise between closed (integral) and open (modular) architecture*, as it has the advantages of coherence and flexibility deriving from the two types of architecture, respectively. Hence, there is a platform/super-module that forms the *integral zone* of a product, whereas the modules produced outside the platform form the *modular zone* of the product. The latter modules can vary in time so as to add other functions to meet the demands of the market (Meyer and Lehnerd 1997).

The use of platforms makes it possible to limit the internal resources devoted to product development, while increasing development projects; moreover, the time-to-market for each project can be reduced, and since a platform remains stable for a relatively long time, costs can be reduced. Finally, if the platform is defined before the derivatives, the manufacturing processes can also be better defined to the advantage of manufacturability, quality and the organization and scheduling of production.

References

Bowen, H. K., Clark, K. B., Holloway, C. A., & Wheelwright, S. C. (Eds.). (1994). *The perpetual enterprise machine: Seven keys to corporate renewal through successful product and process development*. New York: Oxford University Press.

Brown, S., & Eisenhardt, K. M. (1995). Product development: Past research, present findings and future directions. *Academy of Management Review, 20*(2), 343–378.

Clark, K. B., & Fujimoto, T. (1990, November/December). The power of product integrity. *Harvard Business Review*, 107–118.

Clark, K. B., & Fujimoto, T. (1991). *Product development performance: Strategy, organization and management in the world auto industry*. Boston, MA: Harvard Business School Press.

Cusumano, M. A., & Nobeoka, K. (1998). *Thinking beyond lean: How multi-project management is transforming product development at Toyota and other companies*. New York: Free Press.

Henderson, R. M., & Clark, K. B. (1990). Architectural innovation: The reconfiguration of existing product technologies and the failure of established firms. *Administrative Science Quarterly, 35*, 9–30.

Meyer, M. H. (1997). Revitalize your product lines through continuous platform renewal. *Research Technology Management, 40*(2), 17–28.

Meyer, M. H., & Lehnerd, A. P. (1997). *The power of product platforms*. New York: Free Press.

Meyer, M. H., & Utterback, J. M. (1993). The product family and the dynamics of core capabilities. *Sloan Management Review, 34*(3), 29–47.

Tatikonda, M. V. (1999). An empirical study of platform and derivative product development projects. *Journal of Product Innovation Management, 16*, 3–26.

Tonchia, S., De Toni, A., & Nassimbeni, G. (1998). Innovation in product development within the electronics industry. *Technovation – The International Journal of Technical Innovation & Entrepreneurship, 19*, 71–80.

Ulrich, K., & Eppinger, S. D. (1994). *Product design and development*. New York: McGraw-Hill.

Wheelwright, S. C., & Clark, K. B. (1992). *Revolutionazing product development: Quantum leaps in speed, efficiency and quality*. Free Press: New York.

Womack, J. P., Jones, D. T., & Roos, D. (1990). *The machine that changed the world*. New York: Rawson Associates.

Service Design

6

6.1 The Quality of Service

Despite the number and variety of opinions concerning *quality* expressed both in the literature and at a consultancy level, all agree on four issues:

1. Quality involves the *entire* company, management included, which must endorse quality policies and ensure the necessary *commitment*.
2. The focus must be set on *customer satisfaction* (rather than mere compliance with the standards), a customer who can also be *internal* (i.e. inside the company).
3. The importance of *continuous improvement* (a company must also improve during the period between the issuing of two official specifications).
4. The reduction/elimination of variance in the manufacturing processes is a source of quality (the qualitative level depending on the design process, which should have released specifications ensuring conformity of quality).

Process management can be considered pivotal for quality management in the new production contexts, and acts as a powerful catalyst for the deployment of quality programmes.

Figure 6.1 illustrates how the concept of quality has evolved, and the current relationship existing between quality and process management:

- From quality based on standards, we have passed to quality intended as customer satisfaction (quality as a means of meeting customer requirements, by supplying an adequate level of service too: quality coincides with *value*, which in turn is decreed by the customer). All *processes serve to add value*, and must consistently focus on the customer (both final = external, and internal).
- From quality control alone, we have reached a *trilogy* consisting of *planning* (which must meet increasingly severe standards), *control* (along the entire production line, carried out in real time and using sophisticated statistical procedures)

S. Tonchia, *Industrial Project Management*, Management for Professionals,
https://doi.org/10.1007/978-3-662-56328-1_6

Fig. 6.1 Quality and process management

and *improvement* (by training, coaching and accepting suggestions). Improvement by learning can indeed be considered as a key objective of these processes.

- Functional responsibility has become *shared* responsibility, integrating quality management even during the first stages of design and development (hence, the concept of *upstream quality management*). In process management, the existence of internal customer/supplier relationships and teamwork is at the root of shared responsibility.
- The *inspection* approach, featuring measurements, controls and reprocessing, with all its costs, has been replaced by a *preventive* approach, aimed at achieving *first-pass quality* and reducing/eliminating the aforementioned costs before they actually occur.

To manage, it is necessary to measure, but in order to measure, it is essential to identify the *numerous* performance dimensions of quality (Tonchia et al. 1995). In the case of *services*, a particularly useful tool is *ServQual* (Zeithaml et al. 1990), which decomposes the perception of quality service into five constructs, two of which can be further divided:

1. *Tangibles* (physical facilities, equipment, staff appearance)
2. *Reliability* (the ability to perform services as promised or described)
3. *Responsiveness* (willingness to help and respond to customer needs)
4. *Assurance*, comprising the following:
   (a) *Competency* (having the necessary knowledge and ability to deliver the service)
   (b) *Courtesy* (pleasant front office staff, who must be polite and show respect and consideration towards the customers)

(c) *Credibility* (both of the company—in terms of image/reputation—and of the front office staff)
(d) *Security* (the customer perceives no risk, and can also count on a confidential relationship)

5. *Empathy, described as the attention devoted by the company to assisting its clients by means of the following:*
   (a) *Accessibility* (the ease with which the front office can be contacted or reached, which includes where it is placed, opening hours and waiting times)
   (b) *Communication* (constant, complete, understandable, suitable for all types of customers)
   (c) *Understanding the client* (i.e. doing one's best to understand the specific needs and expectations of the customers, as well as their habits)

6.2 Designing Service Content

When designing the contents of a service, it is of utmost importance to keep in mind the key features of a service, which distinguish it from all other products:

- *Intangibility* (the immaterialness of the product delivered)
- *Simultaneousness* between production and consumption
- The *participation* of the customer in service production/delivery

These characteristics imply specific *problems* in the management of the service, which must be taken into account *by the firm* from the very start. A few considerations are as follows:

1. Services cannot be stored, and it is therefore impossible to use warehouses or similar storage facilities to uncouple demand and supply.
2. It is more complicated to quantify production costs and define a price policy, given the few materials required and the variability in the customers' demands, to which the enterprise must readily adapt.
3. Services are difficult to standardise, so it is more arduous to achieve economies of scale.
4. The front office staff plays a key role.
5. There are problems linked to service delivery, which requires closeness to the customer.
6. It is complicated and expensive to make the characteristics of the service known.

On the other hand, also the *customer/user* faces various problems, which however may prove to be an advantage for the enterprise:

1. The specific features of a service increase the customer's trust in the company supplying the service (unlike material goods, comparisons and evaluations can only be made ex-post).
2. It is difficult to remember/describe the service received (one tends to focus on the details only).
3. The customers are in the hands of the supplier, and because they play an active role in the process, if they were to change supplier they would have to make a huge effort to adapt to the new services.

When designing a service, it is necessary to take into account the constructs or dimensions mentioned in the previous section, considering the following:

- The *needs* driving the demand for a service, which should be expressed and described in the customer's language
- The *factors that influence customer expectations*, which in turn depend on the following variables:
  - Individual ones (personal needs and emotional variables, which are affected by the client's personality, past experiences and how they have been judged); emotional variables include rational considerations (pertaining to the logical sphere of the individual), motivations, attitudes, behaviours, values and beliefs
  - Environmental variables (social and cultural aspects, *word of mouth*, competitors' services, image and communication, etc.)
  - Company-related variables (communication with the customer, image, price)
- The so-called perception filters (perceiving risks or, instead, having the feeling of being in control; the company's reputation), which can influence the opinion concerning a service

Some *psychological* rules must also be kept in mind:

- Satisfaction is the sum of numerous, minor aspects.
- Failures or mistakes have a much greater impact than positive outputs.
- Unlike the case of material goods, each customer is interested in a different aspect of the service (*every fish has its bait*).

6.3 Designing Service Delivery

Once the contents of a service have been defined, it is necessary to design service delivery, defining how it is supplied, as well as the necessary resources.

The *ways* through which a service is rendered are illustrated in detailed flowcharts, describing the preparatory, delivery and post-delivery stages. The first stage refers to the activities carried out by the customer to retrieve information and book the service (if needed), and also includes waiting times; the second one is that during which the actual needs of the customer are fulfilled; the last stage can include the establishment of a stable relationship with the client, receiving feedbacks and managing disservice.

The *resources* needed to deliver a service are human ones, infrastructures (such as offices), means of transportation/tools (vectors, dispensers, testing equipment), and other equipment (computers, etc.). Human resources are either allocated to the *front office* (hence, in direct contact with the customer) or the *back office* (which is invisible to the customer).

As mentioned previously, the front office plays a key role, due to the simultaneousness of service production and consumption and the active participation of the customer. It is therefore necessary to study and define the following types of *interactions*:

- *Primary* interactions between the front office staff and customers, as well as with the physical facilities
- *Concomitant* ones, among the customers using the service
- Internal, i.e. both inside the branch and with the company's head office

Also in the case of services, it is necessary to define the *times and methods*, which should do the following:

- Determine the *suitable time* (considering the statistically minimum time, the average adequate time, etc.)
- Separate standard operations from those having uncertain duration
- Distinguish between individual (active) time and collective (*passive*) time spent by the customer either retrieving information or waiting for a service

When designing, it is also necessary to take into account the time required to deal with possible disservices, considering how rapidly the company responds to the complaints and how long it takes to solve the problem and recover the customer (Fig. 6.2).

6.4 Process-Based Service Improvement

The Customer Satisfaction Index (CSI)—which is the weighed sum of how service is judged—measures customer satisfaction according to how the perception of the delivered service matches the expectations: the smaller the gap, the greater the satisfaction of the customer.

To improve services, it is advisable to study this gap. This type of analysis proceeds backwards to identify the causes, which in turn correspond to other gaps (or mismatches).

There are eight different types of gaps (Fig. 6.3), each of which should be analysed in relation to service dimensions (Tonchia and De Toni 2004—modified from Zeithaml et al. 1990) and detailed through cause–effect diagrams, similar to those used for quality management:

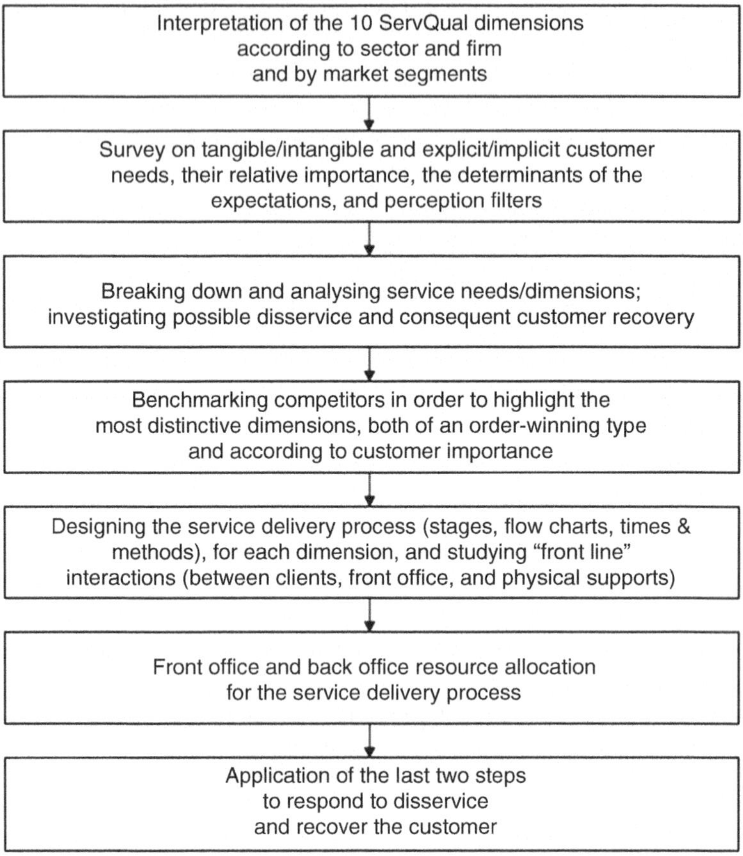

Fig. 6.2 Methodological steps for service design

1. *Gap 1*—between customer's expectations and the perception of customer needs by the management
2. *Gap 2*—between management perceptions and service specifications
3. *Gap 3*—between service specifications and service delivery
4. *Gap 4*—between service delivery and external communication
5. *Gap 5*—between service delivery and how it is perceived
6. *Gap 6*—between perceived service and customer's expectations
7. *Gap 7*—between how the service is perceived and the minimum acceptable service for the customer
8. *Gap 8*—between expectation determinants and expected service (or minimum acceptable service)

To bridge these gaps, it is important to keep in mind that service is a typical organizational process (Tonchia and Tramontano 2004), consisting of activities that are carried out by resources belonging to various functional units of the company (Fig. 6.4).

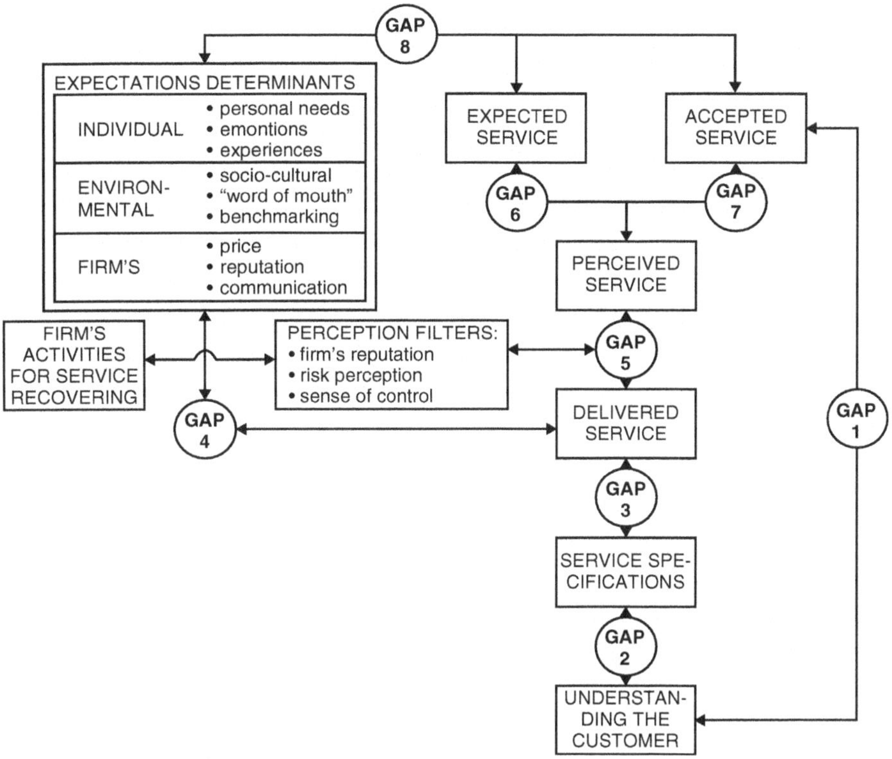

Fig. 6.3 Gaps for customer satisfaction analysis and the design of service improvement (Modified from Zeithaml et al. 1990)

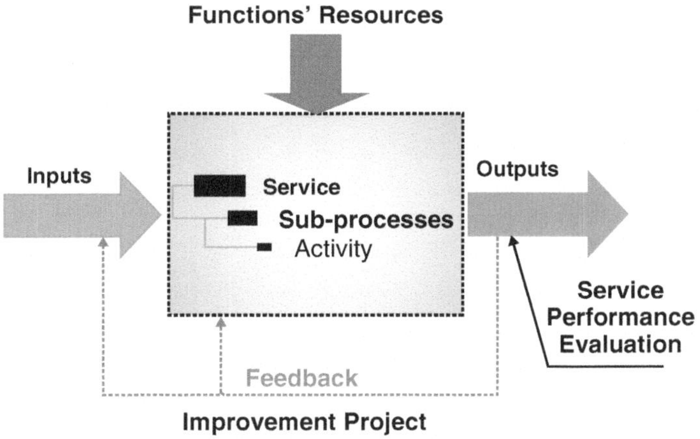

Fig. 6.4 The process-based improvement of service (contents + delivery)

These activities, exploiting certain inputs (material, but mostly information), produce an intangible output known as service; the performance evaluation of this output as well as the gap are the starting points for service improvement or, better, for defining the output performance of the new service.

Considering the service as an organizational process makes it possible to find the causes, which can be attributed to wrong choices made when designing service content and delivery, and redesign the process so as to achieve the targeted performances. This means revising the activities and/or the involvement of the resources and/or the amount and type of input, using a feedback logic.

The output of the project will be a new or modified specification relative to service content and/or delivery.

References

Tonchia, S., & De Toni, A. (2004). Measuring and managing the after-sales service. *International Journal of Services Technology & Management, 5*(4), 385–393.

Tonchia, S., & Tramontano, A. (2004). *Process management for the extended enterprise*. Berlin: Springer.

Tonchia, S., De Toni, A., & Nassimbeni, G. (1995). An instrument for quality performance measurement. *International Journal of Production Economics, 38*, 199–207.

Zeithaml, V. A., Parasuraman, A., & Berry, L. L. (1990). *Delivering quality service*. New York: Free Press.

Part III

Project Scope and Structuring

7 Managing a Project

7.1 Project Charter

The Project Charter is 'the document that formally authorizes the existence of a project and provides the project manager with the authority to apply organizational resources to project activities' (PMI—PMBOK, 6th ed., 2017).

The need for development of the Project Charter resides in documenting basic information for launching the project (start-up stage), so that they may be available at the successive stage (planning stage). In summary, the document must contain the following information:

- Opportunities or commercial needs that led the company to decide to undertake the project
- Description, at the high-level, of the product requirements (or service or process) for which the project was decided
- Project's objectives and KPI (Key Performance Indicators)
- Name of the project manager and range of responsibility/authority
- Environmental factors and organizational assets
- Analysis of influence of major stakeholders.

The reference information for the development of the Project Charter originates from the contract, if the project is a work order (Engineer-To-Order), or from a brief made by the internal sponsor—in the cases of new products/services (commercial opportunities) and for organizational innovations (e.g. Lean transformation). Particularly in the case of Government contracts, it happens that the customer provides the terms of reference as an annex within the tender documents.

In general, the Project Charter should respond for the project to the classic questions: "*What?, How?, Who?, Where?, When?, How Much?*".

Before to approve and release the Project Charter, it should be also asked the opinion of experts from different departments and/or external consultants, exploiting their specialist skills.

S. Tonchia, *Industrial Project Management*, Management for Professionals,
https://doi.org/10.1007/978-3-662-56328-1_7

7.2 Project Scope

Associated with the Project Charter, there is the *Scope Management*. It aims to ensure that the project includes all the activities required for its effective and efficient execution.

Defining and managing the scope of the project greatly affects its overall success.

It must define in particular the project's outputs—called "deliverables"—and the work required to obtain them. It is also crucial to clarify the project's goals (general and detail) i.e. the Key Performance Indicators (KPIs) in terms of time, cost, and quality. These should be *SMART* = *Specific* (target a specific area or fact) + *Measurable* (quantify an indicator) + *Assignable* (specify who will perform it) + *Realistic* (state what results can realistically be achieved, given available resources) + *Time-related* (specify when the result(s) are to be achieved).

External and internal constraints, that delimit the scope of action, should also be highlighted. External constraints are determined by all aspects specified in the contract or suggested by market dynamics. Internal constraints are primarily determined by the resources and competencies of the organization and through the workloads already assigned to other works.

Furthermore it is necessary to consider all the stakeholders involved at the various stages of the project and for the overall results. The stakeholders' needs and expectations are to be analyzed and converted into *requirements*.

Starting from and integrating the Project Charter, a documentation should be developed that specifies the features and limits of the project and the associated products and services, as well as evaluation and control terms.

Preparing a detailed description of the scope of the project is critical for the success of the project itself. To do this, a "*kick-off meeting*" should be organized involving the customer or market representatives, key stakeholders, the project's sponsor and management team. In this meeting, you should usually arrive at key decisions, and also link project's activities to on-going work within the organization; precisely:

- Requirements for the approval of the product, service or process related to the project, and main activities required
- Timing of the activities
- "Make-or-buy" decisions
- Preliminary definition of supply elements, possible variants and economic margins
- Definition of the project organisation
- Assignment of tasks and resources among different departments
- Funding limit, cost estimate and project budget.

7.3 Breakdown Structures: WBS

A *rational approach* to projects requires the use of logical hierarchical structures or *Breakdown Structures* that replicate the way the human mind reacts when facing a problem, namely, breaking it down into sub-problems, in order to analyse them at a greater level of detail.

In general, Project Management considers six breakdown structures, all interconnected (Fig. 7.1), which take into account the three managerial variables of a project (Time—T, Resources—R, Costs—C), as well as its functional–technical ones (Quality—Q):

- *Product Function Structure* (PFS)
- *Product Breakdown Structure* (PdBS)
- *Process Breakdown Structure* (PcBS)
- *Work Breakdown Structure* (WBS)
- *Project Organizational Breakdown Structure* (POBS)
- *Project Budget Breakdown Structure* (PBBS)

These structures and their relationships (both internal and external) are described in greater detail in the following pages.

A project is the result of an idea, more or less destructured (*concept idea*), combining a *why* and a *how*, namely the reason why something new should be created (market demand, unsatisfactory performance, compliance with new regulations, etc.) and how to make the required change in the best possible way (technical, organizational and/or management innovation).

The concept idea is the starting point for defining a structure, known as *PFS*, that describes the functions of the project's output (a product, a different internal organisation, etc.) in a clear, hierarchical manner.

If for instance the project is for a new car, it is necessary to define, at an increasing level of detail, the functions it must possess: traction/drive, steering, braking, a

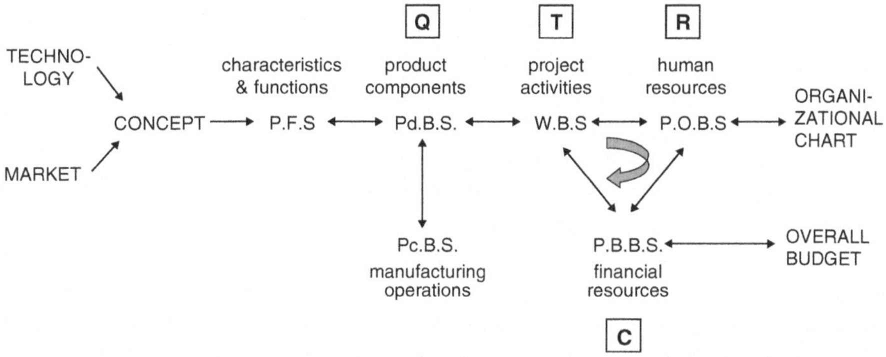

Fig. 7.1 Breakdown structures and project management variables (quality Q, time T, resources R, costs C)

function to dampen the unevenness of the road surface, one to accommodate passengers, hold the baggage, adjust the temperature, commands and controls, etc.

Traction in turn can be divided into various (sub)functions: fuel injection, combustion, the release of exhaust, power transmission and so forth, continuing to divide each sub-function into further (sub)sub-functions.

The PFS generally consists of various levels: level 0 is the overall product; level 1 refers to the functional groups of the product, level 2, the functions of each functional group, and level 3, the sub-functions of each function.

In the case of contract jobs, the functions and performances are specified by the customer and are listed in writing in the contract.

Once the PFS (the tree of functionalities) is defined, it is necessary to identify what parts and sub-assemblies of the product are needed for each function. Hence, a *PdBS* must be made, namely a tree diagram that divides the product into sub-assemblies, which in turn are formed of various parts.

Hence in a car, the function traction will be carried out by the engine, the sub-function fuel injection by the pipes, pump and valves, that termed combustion by the internal combustion chamber and the spark plugs, etc.

The difficulty not only lies in breaking down the product into its functions, but even more in connecting the functions—level by level—to the various items listed in the bill of materials. When proceeding to greater levels of detail, only rarely is there a univocal correspondence between function and part: more often, a given function requires a number of parts, or a certain component may serve for various functions.

If product innovation is accompanied by process innovation, or when a company carries out a contract job, a *PcBS* must also be defined, so as to divide the new manufacturing (or construction) process into stages and operations, and link them to the parts and sub-assemblies that have to be made.

When the manufacturing processes remain unchanged, a correlation should however be made between the PdBS and the (pre-existing) PcBS, so as to link the parts that have been designed to the processes required to produce them.

Once these links have been established, it is possible to identify the activities or *work* needed to develop the product, as illustrated in the bill of materials. These tasks too can be further broken down into groups and sub-groups (the branches of the tree) to reach the maximum level of detail, i.e. the Work Packages or WP, coinciding with the leaves. The project is the trunk of a tree, commonly known as *WBS*, which represents the (unscheduled) deployment of the work needed to complete a given project. Also in this case, the nodes of the family tree must be linked to those of the previous breakdown structures (PdBS and, if present, PcBS).

In its *PMBOK—A Guide to the Project Management Body of Knowledge*, 6th ed., 2017, the Project Management Institute—PMI defines a WBS as 'a deliverable-oriented hierarchical decomposition of the work to be executed by the project team, to accomplish the project objectives and create the required deliverables. The WBS organizes and defines the total scope of the project. The WBS subdivides the project work into smaller, more manageable pieces of work, with each descending level of the WBS representing an increasingly detailed definition of the project work. The planned work contained within the lowest-level

WBS components, which are called *work packages*, can be scheduled, cost estimated, monitored, and controlled'.

So a *work package* is a set of elementary activities having well-identified interactions with other work packages and characterised in a univocal manner by inputs, outputs and internal activities; resources, execution times and ownerships can be linked to these packages, so that they become the basis to plan, budget, schedule and control the advancement of the project.

Although it defines the activities to be carried out, a WBS is usually made using a *mixed logics* approach: in other words, the breakdown process can take into account either the physical structure of the product (PdBS), independently of the future manufacturing stages (PcBS) and how these are planned or, vice versa, it can consider how the process is carried out, independently of the product's structure (Deutsches Institut fur Normung—DIN).

As described in further detail in Sect. 13.3, the Work Packages (WP) of a WBS can be linked to Cost Accounts (CA), using an approach that was originally devised by the US Department of Defence (DoD) and is commonly known as Cost/Schedule Control System Criteria—C/S CSC or *C.Spec.* This procedure however is also referred to as *C.Spec. two-dimensional system*, hence a WBS is also known as a *single C.Spec.* The relevant document "MIL—Military Standard: Work Breakdown Structures for Defense Material Items" poses certain restrictions for the top three levels of the WBS, while the contractor has a greater degree of freedom for the lower levels. According to this document, the top three levels are (1) the project itself, (2) the major elements of the project (e.g., a ship, but also equipment, system test and evaluation, training services, system engineering and the project management itself), (3) elements subordinate to level 2, i.e. sub-elements (e.g. for a ship, its structure, engines, the various plants on board, etc.).

On the other hand, there are also other procedures that break down the project in a different manner. The PRINCE methodology (see the Web sites listed at the end of the book), for instance, considers two levels: the first one refers to the *product*—distinguishing between (1) technical products, consisting of technical documentation, manuals, operational procedures, etc., (2) managerial ones, consisting of plans, description of the work required, management reports, etc. and (3) quality issues, such as the list of specifications, actions, etc.—whereas the second level refers to the *manufacturing stages* and relative milestones. The PRINCE methodology calls this type of WBS a PdBS, a term that in this book—in line with the more accepted meaning—is used instead in reference to the product's BOM.

Following is some advice on how to organize a WBS:

- At the higher levels, a logic of division by project areas should prevail, followed in turn by PdBS and PcBS logics.
- One criterion alone should be used for each level, so as to ensure uniformity of communication among the persons who operate at the same level or at higher/lower ones.
- Breaking down should ensure greater ease in the management of the higher levels, which can be considered as integrated systems to plan and control, having a certain *deliverable* (or output) and measurable cost, time and resource performances.

- The number of levels should ideally range from four to six (not all the branches must necessarily have the same number of levels).
- The sub-levels should define sub-systems that can be managed more or less independently (and which must be tested in a later stage using a *bottom–up* approach, proceeding from the leaves to the roots of the family tree).
- Indirect—i.e. management and control—activities that are needed to deliver the project must also be taken into account (when part of the project is contracted, one or more branches of the WBS are further broken down, giving rise to Contract Work Breakdown Structures—CWBS).

Figure 7.2 depicts the first levels of a WBS for the construction of a new plant; each node of the WBS is tagged with a hierarchical code or number that is equal to that of the levels (level 1—project; level 2—sub-project; level 3—activity; level 4—sub-activity).

WBSs can also refer to organizational or management changes aimed at improving performances: the ISO 9000 certification process is a project, as are Just-in-time (JIT) interventions, internal reorganization, the adoption of a new information system, revising the agents/retailers network, creating partnerships with suppliers, joint ventures, etc. All these projects require careful planning and the definition of the activities to be carried out, which are illustrated by the WBS.

Having defined the activities, it is necessary to allocate the human and financial resources needed for their execution.

As regards human resources, it is first of all necessary to define the competencies needed to carry out the activities defined in the WBS. The structure thus created (*POBS*) illustrates the tasks and responsibilities of the various parties, and indicates who the project manager is. As described in further detail in Chap. 11, devoted to organization, there must not necessarily be a hierarchy (and consequent power to punish or reward), but only a series of *ownerships* that are established within the framework of the project. Given the nature of projects, the POBS is obviously only a temporary structure.

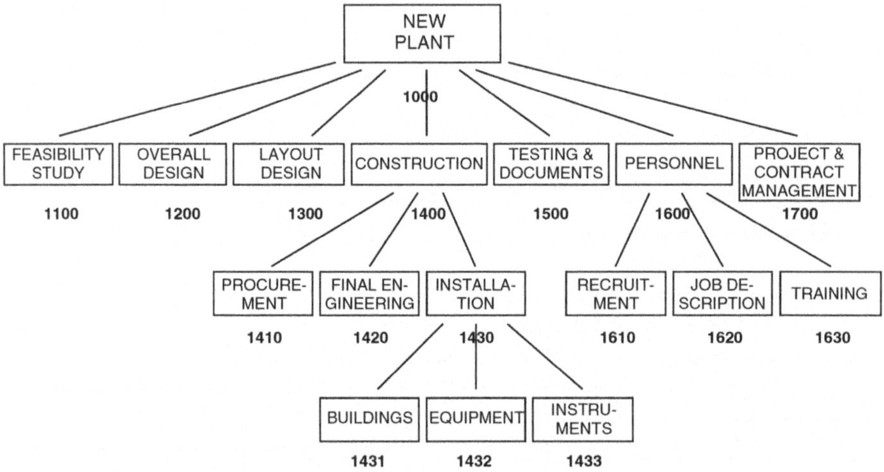

Fig. 7.2 Example of Work Breakdown Structure (WBS) and relative coding

The POBS of the various projects are linked to the *organizational chart* of the company. Each employee can feature in a number of POBS, referring to different projects, or in different branches of the one structure. Obviously, the *call by reason of competency* must be made in such a manner as to ensure that the persons are not overloaded by work.

Although the nodes or elements of a POBS usually comprise resources coming from the Design and Engineering offices, they can, and indeed should, also feature employees working in other functional units, such as marketing, purchasing, production, etc.

If the human resources employed in the projects require valuable equipment or tools whose availability is limited, the POBS should also include the use of these (instrumental) resources, particularly when there is more than one project carried out at the same time.

Finally, the above resources—both human and instrumental (if relevant, namely if it is necessary to enter them as cost items in the framework of a project, because they cannot be considered as routine operations)—require funds, which are provided by the project budget, part of the *overall budget* of the company. The *PBBS* breaks down this budget (which can also be the value of the contract job, minus the margin of profit) into cost accounts that, in a variable measure, must be linked both to the resources used (POBS) and the activities (WBS) that they will carry out. It is a question of costing (PBBS) *who does* (POBS) *what* (WBS).

In brief, the six Breakdown Structures differ in their *nodes,* which are as follows:

- Functions and characteristics in the PFS
- Parts and sub-assemblies in the PdBS
- Stages and operations of the manufacturing cycle in the PcBS
- Activities required for the execution of the project in the WBS
- Project resources in the POBS
- Project costs in the PBBS

7.4 Project Planning & Control

To manage, it is necessary to measure. More specifically, it is necessary to measure operational macro-variables linked to performances, which are namely Quality (technical characteristics), Time and Costs. Resources must also be added to the list because, even though they are not performances, their availability, capacity and saturation levels must also be taken into account.

Project management consists of two stages:

1. *Planning* (prior to the start of the activities specified by the WBS)
2. *Control and reporting (from the start to the conclusion of the project)*

The stage of *planning* defines all the operational variables. The objectives of quality, time and costs (established by PdBS/PcBS, WBS and PBBS, respectively) are specified in detail, and the necessary resources are quantified and made available (as described in the POBS).

The necessary activities are worked out for each class of variables, and are listed here according to priority, rather than in a chronological order:

1. *Budgeting* (definition of the budgeted expenses and successive allocation of these costs according to the PBBS, taking into account the other breakdown structures depicted in Fig. 7.1—in particular the costs arising from the use of resources during the activities)
2. *Scheduling* (defining the start and finish dates for all the activities specified in the WBS)
3. *Resource allocation* (allocation of specific POBS resources to the WBS)

These activities are not carried out in succession, but in loops of greater detail and refinement: the forecasted costs are always better defined after analysing and scheduling each activity; scheduling in turn can be revised after allocating the resources, etc.

In the technical part, which is mainly associated with the quality performance of the project, project strategies are planned and appropriate procedures adopted.

Control and *reporting* take place when the project is already under way. Their aim is to bring it to a successful conclusion within the terms laid down at the time of planning, intervening if variances occur.

Project Strategy Management

8

8.1 Strategic Planning

Strategy can be considered in terms of *content* and *process*. The content refers to both *competitive priorities* (performance-related macroobjectives that can become Critical Success Factors—CSF) and the *interventions* made to achieve them. The latter can deal with *technological* levers (relative to the product/process and to information/communication), internal *organisational* choices and those to interface with suppliers/customers, and *managerial* levers (for example, Just-in-Time, Total Quality Management and Concurrent Engineering). The process, on the other hand, refers to the *formulation* and *implementation* of the strategy.

Traditionally, corporate strategy relies on three options: cost leadership, differentiation and segmentation (Porter 1980). The current trend of overcoming performance trade-offs questions this distinction (Filippini et al. 1998): the aim is the joint achievement of multiple performances, giving them different importance throughout time (the *sand-cone model*—Ferdows and De Meyer 1990).

Nowadays, strategy is mostly about defining priorities, which are not standard, but more of an *order-winning* and *qualifying* type (Hill 1989): the former ones make it possible to steal customers from the competitors; the latter simply allow the company to enter into a certain competitive arena. Both are equally important, and when analysing the scenario, a company must decide whether to invest in one or the other type of strategy.

Recently, Porter (1996) affirmed that 'the *essence of strategy* is deciding to carry out activities in a *different* manner from the competitors'. This is made possible by the so-called *core competencies*, which in recent years have been the object of much debate aimed at renovating the traditional *structure–conduct–performance* strategic scheme, typical of the Industrial Organisation.

The Industrial Organisation scheme has been hugely criticised because of the differences in profitability found between companies operating in the same industry/sector; furthermore, the classic five factors (intensity of competition, potential new entries into the industry, threats deriving from substitute products, contractual power

S. Tonchia, *Industrial Project Management*, Management for Professionals,
https://doi.org/10.1007/978-3-662-56328-1_8

of the suppliers and the customers) have often failed to explain the sustainability of the competitive advantage.

The criticism is moved by supporters of the theory according to which *resources and competencies* are the real source of advantage for a company (and not an adequate strategic behaviour for the industry, as maintained by the traditional theory of the Industrial Organisation). Theories such as Resource-Based View and Competence-Based Competition—which share some of the concepts expressed in the Enterprise Evolutionary Theory and Organisational Behaviour Theory—are approaches that have many features and principles in common, so that it is possible to speak of a *Competency Theory* (Tonchia and De Toni 2003).

According to the Competency Theory (Grant 1996), strategic analysis can be schematically described by the sequence: analysis of the resources/competencies possessed, assessment of their profitability potential, consequent definition of a suitable strategy for exploiting, valorising and consolidating them and strategy implementation by means of appropriate policies of resources management.

Hence, the process of *strategic planning* should take into account both the sequence of the analysis typical of the Industrial Organisation, and that characterising the Competency Theory, identifying the principal point of contact when comparing *competitive priorities* on one hand and *resources/competencies* on the other (shaded in Fig. 8.1 to highlight this link). Competitive priorities emerging from an analysis of the industry are insufficient: the resources/competencies possessed must also be taken into account, after analysing their potential of profitability. Moreover, it must be kept in mind that the typical resources/competencies of a company have a value *in se*, when compared to the outside world (the environment and the industry, this latter sector being increasingly difficult to identify with precision nowadays), but also in relation to the company's competitive priorities.

The link between the internal perspective, typical of the Competency Theory (inside-out approach), and the external one (i.e. the market/competitors), typical of the Industrial Organisation (outside-in approach), can be summarised by the so-called *SWOT analysis* (analysis of Strengths and Weaknesses + Opportunities and Threats).

SWOT analysis is usually carried out after the management has made its mission and vision clear. A company's *mission* is the answer to the questions: Who are we? What do we do? How? Why and for whom? Conversely, the company's *vision* answers questions regarding the future: Where will we be? What will we be? Mission and vision should be communicated both inside and outside the company.

Mission, vision and SWOT analysis lead to the definition of a strategy (depicted in Fig. 8.1 by dashed lines), in terms of competitive priorities, intervention levers and resource management policies. The latter policies affect the competencies, which are placed at a higher level of aggregation than the resources, and assess the capacity of an appropriately managed set of resources to carry out an activity or pursue an objective: competencies can explain how two companies, having the same aims and resources, can perform differently, or have the same performances when possessing different resources.

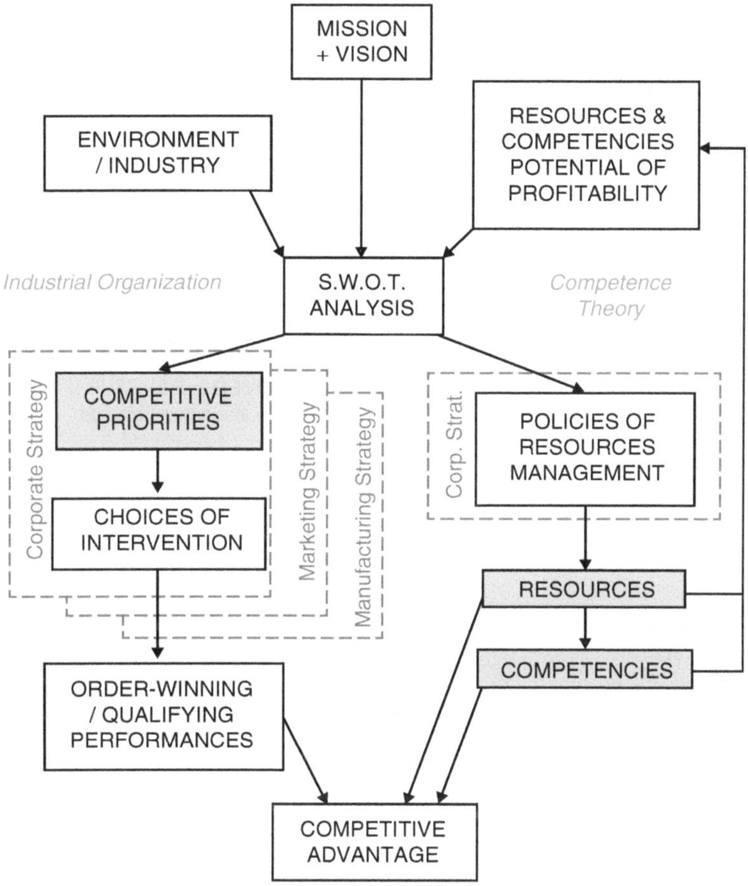

Fig. 8.1 Strategic planning according to industrial organisation and competence theory

The strategic decisions concerning competitive priorities, interventions and resources management are made at a corporate level (*corporate strategy*), but are then further detailed for business units and departments, usually proceeding from the trading area (*marketing strategy*) to the operational functions: *manufacturing strategy* and *design strategy*. In the latter case, specific project management (design) strategies should be developed.

8.2 Project Management Strategies

Clark and Fujimoto (1991) identify three types of strategic decisions regarding projects (Fig. 8.2).

First of all, it is necessary to define the *degree of complexity* to be assigned to the project, and reduce it if the analyses performed—using, for example, breakdown

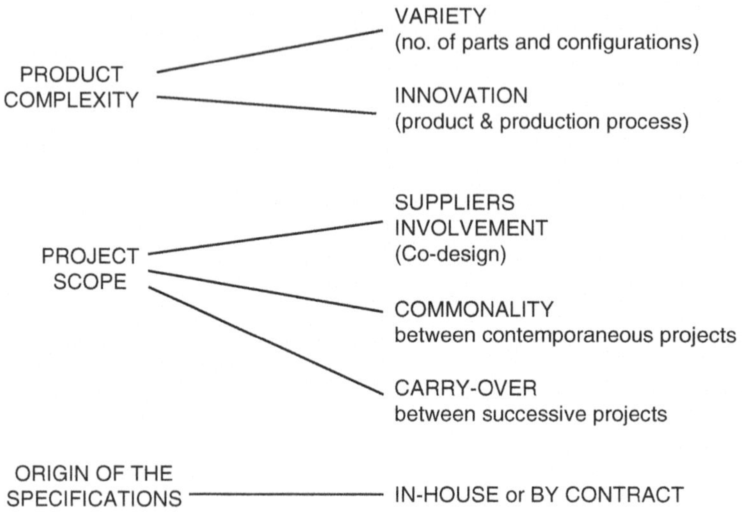

Fig. 8.2 Preliminary strategic decisions regarding projects

structures—reveal a complexity that cannot be dealt with and managed with the available means.

The degree of complexity depends on two factors: (1) the *variety* of the project, namely the number of parts forming it and its various configurations, (2) the level of *innovation* required to develop these parts or carry out the necessary manufacturing/ production processes needed to deliver the project.

Both factors concur in making a project more or less complex: a project cannot be considered complex only because it consists of numerous, variable parts, but innovation must also be taken into account, in other words the novelties (with relative costs and risks) featuring in the project. The difference between large and small projects, or more or less complex ones, can be compared to that existing between an orchestra and a string quartet: both aim at performing well, but the specialisation, coordination, communication and direction needed are utterly different.

A second type of decisions concerns the *size* (*scope*) of the project and possible ways to reduce/limit it. A project's size is defined by the amount of work, effort and money put into it. Size also depends on other factors, such as (1) the involvement of suppliers during the stages of design (in this case, *co-design*), (2) commonality, i.e. the design parts shared by various projects carried out at the same time by the company, (3) the carry-over, namely the number of parts that are inherited from pre-existing projects.

The *involvement of the suppliers,* not only in terms of delivery of materials, parts and sub-assemblies, but also as regards *cooperative design*, especially of those parts that are not considered strategic and for which the company does not possess the necessary competency, ensures a remarkable reduction in the size of the project. This may mean that the supplier establishes a stronger link and a more stable, long-term relationship with the customer, as well as spreading design costs over a larger pool of customers. Moreover, the skills of the supplier can certainly improve the quality of the product.

Commonality on the other hand ensures that, for the same investment, a wider range of products can be placed on the market: a core project termed *platform* is used as a starting point for responding to all specific demands—including customisation. Consequently, the break-even point for the sale of a certain product is lower, risks are smaller and the consequences of a wrong prediction of sales (unsold stock, costs related to production under- or over-capacity, etc.) are reduced.

Carry-over enables new products to be launched on the market with limited investments and risks. Only marginal innovations are introduced, chiefly of an aesthetic or modular nature (additional functions or replacement of certain parts/ sub-assemblies), but these can have a remarkable impact on how the customer perceives the product, which is often more than proportional to the actual degree of innovation. Moreover, it is also possible to prolong the life cycle of a product, delaying investments for newer products, increasing flexibility and waiting for the competitors' moves or new technological developments. Not always are the pioneers more successful than the followers.

A third type of decision concerns the *origin of project specifications*, which may be internal (as often is the case in product development), or receive indications from a certain client of the company: in the latter case, design and production are made to order, or better, *Engineered-To-Order* (*ETO*), that is the *Engineering and Construction Contracting (ECC)*. The client may list and define requirement in a more or less detailed manner (*detailed controlled*, *black box*, or *supplier proprietary*—see Sect. 9.2).

The earlier macrodecisions concerning strategies for a (*single*) *project*—and linked to its complexity, size and specifications—occupy an intermediate level between strategies for a (*single*) *product* and *multi-project, multi-product* ones (Fig. 8.3).

Product strategies can be placed within the wider framework of marketing strategies. They also include the concept ideas defined when planning product development, and the product lines definition.

The decisions concerning *product concept* are affected by the relative position of each product on the market and their sequence of introduction (and for this reason, they are linked to multi-project strategies). Two alternative strategies may be adopted: *continuous spectrum*, which views the market as a set of stratified, continuous segments, a view that is reflected by the products of a certain line (hence the customer is tempted to pass from a basic model to a superior one), and *discrete mosaic*, where the market is considered as a multiple set of unrelated, discrete segments: in this case, the models are uniquely differentiated and are characterised by a very focused product concept (this is the strategy used by producers who want to compete against leading firms).

To design a *product line*, it is essential to first define the differences between products as well as the identity of the line, in other words the common features possessed by the various products. Also, in this case there are different strategies that

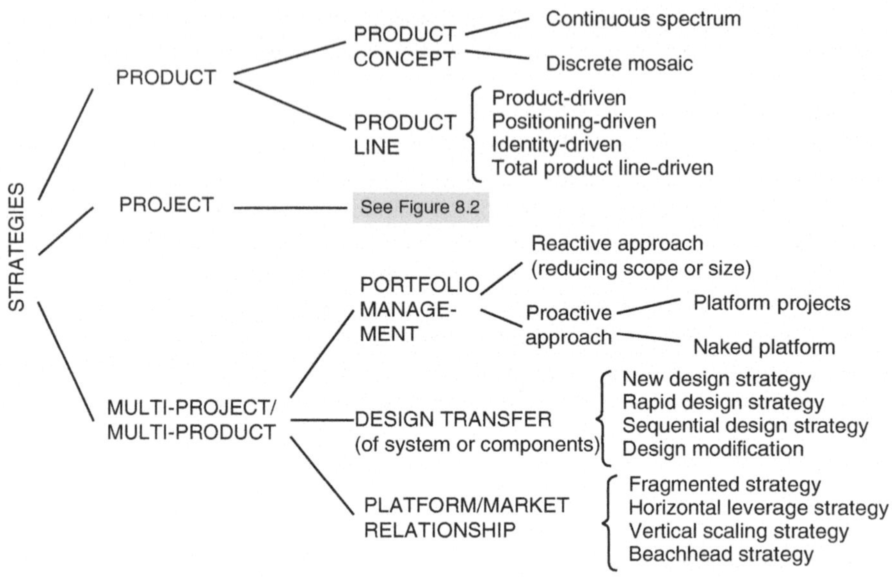

Fig. 8.3 Product, project, and multi-project/product strategies

can be adopted, according to the identity of the line and the importance of product-positioning within the line (Clark et al. 1991).

Multi-project, multi-product strategies are often on the opposite side of single project/product strategies: Sakakibara (1994) observes that although some of the leading companies in Japan have no *star* products, they are extremely successful on the market.

Multi-project, multi-product strategies mostly focus on (1) the management of the project/product portfolio, (2) design transfer, (3) the relationship between platform and market.

Portfolio management can be either reactive or proactive. Traditional policies are based on a *reactive approach*, exploiting commonality and carry-over. When starting a project, an attempt is made to identify the inter-dependence existing with past or current projects, and where possible, ready-developed solutions are adopted. A *proactive approach*, on the other hand, is linked to the concept of *platform* and can offer two alternatives: either develop a new *core design* when preparing a new product, or work in order to set up a platform without, for the time being, launching any product (*naked projects* such as the *concept cars* in the automobile sector that have no immediate commercial aim but in a later stage may also benefit from *off-the-shelf parts*, namely parts taken from *the shelf of innovation*).

Design transfer, namely the transferral of designs and core technologies among projects, requires a three-dimensional analysis, focusing on the type of transfer, the transfer process and the coding of technology. The knowledge transferred can be that relative to the parts or to the entire system (Henderson and Clark 1990). Strategies that provide for the development of a new product with a *core design* that has no significant interaction with other projects developed by the company (a particularly suitable strategy for incorporating state-of-the art design and technological solutions) are defined as *new design strategy* (Nobeoka 1993). There are also other strategies that can be used to transfer knowledge to other projects: (1) in *rapid design transfer,* the base project starts to transfer its core to a new project before it is fully completed, hence great coordination is required, and probably adjustments must be made on both projects; (2) in *sequential design transfer,* the transfer occurs when the base project is completed and the links with the new project are more relevant. There is also another strategy, termed *design modification,* according to which the design of a new product is based on the core design of its predecessor.

If on one hand *new design* is the best strategy in terms of expanding the company's market share, on the other it requires longer development times and considerable resources. The best compromise is a strategy of *rapid design transfer,* which ensures rapid access to the market and a satisfactory productivity of the resources. The time interval between the start of the original project and that of the new one—the rapid transfer time—is known as *platform design age.* Experimental results show that dominant positions on the market are obtained by the frequent introduction of new products, exploiting *young* platforms: the renewal of platforms is a vital issue for today's companies (Meyer 1997).

Platform development can be depicted as a grid whose rows report the levels of price/performances and the columns the market segments. As a consequence, there are four types of *platform strategies* (Meyer and Lehnerd 1997): (1) a *fragmented strategy* if there is a specific platform for each cell (corresponding to a market niche): consequently, there are numerous families of products but also higher costs, because the concept of platform is poorly exploited; (2) a *horizontal leverage strategy* if one platform is exploited by various market niches sharing the same level of price/ performance, thus along various segments of a certain row; (3) a *vertical scaling strategy* if one platform serves different levels of price/performance within a certain market segment, scaling either upwards or downwards; (4) a *beachhead strategy* if the firm prepares a platform for a specific, low-price segment and then extends it to other segments, at higher price levels.

Multi-project strategies based on platforms are part of a wider range of strategies that include joint ventures, globalisation and even the standardisation of platforms and/or architectures among various producers operating in a certain sector (this is often the case of electronic consumables).

Product, design, multi-product and multi-project strategies require the definition of the two factors that characterise all *strategies:* the *scope* (in term of quality, time and costs) and the necessary (technical) *means* to achieve it.

8.3 Project Management Performances

It is first necessary to distinguish between the performances of a single project and those relative to the design process, i.e. all those activities that serve to deliver several projects in a certain period. In a company-wide framework, these activities (engineering, industrialisation, etc.) involve various departments (such as

Table 8.1 Project and design process performances

	(Single) project		Design process (multi-project)		
Quality	External	Total design quality	Internal	Organisational learning	Project flexibility
Time	Internal	Lead time, time-to-market	External	Product replacement and range expansion rate	
Costs	Internal	Project costs	Internal	Costs of the Design Department and of the other resources involved in the design process	
Productivity	Internal	(Technical parameter)/(project costs)	Internal	(No. of projects)/(cost of the design process)	
Profitability	Internal	(Net income from the project)/(project costs)	Internal	(Total revenues)/(cost of the design process)	

Marketing, Production, etc.), and therefore the performances should not only refer to Design function (technical office or similar units).

The performances of a project consider three aspects (Table 8.1):

- The *quality of the project* (how it meets the client's expectations)
- *Lead Time* or *time-to-market* (duration of the project)
- *Productivity cost* (the amount of resources used to produce a certain output)

The quality of a project depends on its output or *deliverable*. Quality is difficult to measure, both because its economic value cannot be immediately determined and because it is, by definition, a multi-dimensional parameter. One of the most popular methods to estimate the quality of a product—that described by Garvin (1988)—considers eight dimensions: (1) key performances/features, (2) other performances/features, (3) conformity with specifications, (4) reliability (the ability to carry out its function for a certain length of time without requiring repairs), (5) duration (linked both to technical obsolescence and economic aspects, i.e. if a certain product is still convenient), (6) aesthetics/design, (7) perceived quality (solidness, etc.) and image (including status symbol), (8) after-sale assistance and services.

The quality of a product, resulting from *total design quality,* can be considered good when (a) the product concept matches the requirements of the customer (*customer–concept matching*) and (b) the product specifications meet the product concept (*concept–design matching*) (Clark and Fujimoto 1991).

Project times are becoming increasingly important, especially in these days of Time-Based Competition (Blackburn 1991), when projects are a key factor of success, especially in the case of *premium price*, usually resulting from being the first company to present a new product on the market (Smith and Reinertsen 1991).

The *lead time* or *time-to-market* of a project is an internal time performance; in other words, it cannot be directly perceived by the customer, who is only aware of the *product replacement* and the *range expansion rates*, namely the rates at which new products appear on the market (either to replace pre-existing ones or to expand the range), or the ratio between the number of new products launched and the average number of products offered in a certain period of time (*New Product Introduction Rate—NPIR*).

Only in the case of ETO, the lead time of a project/contract job is an external performance, since it is perceived by the client, who knows when the project is due to start (usually when the contract is signed, although the company could have started work already, making surveys, preparing estimates, etc.).

The *costs* of a project refer to the resources (human, material and equipment) exploited for the various activities. Also *productivity* is an exclusively internal performance that cannot be perceived by the customer if not indirectly, when assessing the price. Productivity is the ratio between output and input, and—with reference to specific standards—is more accurately defined as *efficiency* (whereas *efficacy* is the ratio between actual output and targeted output, and is therefore independent of the input). Costs and productivity are thus closely linked, but despite this, they can give rise to discordant results: for example, there can be considerable expenses linked to the project but these may have a positive outcome in terms of productivity (with technical results well beyond expectations).

The output and input can be expressed either as amounts or in money value (in this case however there is the problem of price fluctuations, which can be solved by using deflators or the concept of *equivalent units*); generally, the productivity of a project is calculated as the ratio between output (a technical quality parameter) and input (the costs that can be ascribed to the project). When the project is considered as an investment also, the incoming cash flows generated by the project are taken into account, and the value of the numerator is equal to the sum of the profits that can be ascribed to that project. In this case, it would be more correct to describe this value as the *profitability* (or financial productivity) of the project, at a given level of cost.

Cost and productivity are considered as *cost* performances not only for the unit of measurement used to express them, but mostly because they are directly linked—using specific formulas—to the company's financial results; time and quality, on the other hand, are considered as *non-cost* performances because, although they affect these results, they are only indirectly linked to the costs (Tonchia and De Toni 1996, 2001).

Links can be established between product performances listed previously. Clark and Fujimoto (1991) observed an inversely proportional relationship between customer–concept matching and the time needed to develop a project, while there appears to be a directly proportional relationship between concept–design matching and development time; hence, *total design quality* has an optimal development time,

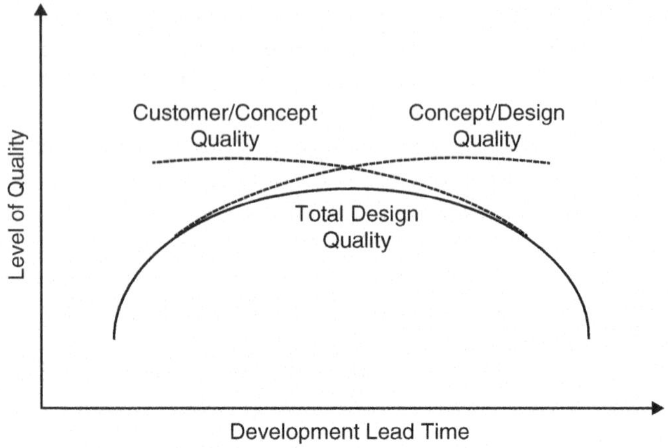

Fig. 8.4 The optimal development lead time is the maximum level of total design quality (Source: Clark and Fujimoto 1991)

corresponding to the peak of the curve obtained by summing the curves of *customer–concept quality* and *concept–design quality* (Fig. 8.4).

There is also a relationship between (project) quality and productivity: when the latter increases, the quality of a specific project may decrease; on the other hand, greater productivity may determine a greater variety, thus increasing quality in terms of customer satisfaction (*fitness for use*). Also, in this case, it is necessary to define an appropriate (minimum) level of productivity.

Wheelwright and Clark (1992) describe a relationship between project performances and organisation-by-projects. Despite some difference among countries (particularly between the Eastern and Western world), when classifying firms according to three different types of organisation (functional specialisation, internal integration of the units, external integration with suppliers and customer), higher levels of productivity and shorter lead times can be observed in the case of internal integration, and higher total product quality in that of external integration.

In a multi-project framework, the overall *design performances* include *time* performances (corresponding to the product replacement and expansion rates), *cost* performances, which refer to the overall cost of the resources, without ascribing them to the single projects (thus without running the risk of a non-objective allocation), *productivity* performances, linked both to the number of projects carried out by the company (or other indicators such as patents, etc.) and their cost, and *profitability* performances, referring to the profits made by the company—thanks to the design activities; in this context, quality can be linked to learning (Nonaka 1991) (Table 8.1).

When operating in a multi-project environment, it is fundamental to take into account the *project's flexibility*, in terms of easiness to modify both pre-existing products (*product flexibility*) and the manufacturing processes (*process flexibility*—not to be confused with *production flexibility*, which refers to changes in the work programme, once the processes/production cycles have been defined). These performances—although unperceived by the customer—determine the range of products, as well as the times and costs needed to expand or replace them. A variation in pre-existing projects (namely, the deliverables of previous design cycles) implies a change in certain characteristics of the product (materials, parts, dimensions, etc.) or its production, such as manufacturing stages and assembly: in other words, a change in quality requires a certain length of time and has a cost (Table 8.1). The product and process flexibility is equal to the ratio between the extent of change (the numerator) and the time and costs required (the denominator) (Tonchia and De Toni 1996, 1998).

Performances, both internal and external, of product development and more in general, of the design process, depend on (1) how the various stages of the process are managed, (2) the practices used, (3) the strategic and organisational context. The most important drivers for these areas are as follows (Griffin 1997):

- The ability to understand in advance the customer's requirements and orient the designing activities so as to satisfy these demands
- Frequent contacts with the clients, so as to listen to the *customer's voice* during all stages of development
- Giving great importance to the first stages of concept definition and product planning; a partial overlapping of the various stages of development, with frequent exchange of information (and relative instruments), is a better alternative to the more traditional, sequential approach
- Frequent *problem-solving* feedback cycles
- The internal involvement of the stakeholders, namely the sponsors (*commitment*) and the management (*leadership*)
- A high level of integration among the functional units (including production, purchasing, trade and marketing, accounting, human resources management, etc.) and widespread *team working*
- A climate oriented towards innovation and learning, inspired by a specific, shared strategy
- Greater emphasis on the architectural and systemic aspects of the project, and on the involvement of suppliers and subcontractors
- The application of innovative techniques to manage variety, modularisation, integrated design, engineering, etc.

References

Blackburn, J. D. (Ed.). (1991). *Time-based competition: The next battleground in American manufacturing*. Homewood, IL: Business One Irwin.

Clark, K. B., & Fujimoto, T. (1991). *Product development performance: Strategy, organization and management in the world auto industry*. Boston, MA: Harvard Business School Press.

Clark, K. B., Fujimoto, T., & Aoshima, Y. (1991). *Managing the product line: A case of the automobile industry*. New York: Mimeo.

Ferdows, K., & De Meyer, A. (1990). Lasting improvements in manufacturing performance. *Journal of Operations Management, 9*(2), 168–184.
Filippini, R., Forza, C., & Vinelli, A. (1998). Trade-off and compatibility between performance: Definitions and empirical evidence. *International Journal of Production Research, 36*(12), 3379–3406.
Garvin, D. A. (1988). *Managing quality: The strategic and competitive edge*. New York: Free Press.
Grant, R. M. (1996). Toward a knowledge-based theory of the firm. *Strategic Management Journal, 17*, 109–122.
Griffin, A. (1997). PDMA research on new product development practices: Updating trends and benchmarking best practices. *Journal of Product Innovation Management, 14*(6), 429–458.
Henderson, R. M., & Clark, K. B. (1990). Architectural innovation: The reconfiguration of existing product technologies and the failure of established firms. *Administrative Science Quarterly, 35*, 9–30.
Hill, T. (1989). *Manufacturing strategy: Text and cases*. Homewood, IL: Irwin.
Meyer, M. H. (1997). Revitalize your product lines through continuous platform renewal. *Research Technology Management, 40*(2), 17–28.
Meyer, M. H., & Lehnerd, A. P. (1997). *The power of product platforms*. New York: Free Press.
Nobeoka, K. (1993, September). Multi-project management: Strategy and organization in automobile product development, Ph.D. Dissertation, MIT, Boston, MA.
Nonaka, I. (1991, November/December). The knowledge creating company. *Harvard Business Review*, 96–104.
Porter, M. E. (1980). *Competitive strategy: Techniques for analyzing industries and competitors*. New York: Free Press.
Porter, M. E. (1996, November/December). What is strategy? *Harvard Business Review*, 61–78.
Sakakibara, K. (1994). *A conceptual note on product innovation management: Issues related to the unit of analysis*. New York: Mimeo.
Smith, P. G., & Reinertsen, D. G. (1991). *Developing products in half the time*. New York: Van Nostrand Reinhold.
Tonchia, S., & De Toni, A. (1996). Lean organization, management-by-process and performance measurement. *International Journal of Operations & Production Management, 16*(2), 221–236.
Tonchia, S., & De Toni, A. (1998). Manufacturing flexibility: A literature review. *International Journal of Production Research, 36*(6), 1587–1617.
Tonchia, S., & De Toni, A. (2001). Performance measurement systems: Models, characteristics and measures. *International Journal of Operations & Production Management, 21*, 1–2.
Tonchia, S., & De Toni, A. (2003). Strategic planning and firm's competencies: Traditional approaches and new perspectives. *International Journal of Operations & Production Management, 23*(9), 947–976.
Wheelwright, S. C., & Clark, K. B. (1992). *Revolutionazing product development: Quantum leaps in speed, efficiency and quality*. Free Press: New York.

Project Quality Management 9

9.1 Designing with the Customers

The most important design techniques aimed at satisfying customer requirements (i.e. the modern definition of quality) include Quality Function Deployment (*QFD*), Robust Design, Value Analysis and Value Engineering.

QFD, a method created in Japan in the 1960s, is aimed at translating the customer's desires into technical details, using a scale of priorities that is also dictated by comparison with similar products made by competitors. Applied for the first time at the Mitsubishi shipyards in Kobe (Japan) in 1972 and later in Toyota, it owes its formalisation to Professor Yoji Akao (who first described it in 1966, although his most famous and complete work is that published in 1990). QFD became popular in the Western world, thanks to Don Clausing, professor at the MIT in Boston, and was soon adopted by General Electric and Ford, where it is now one of the most used methods for product development (www.qfdi.org).

A first, partial presentation of QFD can be made through its *what–how* matrix, illustrating the following (Fig. 9.1):

- Customers' desires (the *whats* to produce) and their priorities, in the rows
- Technical–operational prerequisites (the *hows*, i.e. how to translate the customer's desires into practice) in the columns
- Customer needs versus Technical–operational prerequisites (inside the matrix) and a diagonal correlation between the *hows* (the *roof* of the top of the figure: this is why the matrix is also known as *House of Quality—HoQ*, a definition coined by Hauser and Clausing in their famous paper published in 1988)
- An assessment of the importance of the technical–operational prerequisites and their objective numerical values, in the *basement*
- Evaluation of the competition, considering how the products respond to the demands of the customers (on the right, in the central matrix) and their technical level (at the bottom)

S. Tonchia, *Industrial Project Management*, Management for Professionals,
https://doi.org/10.1007/978-3-662-56328-1_9

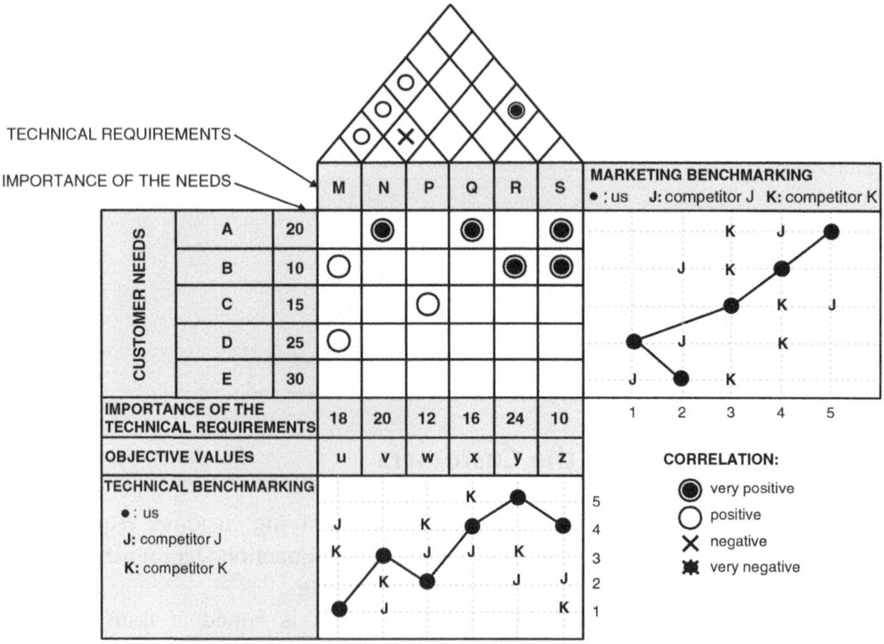

Fig. 9.1 The *whats–hows* QFD (Quality Function Deployment) matrix

The customer's demands (on the rows) are established by marketing research (through interviews, questionnaires, etc.), and it is often necessary to group these requirements into categories, using techniques of Hierarchical Cluster Analysis. In practice, a chart is made (the so-called *Demanded Quality Deployment Chart*), which is organized by rows and develops these demands horizontally by increasing levels of detail (generally, three): the most detailed level will be linked to the technical–operational prerequisites (described in the columns).

Associated with the Demanded Quality Deployment Chart and preliminary to the HoQ matrix is the *Quality Pre-Plan*, aimed at calculating the priorities given by the customers (Mizuno and Akao 1994).

Also the Quality Pre-Plan has a matrix structure, with rows expressing customer needs and columns describing (1) the score given by the customers to the goods produced both by the company and its competitors, according to how these match their requirements, (2) the objective score that the company intends to achieve by developing the new product, (3) the improvement rate deriving from the comparison between past scores and the objective score, (4) the *sales point scores*, namely factors of 1.2 and 1.5, respectively, linked to requirements that, if satisfied, can make a difference in terms of image and sales success, (5) the *overall score* resulting from the score given by the customers, the rate of improvement and the sales point score, (6) the *demanded quality weight* or *relative weight*, namely the value of the previous score expressed as a percentage: this value is then written down in a column of the *HoQ* matrix (Fig. 9.1).

The technical–operational prerequisites listed in the columns of the HoQ represent the technical features of the product and are also known as the *voice of the engineer*. These prerequisites—indicated as w_j—are also weighed, using factors obtained by multiplying the demanded quality weight d_i (see earlier) by the crossed r_{ij} values of the matrix correlating the *whats* and the *hows*; the latter values are converted from rankings (like those shown in Fig. 9.1) into numerical values using techniques of Analytic Hierarchy Process: four alternative rankings usually become −2, −1, +1, +2; three alternatives can become 1, 3, 9 (or 1, 3, 5 or even 1, 5, 9).

Thus

$$w_j = \sum_{i=1}^{n} d_j{}^* r_{ij}$$

or, in vectors: $w_j = \bar{d} \cdot \bar{R}_j$ (where $\bar{R}_j$ is column j in the what–how correlation matrix).

The relative importance of each technical feature (i.e. the columns in the HoQ) can then be determined, so as to obtain a list of priorities that the engineers must keep in mind when designing the product:

$$\tilde{w}_j = \frac{w_j}{\sum_{j=1}^{m} w_j}$$

As well as the HoQ (or Product Planning matrix), the QFD method also includes other types of matrices (Mazur 1990), such as the *Part/Sub-System Planning matrix*, the *Process Planning matrix* and the *Production Control matrix*, which are linked to each other as shown in Fig. 9.2.

Moreover, by extending the method, it is also possible to prepare other charts for the deployment of various needs: the technology required (Technology Deployment Chart), the technical mechanisms/solutions to adopt (Mechanism Deployment Chart), product reliability (Reliability Deployment Chart) and costs (Cost Deployment Chart), the latter two being linked to FMEA and Value Analysis, respectively.

QFD has now become a common practice, linking design to quality management. The driving force is the customer, since quality—as established by norm ISO 8402—is the satisfaction of market needs and desires. Quality therefore starts from the stage of design.

QFD can be successfully applied not only to the manufacturing industry but also to process industries, construction firms and software or service companies. There are also other approaches similar to QFD, such as FAST (Function Analysis and System Technique—Pugh 1991).

Robust Design (*RD*) is aimed at designing a product in such a way as to minimise loss of quality, which would mean failing to satisfy customer needs. Loss of quality is caused by a variance from the identified targets, and according to Professor

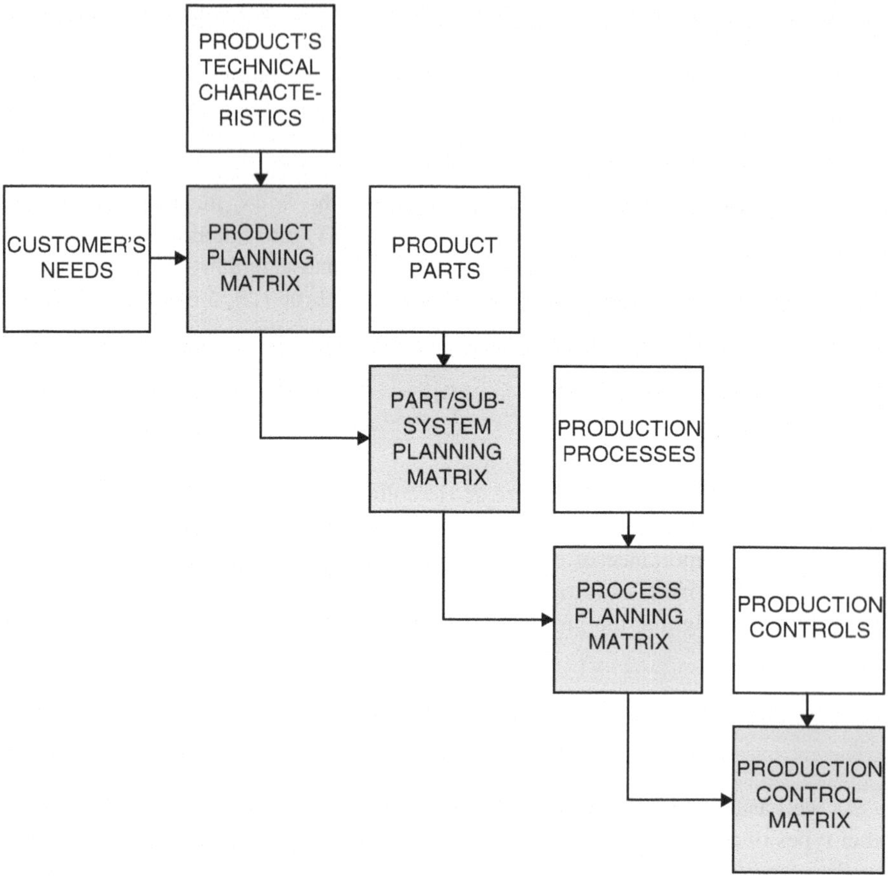

Fig. 9.2 The four QFD matrices

Taguchi's "loss function", the failure to achieve these targets determines the onset of costs that are proportional to the squared value of the variance (Barkley and Saylor 1994). Taguchi's approach considers three stages of design, relative to (1) the *system*, which defines the architecture of the parts, the shape of the product, the materials used, etc., (2) the *parameters* of the product, which must ensure that the performances are the least sensitive possible to variations from the target values, (3) the admitted *tolerance*.

Value Analysis (*VA*) consists in studying the relationships between functions and performances on one hand, and the cost of the product on the other (an example is the classic quality/price ratio); the analysis, when referred to the product, determines an analysis of its parts and sub-assemblies (Fowler 1990; Miles 1992). It is also known as Functional Cost Analysis (Michaels and Wood 1989).

The *value* is defined as the ratio between the weighed sum of benefits and costs:

$$\text{Value} = \left(\sum_i p_i B_i\right) / \left(\sum_i p_i C_i\right)$$

This value can be as follows:

- A *use value* or *functional worth*
- A *cost value* (summing the costs of materials, work and overheads)
- An *esteem value* (or prestige value, for which one is willing to pay an extra)
- An *exchange value*

The benefits considered to calculate this value can be functions, performances or both, and are ranked on different scales.

It is important to distinguish between primary functions (those responding to the primary scope for which the product is manufactured—e.g., in the case of a screwdriver, to transfer torsion) and secondary ones (referring for example to how the primary functions are carried out—in the earlier case, it could be to provide electric insulation).

It is also necessary to distinguish between (a) configuration parameters (e.g. the engine displacement in a car), (b) performance parameters (horsepower or fuel consumption), (c) performance indexes (the ratio between the earlier parameters, such as power/engine displacement and speed/fuel consumption).

VA considers all the primary and secondary functions of a product, assessing their usefulness and the level of appreciation by the customer. The key concept in this analysis is that the consumer is willing to pay an extra sum for some accessory functions. The scope is to create a product of lower price, supplied with all the functions appreciated by the customer and lacking all those that the customer is unwilling to pay for.

Value Engineering (*VE*) was applied for the first time in General Electric during World War II. It considers the materials and manufacturing processes needed for every function and component, selecting those that are both the cheapest and most efficient to perform the desired functions (De Marle and Shillito 1992).

9.2 Designing with the Suppliers

The relationship with the suppliers has gained increasing importance over the last few years, as it became more evident that working together during the stages of product design/development and indeed during the entire production cycle could greatly improve time, cost and quality performances; another factor promoting collaboration is the increasing weight of purchase costs on the price of the end product (Clark 1989).

The main novelties concerning the evolution of the customer–supplier bond refer to the *models* and *types* of relationship.

With reference to the *models*, these mainly consist in (Tonchia et al. 1994a) the following:

- *Revising the traditional antagonistic model of the customer–supplier relationship.* The closer interdependence between the various units of the supply chain has transformed the transaction of goods into a relationship of cooperation, which has also become more exclusive, reducing the number of suppliers. The evolution of the relationship has led to *single sourcing*, *dual sourcing* and *parallel sourcing*, featuring, respectively, one supplier per component, two in competition with each other, or again, only one, but with the possibility of changing the supplier in favour of someone who produces a similar part and/or one that can be used for another model.
- A *reconfiguration and integrated management of the supply chain.* The broader area of customer–supplier interaction redefines the profile and role of the upstream interlocutor. In particular, privileged suppliers are those who provide sub-assemblies that are not critical for the purchasing company, thus allowing the latter to reduce the relational effort arising from a larger number of supplying interlocutors. In other words, the company interacts with a smaller number of suppliers than before; these are often tier-1 suppliers managing tier-2 and tier-3 suppliers, who in the past would have interacted directly with the purchasing company (Cusumano and Takeishi 1991).
- An *extension of the traditional geographical area for procurement.* These changes arise from the world-wide competition that exists in certain sectors, and the need for a company to acquire distinctive features on the global market.

As regards the *types* of relationship with the suppliers, it is possible to distinguish between the following:

- A *traditional relationship*, featuring negotiation, inspections of the incoming products and safety stock
- An *operational integration*, with medium- or long-term relationships, the possibility of changing prices and volumes, guaranteed and certified quality, responsibility for the modules or sub-assemblies provided, frequent supplies in small lots (when operating in an *open order* mode) or according to the needs of the manufacturing departments (i.e. the *pulled* supply of Just-in-Time Purchasing), continuous quality improvement and cost reduction, offers of *service packages* (Tonchia et al. 1994b), consultancy and training activities
- *Partnership*, which results in cooperation right from the stages of design (*co-design*—Hartley et al. 1997; Ragatz et al. 1997), common investments in technology, research and development, a constant exchange of information, and in general, a coherence of strategies

Delegating the suppliers to partly or wholly design and develop the components is the last step towards establishing a partnership. While a medium- to long-term contract with an *integrated* supplier guarantees certified quality and just-in-time delivery (i.e. in small, frequent lots), a relationship with a *partner* also ensures cooperation in component design/development and, thanks to shared investments in research and technology, an innovation in the manufacturing/assembling processes. The companies can finally decide to establish solid builder/supplier groups that in Japan are known as *keiretsu*.

There are two types of involvement:

- Spot, when the supplier is occasionally invited to meetings with the engineering team of the company so as to discuss certain issues and design solutions
- Continuative, when the supplier's team works with the company for the entire duration of the designing activity

In the latter case, the involvement ranges from *detailed controlled* (when all specifications are provided by the customer, and the supplier must only carry out engineering and production) to *black box* (when the client specifies the basic features—performances, external shape, interface, and so forth, which arise from a rough, preliminary design—leaving the details to the supplier), to *supplier proprietary* (where the specifications, although congruent with the project of the client, are entirely defined by the supplier) (Clark and Fujimoto 1991). The second case is becoming increasingly widespread, because the company can exploit the technical and engineering know-how of the supplier while preserving control of the product's architecture; the advantages are however counterbalanced by the risk of depending on one supplier only and being spied on by the competitors. Usually, the smaller the interdependence between product parts, the more the black box policy is used. The final drawings are either *approved* or *consigned*, according to whether the supplier keeps the property of the drawings (and therefore the client only approves their use) or not (usually when a large number of specifications, although not detailed, are defined by the customer).

The involvement of suppliers largely depends on their *designing ability*, and its two parameters: breadth, i.e. the number of black box transactions, and depth, namely the different types of design activities that the supplier can carry out (basic specifications, detailed design, assemblies, prototypes, etc.). The criteria used to assess the supplier's designing capability consider various areas: (1) devising and designing the part and/or sub-assembly, (2) planning and engineering it, (3) planning and engineering its manufacturing process.

Nowadays it is a common practice to involve the suppliers in the design; in the automotive sector, for example (Cusumano and Nobeoka 1998), producers are reducing their design efforts and are moving towards the development of an overall design solution (*modular platforms*); at the same time, the suppliers of sub-assemblies or systems (first level suppliers) are gaining more space in designing activities (illustrated by the larger area in the lower part of Fig. 9.3), although they too are moving towards more *systemic* solutions.

The automotive group Fiat, for example, operates a *strategic segmentation* of the suppliers, grouping them into five categories: (1) *co-design A* (for parts that have a strong impact on the overall style and/or performance of the vehicle, hence the suppliers are involved from the stage of concept development), (2) *co-design B* (for components that are influenced by the style of the vehicle, such as the parts inserted into the dashboard or the wipers, supplier involvement occurs at a later stage), (3) *simultaneous* suppliers (who provide the major metal parts; simultaneousness

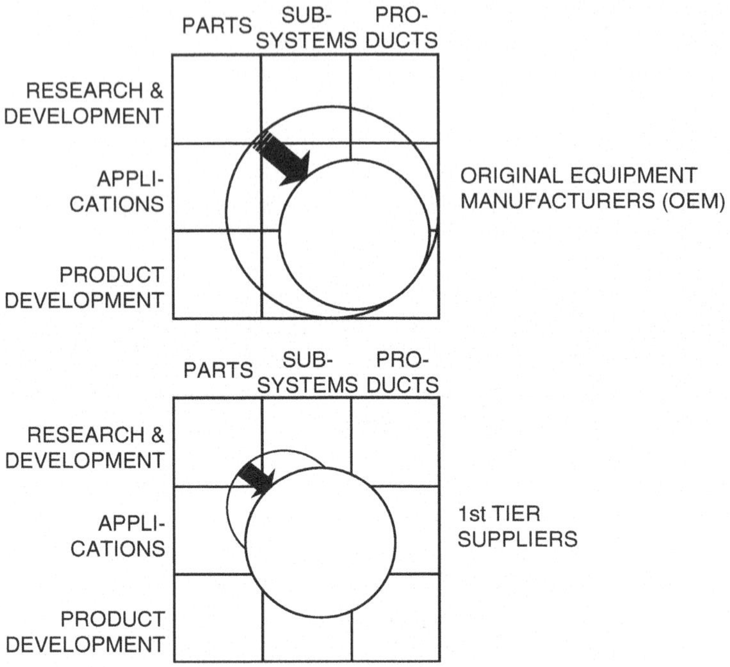

Fig. 9.3 Reconfiguration of the design activities of the OEM (*above*) and greater design activities by the first level suppliers (*below*)

is fundamental, given the long lead times), (4) *detailed controlled* (for smaller metal or plastic parts), (5) *off-the-shelf* or *commodity* suppliers (for other components such as sparking plugs).

9.3 Design for Manufacturability

Design for Manufacturability (*DFM*) considers the effects of the product's structure on manufacturing costs and, thanks to cooperation between the design and production units, aims at simplifying the manufacturing processes, while preserving the characteristics and performances of the product (Niebel and Liu 1992; Youssef 1994).

Stoll (1988) lists a series of key points ensuring a good DFM, which include assessing the *manufacturability* of the various parts, so as to choose design solutions ensuring a high degree of conformity; evaluating the complexity of the manufacturing process required by certain design solutions and choosing the easiest ones among those ensuring the functionality of the project; assessing the variety of machinery setups required by a certain solution.

It is also possible to create two types of correlation matrices, known as *Houses of Producibility*: one correlates the design parameters of the product with the process, and the other the process design parameters with the production performances (Wheelwright and Clark 1992).

Like DFM, *Design for Assembly* (*DFA*) is aimed at limiting assembly costs while preserving a high level of quality, by choosing the most appropriate assembling methods, reducing handling and change of direction, inserting and connecting the components according to their shape, the material they are made of, technology, etc. (Boothroyd and Dewhurst 1987).

Since it considers the impact that certain choices of design have on production (manufacturing and assembling), DFM/DFA is also known as *Design for Operations—DFO* (Schonberger 1990). Given here is a list of the main rules of DFO:

- Limit the number of components
- Reduce component variety as much as possible
- Design components that can perform multiple functions
- Design components that are suitable for various uses
- Design components that are easy to manufacture
- Maximise conformity, so as to make it easier to comply with the specifications
- Reduce handling during production
- Design so as to make assembly and disassembly more simple
- Choose if possible only one direction for assembly, keeping in mind that top to bottom is better than vice versa
- Avoid separate fixing/connecting devices
- Eliminate or reduce changes while assembling

Applying these rules and keeping in mind past experiences simplifies the following stage of engineering, making it possible to limit the number of tests and trials (pilot runs or prototypes).

While DFM/DFA lists a series of principles that, if applied, should simplify manufacturability, *Integrated Product–Process Design* (*IPPD*) aims at integrating product design with production design (Ettlie 1997). To do so, it is necessary to find points of similarity between product and process, on the grounds of the following (De Toni and Zipponi 1991):

- *Product repetitiveness*, which can be analysed by considering the production volume (i.e. the total number of items per their average volume)
- *Process repetitiveness*, which can be analysed by considering the production volume as the product of the number of families (i.e. a group of products characterised by similar component items produced by groups of similar machines—*manufacturing cells*), per the average number of items per family, and the average volume of each item

Interventions of IPPD, aimed at rationalising the product and simplifying the manufacturing processes, can occur at three different levels:

1. Level 1, the lowest (level of *single component* as regards the product, and *single operation* in the manufacturing process)
2. Level 2, intermediate (level of *sub-assembly or functional group* in the case of the product, and *functional unit* in that of the process)
3. Level 3, the top level (level of *end product*, and *production process*)

9.4 Reducing Variety

The Variety Reduction Program (VRP) is a technique, theorised by Koudate and Suzue (1992), that is aimed at reducing the cost of design and product development by limiting the number of components and processes needed to manufacture a product, while responding at the same time to the market demand for an ample variety of products. It may seem a paradox to reduce variety when the market requires a wide range of models and customization: however, the technique ensures a reduction of *internal* variety (which, for a company, means costs), while preserving *external* variety and therefore the number of goods offered to the customers (it is this type of variety that contributes to customer satisfaction).

VRP considers three different types of costs, which differ extensively from the traditional ones examined in cost management: the latter costs can be easily classified according to their nature (e.g. those relative to the materials, machines/plants, staff, etc.), their variability (fixed costs, variable costs, semi-variable, etc.), their attribution to a cost centre or a product (direct/indirect) or in relation to time (predicted, actual, final), etc. The costs considered in the VRP are as follows:

1. *Functional costs*, linked to specific performances/functions that the product must possess (they indicate the cost needed to ensure a certain performance of a given function added to the product, which implies additional materials, components, manufacturing processes, etc.). In other words, the focus is no longer on the traditional bill of materials (Product Breakdown Structure—PdBS) but on the product's functions (Product Function Structure—PFS). The relationship between performances/functions and costs is determined by its Value Analysis (VA).
2. *The cost of variety*, due to the fact that if a wide range of products is appreciated by the market, it requires different processes and therefore different types of machinery and equipment.
3. *The cost of control*, namely assessing the economic impact of complex management.

These costs evolve differently in relation to variety, with the latter two rising as product and process variety increase, whilst, under the same conditions, functional costs decrease (Fig. 9.4). A greater variety allows the latter costs to be distributed over a wider range of products: accessories, for instance, can be applied to different families of products. The VRP technique helps identify the level of variety needed to

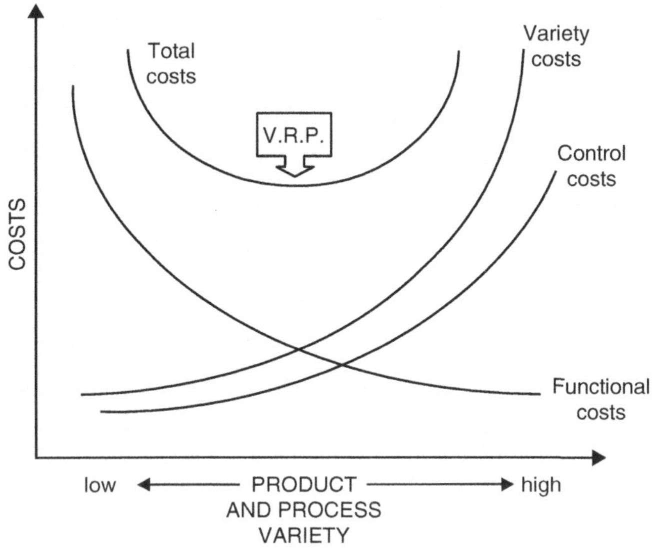

Fig. 9.4 The optimal degree of variety is the minimum of Variety Reduction Program (VRP) total costs (Source: Koudate and Suzue 1992)

minimise the three types of cost. It is an innovative approach, differing extensively from the more conventional methods used to examine costs.

External variety coincides with the *breadth* of a given range of products, while the *depth* measures the number of options for a certain family (for instance, in a car: three or five doors, with or without air conditioning, etc.).

Internal variety can be assessed by means of various *variety indexes*, such as the following:

- The *parts index*, based on the type and number of parts forming a product
- The *production process index*, based on the different types of processes needed to manufacture a product and the number of machines involved in each process

An example of calculation used to determine the parts index is shown in Fig. 9.5: A, B and C are three different models, X, Y and Z are their respective components, which differ slightly—in size, for instance—and hence there are variants x', x'', x''' for X, etc.; 2* and 4* are the use coefficients.

Figure 9.6 illustrates how the production process index is calculated: the circles are the departments and the triangles are the warehouses; the flows are not necessarily linked to production lines as shown in this example, but could be merely logistic, in other words calculated for executing production.

By using these indexes and the functional, variety and control costs mentioned previously, it is possible to define programmes aimed at reducing variety by combining the different parts and minimising the number of components needed to carry out a certain function. Cost evaluation will help define a level of variety close to the

Fig. 9.5 *Part Index* calculation (Source: Koudate and Suzue 1992)

models / parts	A	B	C	Type of parts
X	x'	2*x"	x"'	3
Y	2*y'	y'	–	1
Z	4*(z'+ +z")	4*(z'+ +z")	4*(z'+ +z")	2
No. of parts	7	7	5	6 / 19

"Part Index"= 6 * 19 = 114

Fig. 9.6 *Production Process Index* calculation (Source: Koudate and Suzue 1992)

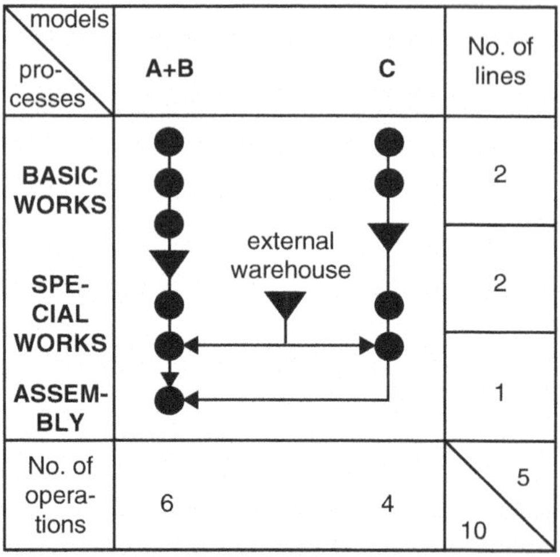

"Production Process Index" = 5 * 10 = 50

lowest point of the U-shaped curve obtained by adding up the three types of cost (Fig. 9.4).

The VRP technique also suggests ways to reduce (internal) variety by applying five principles:

1. Distinguish between *fixed and variable parts*. In any given project, a clear distinction must be made between these two types of parts (according to the customer's need), as well as semi-variable (or similar) ones, and only increase the number of variable parts that respond to the real needs of the customer, reducing superfluous accessories.
2. Use *combinations of parts*, both to amplify the range of products and create higher level functions by combining various lower level ones.
3. Apply *multi-functionality/integration*, so that only one part or sub-assembly performs various functions (e.g. using the engine shaft both to transmit motion and distribute oil or conditioned air), if necessary by integrating various parts into one.
4. Analyse the *range of performances* for a certain part, without having to substitute it with another, so as to reduce the number of parts that must be designed/produced for the entire range of products.
5. Analyse the *series*, namely seek a constant ratio (different for each series) between two dimensions, or two performances, or between size and performance.

Internal variety should also be reduced so as to ensure low production process indexes, at least during the first stages of production. Thus, in order to preserve the same range of products, external variety must be created downstream, towards the end of the manufacturing process (usually when assembling). This, in short, is the *mushroom concept* illustrated in Fig. 9.7: differentiation only occurs during the last stages of production (i.e. the mushroom cap), while the former ones (the stem) remain few and standardised. Hence, according to the mushroom theory, internal variety can be reduced while external variety remains unchanged.

Among the programmes used to reduce variety, *modularisation* or *modular design* (Rajput and Bennett 1989; De Toni and Zipponi 1991) deserves special mention. It helps obtain sufficiently different products while economising on the activities of design, production and management of logistic flows, thanks to the repetitive use of standard modules and parts when designing the product.

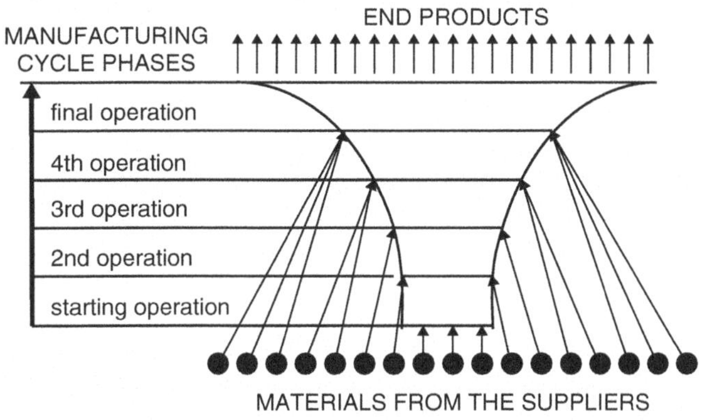

Fig. 9.7 *Mushroom* manufacturing cycle

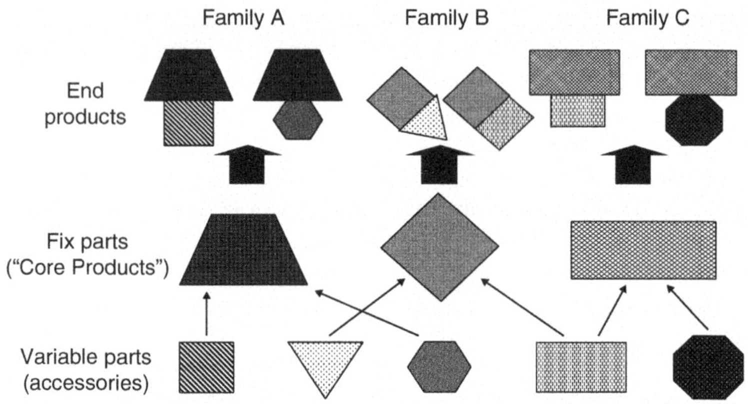

Fig. 9.8 Vertical modularisation

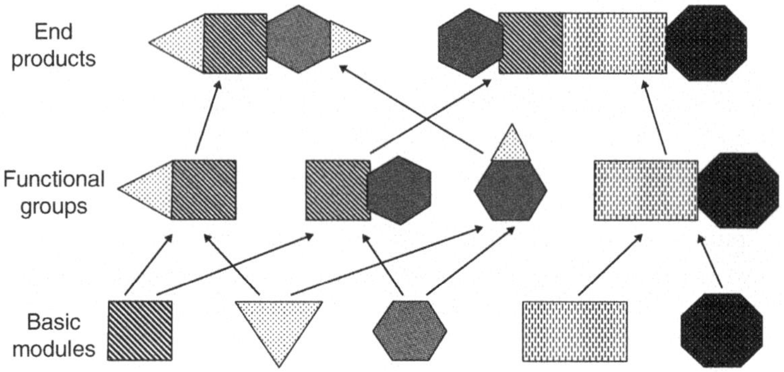

Fig. 9.9 Horizontal modularisation

Modularisation is performed on the bill of materials, and can be of two types, *vertical* and *horizontal* (Baldwin and Clark 1997). In the former case (Fig. 9.8), there is a fixed part that is common to all products (thus known as *core product*), and variety originates when assembling variable parts (such as accessories) onto the core products. Thus, families of end products are created starting from different core products (depicted as a trapeze, rhombus and rectangle in Fig. 9.8). In the latter case (Fig. 9.9), variety originates from the combination of basic modules, leading to the production of intermediate ones (also known as *functional groups*, which in turn determine the end product); this type of modularization is defined as horizontal, since in the product there is no prevalence of a certain part (one of the geometrical shapes illustrated in Fig. 9.9).

From a production viewpoint, the concept of modularisation has given rise to a specific process technology known as *Group Technology* (Burbidge 1975), which, as the name suggests, groups the machinery into *production cells* according to the *families of products* that are manufactured and which features groups of similar parts (Figs. 9.10 and 9.11).

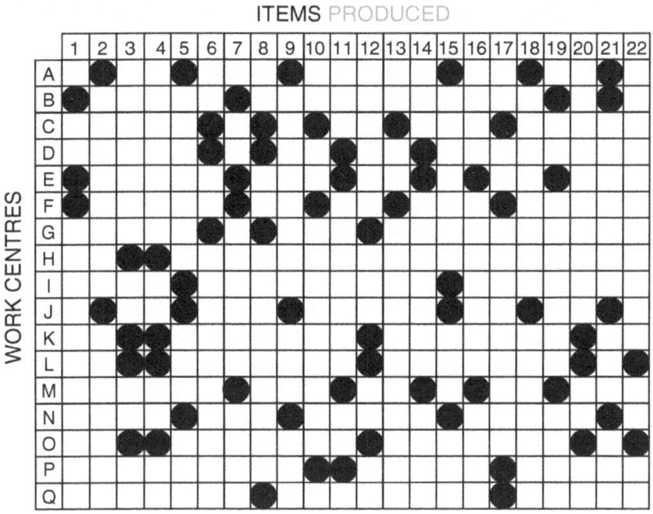

Fig. 9.10 Creation of families of products and production cells through the Group Technology (A)

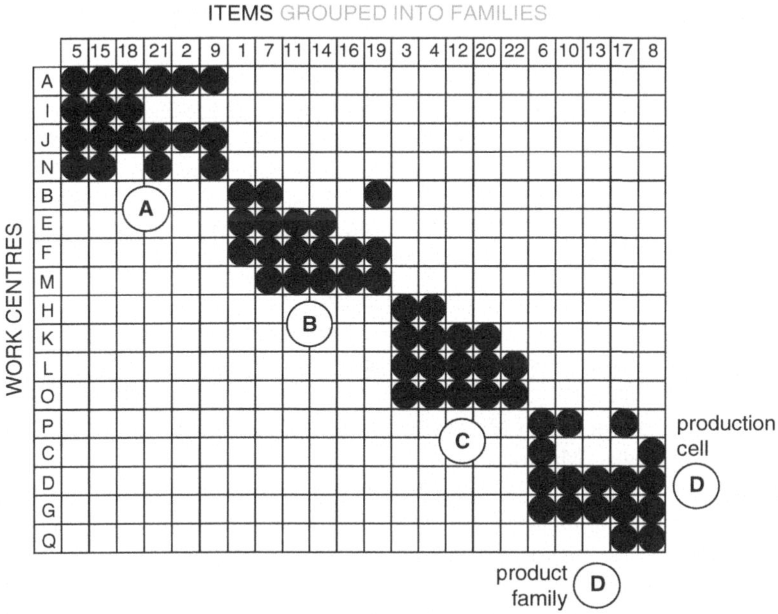

Fig. 9.11 Creation of families of products and production cells through the Group Technology (B)

Finally, certain authors combine the management of carry-over, the principles of VRP and modularisation practices into one technique; Koudate (1991), for instance, calls it *Edited Design* or *Henshu Sekkei* and underlines the importance of distinguishing between fixed and variable parts, the latter ones playing a key role in satisfying market demands.

9.5 Simultaneous Engineering

Simultaneous (or Concurrent) Engineering uses a systematic approach for the simultaneous and integrated design of products and processes, including production and support. Ideally, it should take the place of the traditional, time-consuming pathway of serialised, repetitive adjustments, which determine a net separation between client and designer, as well as between designer and technician or the persons in charge of maintenance and technical assistance (Carter and Stilwell 1992; Koufteros et al. 2001).

In other words, the duration of a project can be shortened, not only by—thanks to—the reduction of each activity, but also by joining the efforts and contributions of the protagonists of each stage. In a more stringent sense, Concurrent/Simultaneous Engineering (C/SE) aims at parallelising the activities, so as to reduce the overall duration of a project.

Some authors distinguish between Concurrent Engineering (joint, parallel interventions so as to shorten times by having overlapping stages of development) and Simultaneous Engineering (intended as contemporaneous design of various products—a family—using a multi-project management approach).

In practice, C/SE aims at integrating and overlapping the activities by increasing the exchange of information, producing interim reports and passing relevant data to the following activity in advance, before the previous one is over—thanks to intensive team working and transversal meetings.

The principle behind C/SE is that a given activity must not necessarily start when the previous one is over, but there can be a certain degree of overlapping, which would promote a bilateral exchange of information and help revise/adapt both activities, so as to reduce not only design times but also problems that could occur in the following stages (Maylor 1997).

Practice has shown however that C/SE may have certain drawbacks, due to an excessive amount of parallel modules and a more complex management and technical integration.

Rapid Prototyping (*RP*) makes it possible to pass directly from 3D CAD drawings to prototypes made either of special resins or the actual materials used for the end product. The prototype is constructed by means of stereolithography (which hardens a photopolymer—thanks to the selective action of a laser beam) or laser syntherisation (in this case, the models are obtained by the thermo-fusion of powders).

Not only can these procedures be used to create the prototypes of parts and products, but also to test production equipment such as tools and moulds (*Rapid*

Tooling). The Research Centre in Fiat—CRF—has defined a process to build moulds for the pre-series, consisting of four elements: (1) a hull derived from the CAD model, made by electro-deposition or *metal spray*, which is used to create the shape, (2) its lining with carbides, ensuring resistance to wear, (3) its resin support, ensuring mechanical resistance, (4) a metal sheet container, which connects the mould to the press.

It is also possible to carry out *Virtual Prototyping* (*VP*), which helps study the functionality of a product or its manufacturing process *without* actually creating any physical artefact. Virtual technology, originally applied for space exploration and defence, then expanded to the field of entertainment and has now become popular in the industrial sector too. The computer recognises the virtual product and the context where it is used; the model defined by the CAD is transferred into a virtual world so as to visualise/animate it and simulate functionality tests. There are three main systems currently used, listed according to their complexity and costs: (1) the *CAD viewer*, where the engineer is not immersed in the virtual world but simply observes it through special glasses and can use virtual dummies to test the ergonomics of a product, (2) *third-party* systems, where the engineer is shot, and the images are reproduced by the computer and then immersed into the virtual world, so that the engineer can interact with the product, (3) *immersion systems*, where (thanks to special devices such as gloves, helmets, etc.) the engineer is de facto in a virtual world.

9.6 Reverse Engineering

Techniques of *Reverse* (or Feedback) *Engineering—RE—*include Design of Experiments (DOE), Rapid Prototyping (RP) and Failure Mode and Effect Analysis (FMEA). These tools help assess and approve the choices already during the stages of design/engineering, hence before the product is actually made, so that it is unnecessary to use the manufacturing plants and start production.

Design of Experiments (*DOE*) helps identify the physical and operational parameters that have the greatest influence on the features or performances of the product: in this way, the engineers know what tests and trials to carry out when designing and optimising the product (Mason et al. 1989; Wang et al. 1992). It is also known as *design–build test cycle*, to underline that the definition of the tests and trials is an integrating part of the product's design. Experiments are carried out to define models linking the physical/operational parameters of the product to the results, in order to assess their impact as their values change. During this stage, also the trials and experimental designs are defined. From a practical viewpoint, it is necessary to follow the steps listed here:

- Define the feature/performance to achieve or improve, identifying its target value
- Define the physical parameters and the conditions under which it is tested and used

- Choose the experimentation levels for each parameter (number of levels and values)
- Choose the experimental design/procedure
- Interpret the results of experiments
- Repeat the experiments
- Identify the best parameters/solutions

Early Problem Detector Prototyping (*EPDP*) exploits prototypes to identify the problems and defects of a given design as soon as possible. The prototypes may be similar to the end product or only perform some functions, each of which must be tested separately.

Prototyping, when not rapid/virtual, requires *pilot runs* and can be carried out while the previous product is being manufactured, to then gradually extend to the other stages of production (*step-by-step prototyping*).

It is important to detect problems as soon as possible, because the cost of change increases exponentially as product development advances; for this reason, it is advisable to prepare a wide range of solutions during the first stages of design, which should be then screened, to narrow down the options. This procedure is like a funnel, and the wider it is at the top and the sooner it becomes narrow, the less it will cost to revise the project.

A *dilemma* thus arises: whether to make a decision that will have a fundamental impact on the performances of the product/plant at an early stage, relying on uncertain and incomplete data, or postpone the choice, awaiting more detailed information and clearer opportunities.

FMEA, also known as Failure Mode Effect and Criticality Analysis (FMECA), is a technique used to assess the reliability of a product. It was first defined by the US Army to determine equipment *failures* and their consequences on the system (military procedure MIL-P-1629 dated 11 September 1949). FMEA, comprising both a *product FMEA* and a *process FMEA*, became the standard method (and procedure) applied in the field of Advanced Product Quality Planning (APQP), in compliance with TS16949 norms.

The importance of reliability as an intrinsic performance of the product (together with its features and functionality) requires great care during the stages of design, where possible causes of product failure must be identified, studied and eliminated. FMEA considers the possible types of failure (of the whole product, and due to faults arising from wear, incorrect handling, etc.), their effects and causes, verifying whether they can be ascribed to the materials used or the manufacturing processes.

Both product and process FMEA follow the procedure described in Fig. 9.12. The starting point is the identification of the *functions*, i.e. the intended scope of the product or the process, which must be analysed; the following step consists in the identification of failure modes and their effects, their severity being ranked on a scale from 1 to 10 (the latter value being the most severe, i.e. that in which the life of the person—either the user or the process operator—may suddenly be in jeopardy).

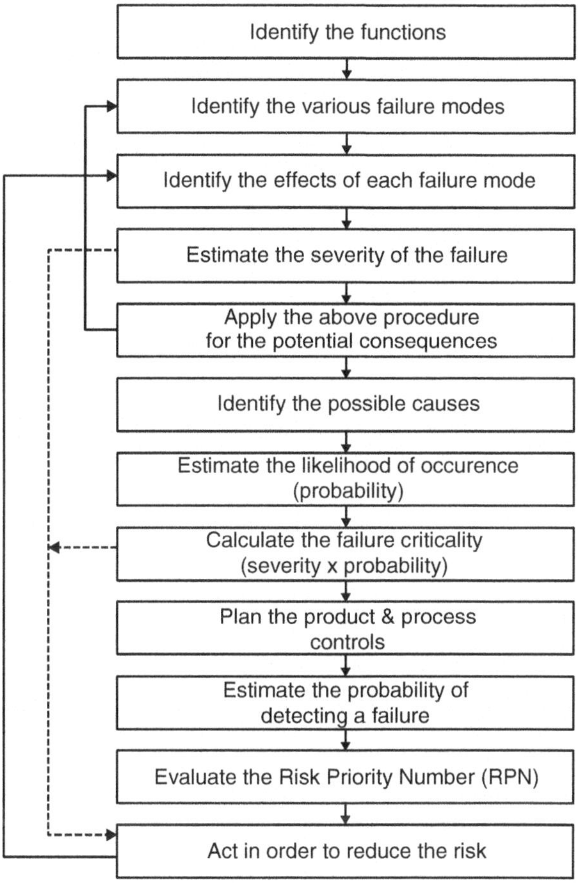

Fig. 9.12 The Failure Mode and Effect Analysis (FMEA) methodology

After identifying possible causes of failure and analysing their potential consequences, it is necessary to calculate the likelihood of failure occurrence, once again on a scale from 1 to 10 (1 being a fault occurring with a probability lower than 1/1,500,000 and 10 with a probability greater than ½).

When its *criticality*, namely the severity of failure multiplied by the likelihood of its occurrence, has been determined, it is time to define the tests to be carried out both on the product and on the process. There are three types of test: (1) *type 1*, to avoid the occurrence of a failure cause or mode, thus reducing the likelihood of failures, (2) *type 2*, to discover the cause of failure and define corrective actions, (3) *type 3*, to reveal failure modes before the product reaches the customer.

It is then necessary to estimate the probability of detecting a fault; the values, in this case, range from 1 (the test will nearly certainly reveal the potential cause of the failure or the consequent failure mode) to 10 (the test will probably fail to reveal the potential cause of failure and the consequent failure mode, or the test is not carried out).

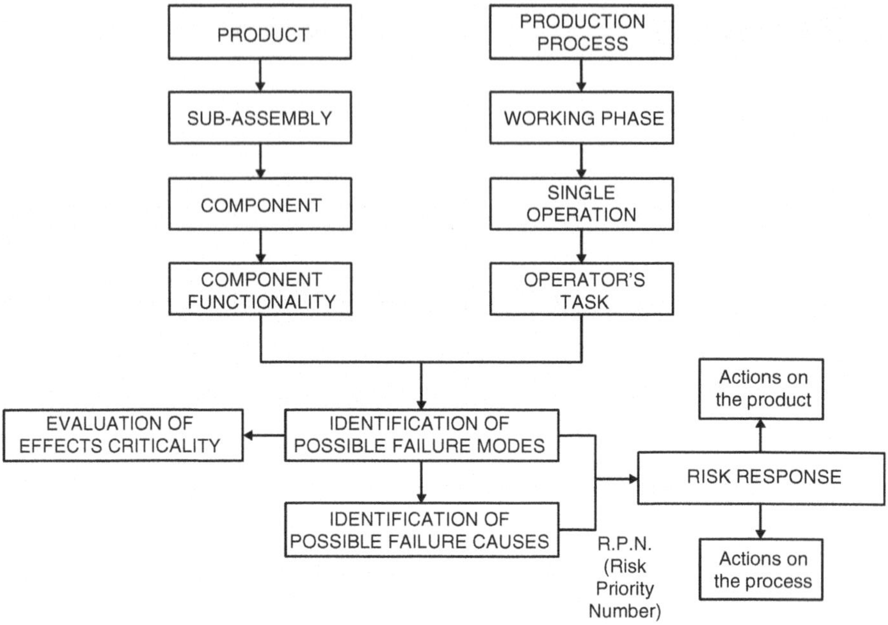

Fig. 9.13 Logical flow for FMEA

The last step is to calculate the so-called *Risk Priority Number* (RPN), which is equal to the *severity* of failure multiplied by the *probability of occurrence* and the *probability of interception*, namely the possibility of identifying the cause of failure. RPN values range from 1 to 1,000: statistical data show that the average value is around 160, whereas only 6% of all indexes are higher than 500 (median: 105).

FMEA should be performed both on the *product* and on the manufacturing *process*, as shown in Fig. 9.13.

FMEA can also be applied in the field of services; in this case, the RPN is calculated by multiplying the likelihood of a disservice by the criticality of its effects and the ability to detect it.

References

Baldwin, C. Y., & Clark, K. B. (1997, September/October). Managing in the age of modularity. *Harvard Business Review*, 84–93.

Barkley, B. T., & Saylor, J. H. (1994). *Customer-driven project management: A new paradigm in total quality implementation*. New York: McGraw-Hill.

Boothroyd, G., & Dewhurst, P. (1987). *Product design for assembly handbook*. Wakefield, RI: Boothroyd & Dewhurst.

Burbidge, J. L. (1975). *The introduction of group technology*. London: Heinemann.

Carter, D. E., & Stilwell, B. (1992). *Concurrent engineering: The product development environment for the 1990s*. Reading, MA: Addison-Wesley.

Clark, K. B. (1989). Project scope and project performance: The effect of parts strategy and supplier involvement on product development. *Management Science, 35*(10), 1247–1263.

Clark, K. B., & Fujimoto, T. (1991). *Product development performance: Strategy, organization and management in the world auto industry*. Boston, MA: Harvard Business School Press.

Cusumano, M. A., & Nobeoka, K. (1998). *Thinking beyond lean: How multi-project management is transforming product development at Toyota and other companies*. New York: Free Press.

Cusumano, M. A., & Takeishi, A. (1991). Supplier relation and management: A survey of Japanese-transplant, and U.S. auto plants. *Strategic Management Journal, 12*, 563–588.

De Marle, D., & Shillito, M. L. (1992). Value engineering. In G. Salvendy (Ed.), *Handbook of industrial engineering*. New York: Wiley.

De Toni, A., & Zipponi, L. (1991). Operating levels in product and process design. *International Journal of Operations & Production Management, 11*(6), 38–54.

Ettlie, J. E. (1997). Integrated design and new product success. *Journal of Operations Management, 15*(1), 33–55.

Fowler, T. C. (1990). *Value analysis in design*. New York: Van Nostrand Reinhold.

Hartley, J. L., Meredith, J. R., McCutcheon, D., & Kamath, R. R. (1997). Suppliers' contribution to product development: An exploratory study. *IEEE Transactions on Engineering Management, 44*(3), 258–267.

Koudate, A. (1991). *Il management della progettazione*. Torino: ISEDI – JMAC.

Koudate, A., & Suzue, T. (1992). *VRP: Variety reduction program*. Torino: ISEDI – JMAC.

Koufteros, X., Vonderembse, M., & Doll, W. (2001). Concurrent engineering and its consequences. *Journal of Operations Management, 19*, 97–115.

Mason, R. L., Gunst, R. F., & Hess, J. L. (1989). *Statistical design & analysis of experiments*. New York: Wiley.

Maylor, H. (1997). Concurrent new product development: An empirical assessment. *International Journal of Operations & Production Management, 17*(12), 1196–1214.

Mazur, G. H. (1990). *Introduction to quality function deployment*. Ann Arbor, MI: Japan Business Consultant.

Michaels, J., & Wood, W. (1989). *Design to cost*. New York: Wiley.

Miles, L. D. (1992). *Techniques of value analysis and engineering*. New York: McGraw-Hill.

Mizuno, S., & Akao, Y. (Eds.). (1994). *QFD: The customer-driven approach to quality planning and development*. Tokyo: Asian Productivity Organization.

Niebel, B. W., & Liu, C. R. (1992). Designing for manufacturing. In G. Salvendy (Ed.), *Handbook of industrial engineering*. New York: Wiley.

Pugh, S. (1991). *Total design: Integrated methods for successful product design*. New York: Addison-Wesley.

Ragatz, G. L., Handfield, R. B., & Scannell, T. V. (1997). Success factors for integrating suppliers into new product development. *Journal of Product Innovation Management, 14*(3), 190–202.

Rajput, S., & Bennett, D. (1989). Modular system design and control for flexible assembly. *International Journal of Operations & Production Management, 9*(7), 17–29.

Schonberger, R. J. (1990). *Building a chain of customers*. New York: Free Press.

Stoll, H. D. (January 1988). Design for manufacture. *Manufacturing Engineering, 100*(1), 67–73.

Tonchia, S., De Toni, A., & Nassimbeni, G. (1994a). New trends in the supply environment. *Logistics Information Management, 7*(4), 41–50.

Tonchia, S., De Toni, A., & Nassimbeni, G. (1994b). Service dimensions in the buyer-supplier relationship. *International Journal of Physical Distribution & Logistics Management, 24*(8), 4–14.

Wang, H. S., Koo, T. Y., & Lu, I. Y. (1992). Design of experiments. In G. Salvendy (Ed.), *Handbook of industrial engineering*. New York: Wiley.

Wheelwright, S. C., & Clark, K. B. (1992). *Revolutionazing product development: Quantum leaps in speed, efficiency and quality*. Free Press: New York.

Youssef, M. A. (1994). Design for manufacturability and time-to-market, Part 1: Theoretical foundations. *International Journal of Operations & Production Management, 14*(12), 6–21.

Part IV

Managing Project's Variables

Project Time Management 10

10.1 Time Representation: Gantt Chart and Network Diagram

Time was the first variable in Project Management to be supported in a coherent manner by specific techniques. Thanks to the great repercussions caused by variations in the costs and delivery times for the high-profile military and civil orders made after the war. But while costs could be ascribed to a number of causes that, although often impossible to predict or control (as in the case of cost fluctuations), could be managed through various techniques (see Chap. 13), in the case of delays, management seemed less complex, and problems easier to solve.

The situation at the time is illustrated by two research projects dating back to the 1950s, one of which was carried out by the Rand Corporation (Marshall and Meckling 1959) and focused on military projects, whereas the other, on civil projects, was carried out by the Harvard Business School (Peck and Scherer 1962). The investigations reveal a cost factor (i.e. the ratio between actual and forecasted cost of the project) of 2.4 and 1.7, respectively, and a time factor (the ratio between actual and forecasted implementation times) of 1.5 and 1.4, respectively. Hence, although the delays were similar, there was a large gap between the two cost factors, which was ascribed to the greater degree of innovation characterising the military projects.

However, even before attempts had been made to improve project management, the first steps in this direction date back to 1911, when Frederick W. Taylor experimented on *scientific* production management (coinciding with today's times and methods) and Henry G. Gantt (1861–1919) was working for the US Army.

Gantt is the father of one of the two time representations still in use today: the activity bar on time axis or *Gantt chart* (the other being the network diagram).

Figure 10.1 depicts a simple Gantt chart:

- The start date and end date for each activity (also known as task) are indicated on the horizontal time axis, while its duration is proportional to the length of the bar.

S. Tonchia, *Industrial Project Management*, Management for Professionals,
https://doi.org/10.1007/978-3-662-56328-1_10

Fig. 10.1 Gantt chart

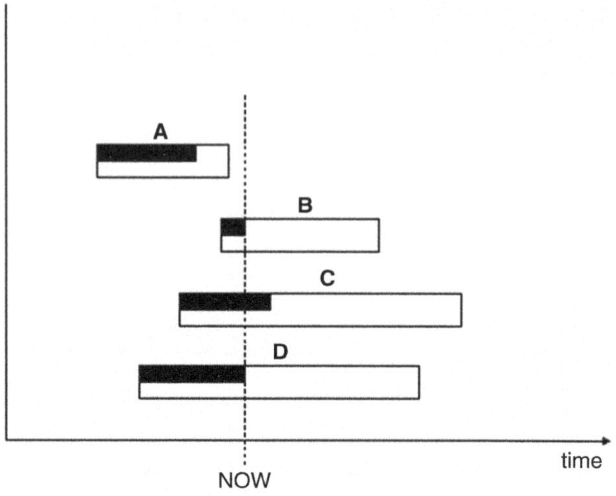

- The state of advancement of the tasks at a given time (usually coinciding with the present) is indicated by the blackening of the bar (in Fig. 10.1, task A is late; B and D are on time; C is early).

The Gantt chart, although easy to read, has various drawbacks. It does not indicate priorities between the activities. The fact that task B starts when A ends does not necessarily imply that task A must be completed before carrying out B. For instance, when building a house, A could indicate plumbing and B wiring (two activities that could be carried out contemporaneously): the fact that one task is performed before the other does not indicate a priority, but could be due for example to the unavailability of the electrician.

To visualise the actual priorities, the bars should be linked with arrows, but in the presence of numerous bars/activities, the complex network of arrows that would arise would make the chart difficult to read. In these cases, it is advisable to use *network diagrams*, which depict activities and their priorities with nodes and arrows.

Typically, a given project can be described by two types of network diagrams (Fig. 10.2), which are known as European and American. In the former diagram, the activities are depicted by nodes, in the latter by arrows. In the latter case, the nodes represent *events*: in Fig. 10.2, node 1 is the event *start of activity A* and node 2 is *end of activity A*. Since the American representation makes the diagram more intricate (e.g. node 5 is necessary to signal the event *activities B and D are concluded* and is a so-called *dummy node*), this type of representation has been almost completely set aside.

Generally, a Project Management network consists of activities (the nodes) and links of precedence (the arcs or arrows). This network is a graph (a combination of nodes and lines) that is interconnected (there are no isolated nodes), oriented (the lines are arrows) and free of circuits (it is impossible to return to a node). To

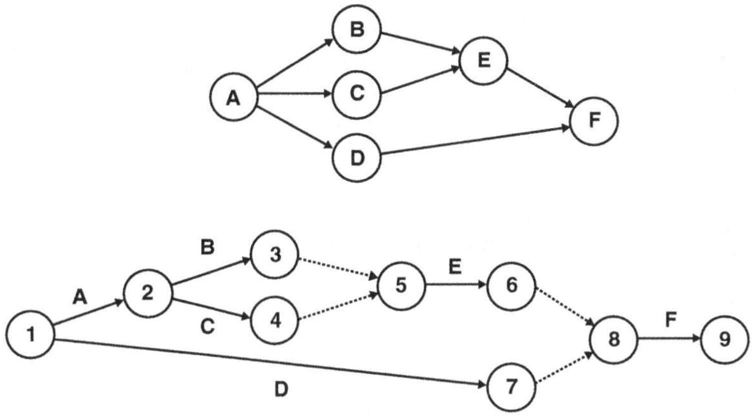

Fig. 10.2 Network diagrams according to the European (above) and the American (below) representation

underline its field of application, it is also known as *Critical Path Method* (*CPM*) *diagram or Program Evaluation and Review Technique* (*PERT*) *diagram*, after the names of the main techniques exploiting it.

Network diagrams can also be hierarchical, according to the principle of Chinese boxes, as shown in Fig. 10.3. It is possible to have either a general description (*above*), or a more detailed one (*below*), where all the links between micro-activities reflect those between macro-activities, and vice versa.

Although the activities are connected to each other by links of precedence/subsequence, some may be concurrent; hence, a given task may start before the previous one is over. The degree of concurrency is measured by the so-called *overlapping ratio*, defined by the formula:

$$IS_i = \frac{\sum_{j=1}^{N} t_{ij}}{T_i}$$

where t_{ij} is the time required by task j of a project i consisting of N activities, and T_i is the overall duration of project i; the total overlapping ratio is then calculated by summing the overlapping ratios for each project i.

If, for instance, we consider, for the sake of simplicity, an identical duration d of the three activities forming a mini-project, and a link of precedence among them, the overlapping index is equal to 1 when all the activities are carried out in a rigid sequence (IS $= (d + d + d)/3d = 1$), whereas it increases, reaching a maximum value of $N = 3$, when they can be carried out simultaneously (IS $= (d + d + d)/d = 3$).

Figure 10.4 illustrates the significance of precedence. A normal link of precedence between a certain activity A and the following one B indicates that A delivers an output that is an input for B. In the event of overlapping, A must be able to supply

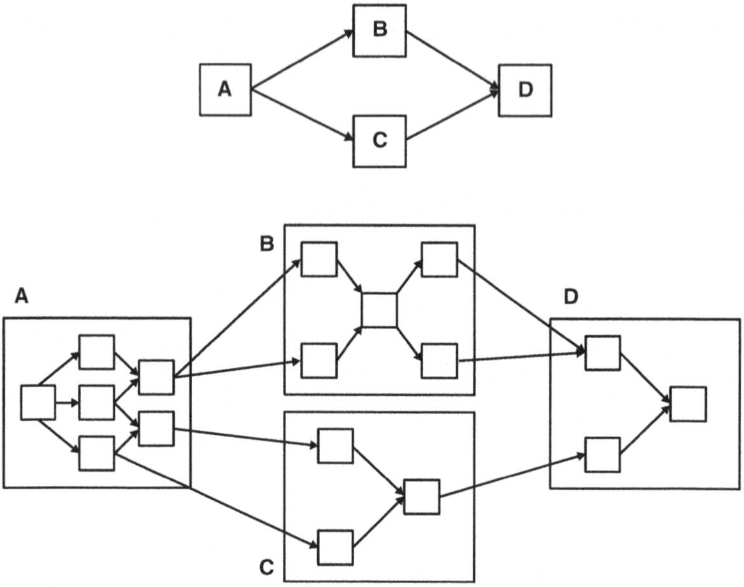

Fig. 10.3 Overall (above) and detailed (below) network diagrams

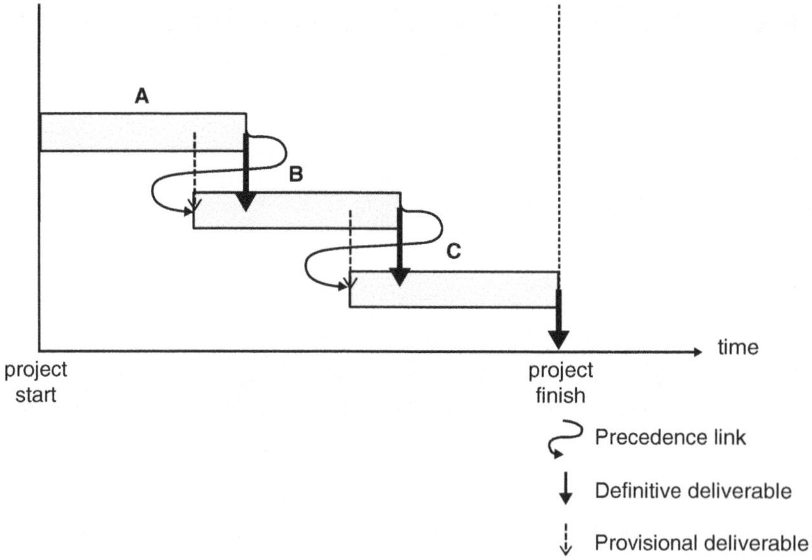

Fig. 10.4 Significance of the precedence link between activities

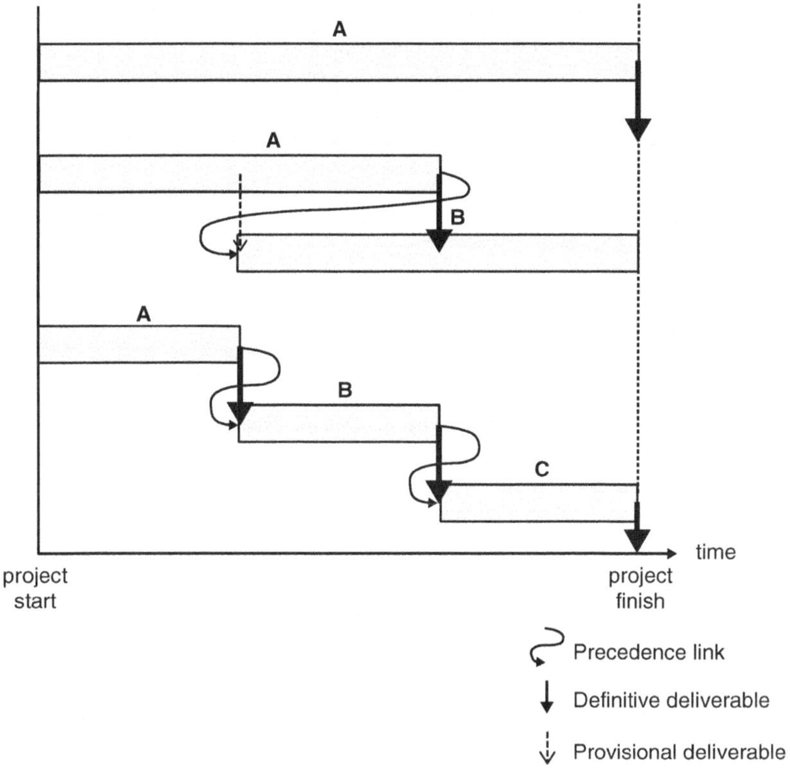

Fig. 10.5 Significance of activity splitting

a provisional deliverable for B to start, and only when it delivers its final output, will activity A end.

Figure 10.5 illustrates the meaning of activity splitting, and the difference that runs between having a large and a small number of activities. The splitting process must be such as to ensure that each sub-task produces a deliverable. If a project has only a small number of activities, there will be very few intermediate deliverables.

10.2 Network Techniques: CPM, MPM, PERT, GERT, VERT

Network techniques are used for managing time; they are called so because they are based on the network diagrams, in this case not intended to illustrate data but to elaborate them, so as to do the following:

– *Schedule activities*, namely define the start and end date of each activity and, consequently, the duration of the entire project
– *Analyse allowable floats* (*or slacks*)—for non-critical activities—that do not increase the duration of the project, which means identifying those activities that, if delayed, would prolong the duration of the project, and are therefore known as *critical*

There are three types of starting data:

1. The activities (task name or identification code)
2. Their duration
3. The links of precedence between the activities

The activities are those described in the Work Breakdown Structure (WBS), whereas duration and precedence are discussed and defined at meetings, and can be either deterministic (certain) or probable (estimated with a certain margin of risk).

There are therefore various network techniques:

- *CPM*, used when the duration of all the activities is considered fixed, and likewise the links of precedence, which are of the Finish-to-Start type (namely, the beginning of a certain activity is linked to the end of a previous one)
- *MPM* (*Metra Potential Method*): it is similar to CPM, but also includes three other types of relationship between the activities (namely, Start-to-Finish, Start-to-Start, Finish-to-Finish: e.g. it can sometimes be useful to have two activities finishing at the same time, and in this case a Finish-to-Finish link is established between them)
- *PERT*, namely a CPM having all durations expressed in a probabilistic manner
- *GERT* (*Graphical Evaluation and Review Technique*), namely a PERT that also has the precedence links expressed in a probabilistic manner (logical gates are therefore used to trace the paths, ensuring feedback among the activities and various types of conclusion)
- *VERT* (*Venture Evaluation and Review Technique*), which takes into account the time, cost, resources and risk variables contemporaneously, and is particularly suited in *what-if* scenario analyses and when evaluating and reviewing new businesses or strategic initiatives (as the name suggests)

CPM and MPM can thus be considered deterministic techniques, whereas PERT, GERT and VERT are probabilistic ones.

Defining the likely duration of the activities is not easy: it is insufficient to say that, for instance, in 80% of the cases a certain activity has duration *d*, since this statement does not explain how long it lasts in the remaining 20% of the cases. It is therefore necessary to consider the *distribution* of duration probability, using for example a Gaussian curve that not only indicates the average duration of a certain task, but also the variance (the distribution of values around the mean). However, a new problem arises, namely the huge amount of calculations required to define the overall duration of a project, which must take into account the *sequence* of probabilities. To simplify calculations, it is a common practice to use simulation methods (such as Monte Carlo) or a formula (derived from the Beta distribution) to estimate the duration of each activity.

In the first case, for each activity, a duration value is selected at random, using the Monte Carlo Method; the values thus picked are used to calculate the overall duration of the project. By repeating this procedure for a large number of times (generally, a few thousand), it is possible to obtain a significant sample of the time needed to complete the project, and this sample can be used to calculate distribution, mean value and variance.

Beta distribution can also be used to estimate the duration of a project. This function describes the distribution of duration probability, and intersects the x-axis in two points: the first one corresponds to the times required in an optimistic scenario (i.e. the shortest duration), the second to those in a pessimistic one (i.e. the longest duration). From a practical viewpoint, three types of duration are taken into account: low (d_{opt}), base (i.e. the most likely, d_{ml}) and high (d_{pess}). The estimated duration is a weighed sum of the three, giving more weight to the most likely duration:

$$d = \left(d_{\mathrm{pess}} + 4d_{\mathrm{ml}} + d_{\mathrm{opt}}\right)/6$$

Thanks to these methods, it is possible to obtain deterministic values, and analyse them by CPM.

GERT, on the other hand, is even more complex, because not only is it necessary to calculate probabilities, but also determine the types of relationship. These can in fact change from AND/AND (when all the previous activities or predecessors trigger the following ones or successors) to OR/AND, AND/OR, OR/OR; moreover, the OR can be *inclusive* (the activity only starts if at least one of its predecessors starts) or *exclusive* (it starts when only one of the predecessors is carried out). For example, an inclusive OR/AND relationship must be thus interpreted: if at least one of the predecessor activities is executed, all the other subsequent ones can start.

The introduction of a risk variable and the contemporaneous consideration of all the other variables (times, costs and resources) makes VERT the most complex tool of all. It operates in a *what-if* scenario, and requires a certain amount of information before it can provide statistically significant solutions. The final result is given by a function or an index (Return On Investment, Net Present Value, etc.) representing the goals of the venture; each variable is weighed with the likelihood of its occurrence and the trend of the output is determined empirically, analysing the frequency curve obtained as the values of the variables change. The joint identification of the risk profiles makes this a very useful tool for decision-making purposes.

The complexity of PERT, and even more GERT and VERT, is such that in most cases an attempt is made to use CPM (or better, MPM, if there are various types of links): hence this method will be described in greater detail in the following section.

10.3 CPM: Activity Scheduling and Float Analysis

The CPM was developed in 1957 in the United States by the Catalytic Construction Company, thanks to the efforts of Morgan Walker, who, when working on a construction project for the DuPont Corporation, devised a method to simulate various planning alternatives, all of which would have a fixed duration. Soon after, a more complex CPM, called PDM (Precedence Diagramming Method), was defined at the Stanford University. PDM also included start-to-start and finish-to-finish relationships, and can thus be considered the first MPM. The CPM/MPM was further developed and became increasingly popular—thanks to the aid offered by calculators and computers.

PERT, on the other hand, was defined about a year later in Lockheed, with the contribution of Booz, Allen and Hamilton, when designing and building atomic submarines armed with ballistic missiles for the Special Project Office of the US Navy. This was the Polaris project, involving over 250 contractors and 9000 subcontractors. It was impossible to manage such a complex project using conventional methods, especially because of the cold war, which made the project a national priority (it was completed in a record time of 4 years).

GERT was created by NASA for the Apollo project, which brought man to the moon (1969). VERT, on the other hand, was only defined in the late 1970s and is still evolving.

These projects, formalising the various network techniques, were preceded by other ones, exploiting a series of more or less defined procedures that form the core of the theory and practice of Project Management. Among these is the *Manhattan* project, which led to the first A-bomb experiments in Los Alamos in 1943. Presumably however, some sort of management was required for other complex projects carried out in the past, such as the construction of the pyramids in ancient Egypt.

CPM considers a project as a set of activities, having a certain duration and clear links of precedence. Figure 10.6 is an example, depicting the activities, the names and duration of which are indicated inside the rectangles, and the relationships, indicated by the arrows. Two questions arise: how much time passes from the start of the first activity (A) to the end of the last one (I), in order words what is the duration of the project? In addition, what is the schedule of *each* activity, namely its start and end date?

To answer the first question, we must verify *all* the possible paths leading from the first activity to the final one, and consider the (minimum) duration of the project as the time required to complete the longest path (in the example: A–D–F–H–I, for a total duration of 60 time units).

This pathway, dictating the duration of the project, is the *critical* one, since a delay on this route determines a delay in the delivery time of the project. A certain slack can be tolerated in activities that are *not on the critical path*; for instance, both B and C (together) can slack up to a maximum of 10 time units, without causing any delay in the conclusion of the project (whose minimum duration is, as mentioned previously, of 60 time units).

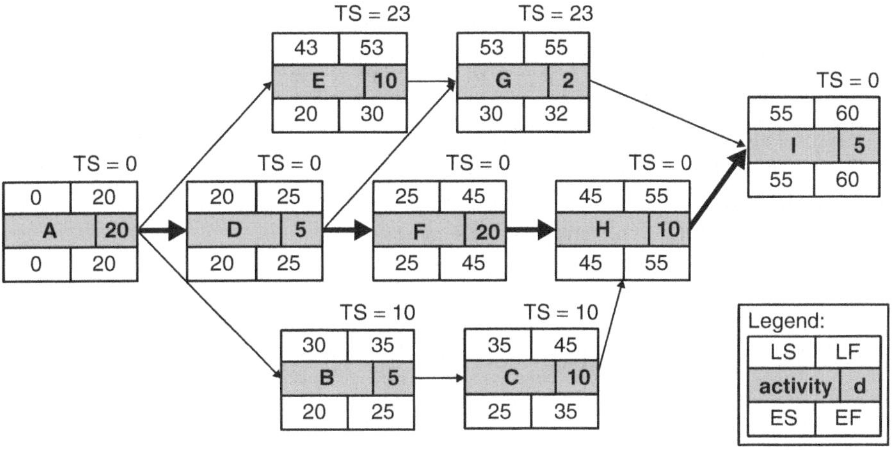

Fig. 10.6 CPM—in bold, the critical path, where TS = 0 (*d* activity duration, *ES* Early Start, *EF* Early Finish, *LS* Late Start, *LF* Late Finish, *TS* Total Slack)

We must then ask ourselves what are the start (and end) dates for each activity. Obviously, the end date depends both on the start date and on the duration of the activity.

In certain cases (i.e. for those activities that are outside the critical path) there is no definite start date, but there is instead a time interval given by the earliest and the latest date when the activity can start (*Early Start date* and *Late Start date*, respectively). For instance, activity B cannot start before time 20, or after time 30, else the project would not be completed in time.

This is obviously the maximum possible interval, and may not be completely available. The start date, for instance, may depend on that of the other activities that are directly linked to it; in the example, it is clear that C can start between time 25 and 35, but only if B starts between 20 and 30.

The CPM provides the algorithms to calculate these dates. By representing the Early and Late—Start and End dates, as shown in the caption of Fig. 10.6, these algorithms help schedule the activities depicted in the figure on a clear background, and calculate the overall delay or *Total Slack* (*TS*), which is the difference between the earliest and latest date (it is indifferent if start or finish, since the latter date is equal to the former plus the duration of the activity).

All the *critical activities* (i.e. the activities on the critical path) have a TS = 0, whereas the others have a TS $\neq$ 0. Therefore, all critical activities can be identified both by testing all possible paths—as illustrated at the beginning of this section—or by verifying whether early and late dates coincide. It is fundamental to calculate TS, since it identifies the criticalities (from a time viewpoint) for each activity: in the example, E and G are the less critical activities.

Given here are the formulas used for CPM algorithms (MPM formulas are more complex, since they must consider all types of relationship between any two activities: finish-to-start, start-to-finish, start-to-start, finish-to-finish): *d* is the duration of the activity, and activity *j* is preceded by *i* and followed by *k*:

1. *Initial activities*
   Early Start date $\mathrm{ES}(j) = 0$
   Early Finish date $\mathrm{EF}(j) = \mathrm{ES}(j) + d(j) = d(j)$
2. *Other activities (forward pass)*
   Early Start date $\mathrm{ES}(j) = \max_i \mathrm{EF}(i)$
   Early Finish date $\mathrm{EF}(j) = \mathrm{ES}(j) + d(j) = d(j)$
3. *Final activities*
   Late Finish date $\mathrm{LF}(j) = \max_i \mathrm{EF}(i) + d(j)$
   Late Start date $\mathrm{LS}(j) = \mathrm{LF}(j) - d(j) = \max_i \mathrm{EF}(i)$
4. *Other activities (reverse pass)*
   Late Finish date $\mathrm{LF}(j) = \min_k \mathrm{LS}(k)$
   Late Start date $\mathrm{LS}(j) = \mathrm{LF}(j) - d(j)$

Early and late dates are calculated using two separate processes (there are four variables, two of which however are not independent, because the end dates, as mentioned previously, are the sum of the start date and duration of the activities):

- The first step consists in calculating the early dates, starting from activity 1 (the first of the project) and trying to start all the subsequent activities as soon as possible (in other terms, we move forward—*forward pass*—along the network). When various activities flow into one, the beginning of this latter activity coincides with the latest (indicated as *max*) among the early finish dates of the previous activities (thus the rule is *the latest among the earliest*).
- Having reached the project's finish date (the end of the last activity), we can calculate the late dates, starting from the end of the network (therefore moving backwards—*reverse pass*). When activities branch, the finish date of the predecessor activity is the earliest (indicated as *min*) among the late start dates (so the rule now is *the earliest among the latest*).

The total slack must also be analysed, so as to verify how and to what extent it depends on upstream or downstream activities. There are various types of slack:

- *Total Slack* (*TS*) is the maximum slack allowable for an activity, when the established start and finish times of the project are respected (it is simply the difference between late and early start dates).
- *Free Slack* (*FS*) is the maximum amount of time for which a given activity, started on its earliest date, can postpone its finish date without causing delays in the early start of all the following activities (it is free in the sense that it does not create restraints downstream).
- *Chained Slack* (*CS*) is the difference between Total Slack and Free Slack.
- *Independent Slack* (*IS*) is the maximum possible slack for an activity carried out under the most unfavourable conditions, namely when all the previous activities are completed on their latest finish date and the following ones have started on their earliest possible date.

Obviously, if TS = 0, there can be nor FS neither IS. If FS $\neq$ 0, IS is not necessarily $\neq$ 0, because of the constraints present both upstream and downstream. In other words, TS $\geq$ FS $\geq$ IS.

Listed here are the formulas used for calculating slacks. The results of these calculations are one of the greatest practical implications of the CPM, because they help clarify the times needed to carry out each activity of a project, analysing in advance time criticalities and reducing the risk of organisational conflicts.

Total Slack = $TS(j) = LF(j) - EF(j) = LS(j) - ES(j)$
Free Slack = $FS(j) = \min_k ES(k) - EF(j)$
Chained Slack = $CS(j) = TS(j) - FS(j)$
Independent Slack = $IS(j) = \max\,[0; \min_k ES(k) - \max_i LF(i) - d(j)]$

Table 10.1 is an exercise on CPM: starting from the given data (the name and duration of the tasks, and their predecessors), calculate the schedule (thus, four dates per activity) and analyse possible slacks (TS, FS, CS and IS). Note that, in this exercise, there are three different initial activities and two different final activities (which could be, for example, a new product and the campaign to launch it). The solution is shown in Table 10.1 from column ES to column IS, and by Fig. 10.7 (network diagram solution).

Both activity scheduling and float analysis can be carried out before the project starts (i.e. planning the project). Once it has begun, it is necessary to control how the project proceeds, monitoring the differences with the scheduled times. These differences may refer to both start dates and duration of the activities, and may cause changes in the finish date of the project.

These variations can be analysed and discussed using various tools—typical of quality control management—such as the *fishbone* or Ishikawa diagram. The main

Table 10.1 CPM exercise

Activity	Duration	Predecessor	ES	EF	LS	LF	TS	FS	CS	IS
A	9	–	0	9	6	15	6	0	6	0
B	3	–	0	3	0	3	0	0	0	0
C	7	–	0	7	7	14	7	0	7	0
D	10	A	9	19	15	25	6	0	6	0
E	1	B	3	4	23	24	20	20	0	20
F	21	B	3	24	3	24	0	0	0	0
G	1	B	3	4	13	14	10	3	7	3
H	5	C, G	7	12	14	19	7	0	7	0
I	6	D	19	25	25	31	6	0	6	0
L	3	H	12	15	21	24	9	9	0	2
M	11	H	12	23	19	30	7	0	7	0
N	8	E, F, L	24	32	24	32	0	0	0	0
O	1	I	25	26	31	32	6	6	0	0
P	2	M	23	25	30	32	7	7	0	0
Q	3	M	23	26	31	34	8	8	0	0
R	2	O, N, P	32	34	32	34	0	0	0	0

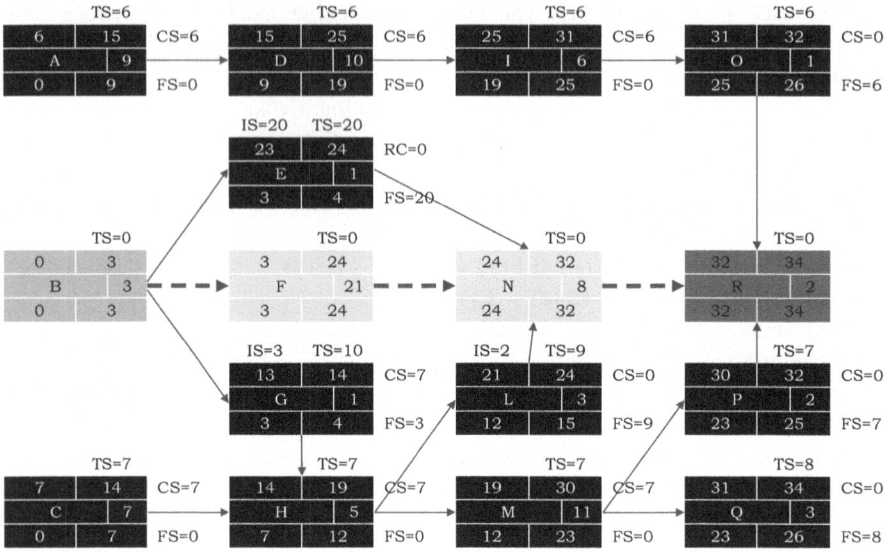

Fig. 10.7 CPM exercise

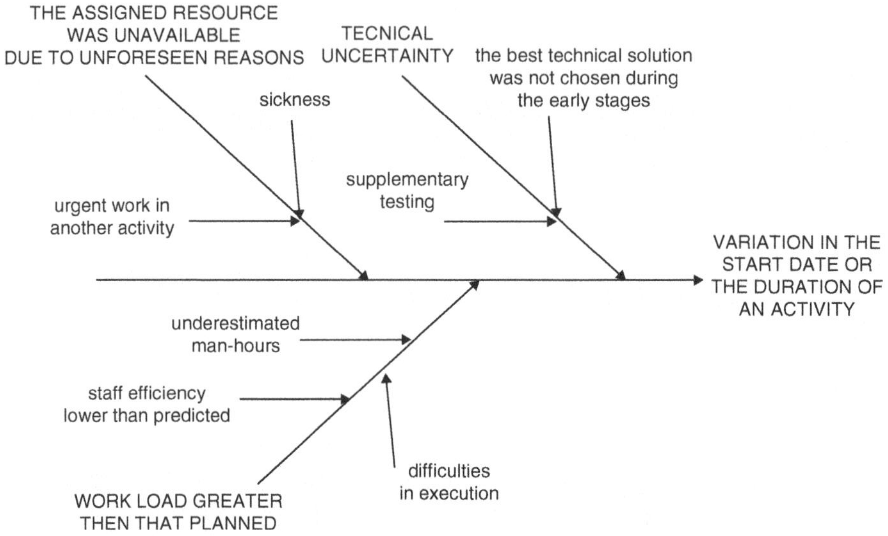

Fig. 10.8 Cause–effect diagram for the analysis of the variation between estimated and actual times

arrows point to a certain effect; on each of these arrows, there can be secondary ones that indicate the possible causes of that effect; these causes in turn will also have their upstream causes, so that a further arrow can be placed on the secondary one, and so forth (Fig. 10.8). The causes can be evaluated in probabilistic terms and/or weighed according to the opinions of the persons involved in the project.

References

Marshall, A. W., & Meckling, W. H. (1959, December). *Predictability of the cost, time, and success of development* (RAND Corp. Report, No. 1821).

Peck, N. J., & Scherer, F. M. (1962). *The weapons acquisition process: An economic analysis*. Boston, MA: Harvard Business School Press.

Project Organization and Resource Management

11

11.1 Company-Wide Project Management

Projects, even the more conventional ones for new products, are created and developed thanks to the contribution of resources belonging to *various* departments (although Design department plays the principal role). In other words, even those functions whose tasks do not include the definition of project specification may and indeed must contribute to the project (the matrix structure—typical of a project, as we shall see later—has the objective of formalising this *call for competencies*).

Figure 11.1 depicts the involvement of three specialisms in a project: Marketing (whose task is to ensure that the output of the project appeals to the market, thus justifying the investments made), Design (whose specific task is to develop and implement the project) and Production/Delivery (in charge of delivering the product/service specified in the project in an efficient manner and with a high level of conformity). However, other functions can be involved, such as Purchasing, the Engineering or Technology Department, Quality Management, Machinery Maintenance, Technical Assistance, Customer Service, and so forth.

Their contribution to a project is not constant throughout time, but varies according to the stage reached by the project, as illustrated in Fig. 11.1. By tracing an upright line, it is possible to see the percentage of involvement of the various functional units at a given time.

Every project has its stakeholders, whose interest goes beyond technical and economic issues, and may include personal satisfaction as well as that of the entire company, a factor that contributes to making a serene working environment. The field of interest can also stretch beyond the boundaries of the firm, to include customer satisfaction, that of the suppliers, who preserve a market outlet, and even of the entire community, when themes of social utility and eco-compatibility are at stake.

Company-Wide Project Management requires integration between functional units, but also between the tasks assigned to single persons. Two types of integration are possible: horizontal (or *job enlargement*, with a larger number of tasks being

S. Tonchia, *Industrial Project Management*, Management for Professionals,
https://doi.org/10.1007/978-3-662-56328-1_11

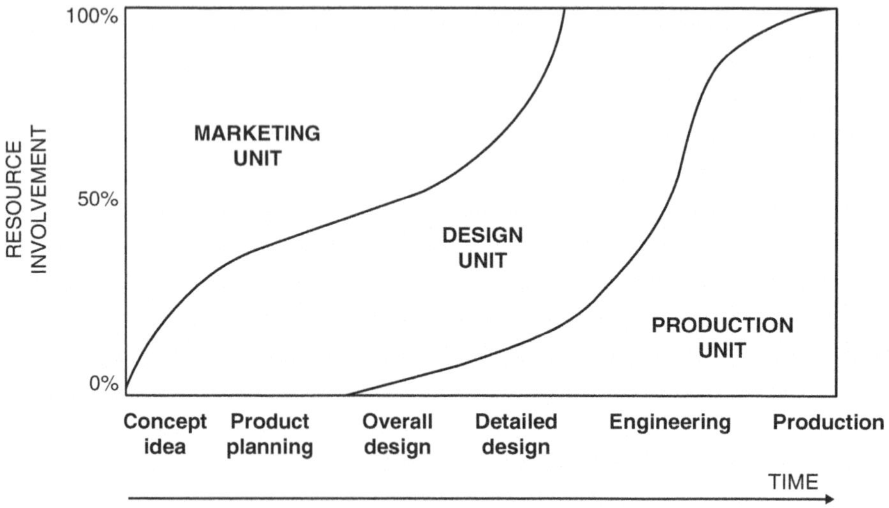

Fig. 11.1 Involvement (percentage) of the organizational units in relation to the project stages

carried out by one person) and vertical (or *job enrichment*, with greater responsibilities, including decision-making). In the former case, the employees become more skilled in a variety of activities, and are therefore able to carry out a wider range of operations, either sequential or similar, to the advantage of coordination and flexibility; motivation and a bent for learning is required, as well as a solid background of knowledge. In the latter case, the employees are encouraged to control and monitor the activities autonomously and solve problems independently, at least those of minor relevance. Job enrichment is linked to *delegation* and *empowerment*, two factors that spur the staff to improve and take on greater responsibilities, to understand more clearly the effect of their actions on the overall performances and constantly seek new solutions.

11.2 Organizational Structures for Project Management

Typically, companies are organized by functional units, i.e. hierarchies of various specialisms, and the human and technical resources are divided among them. The principle of dividing work to obtain efficiency used in this type of organizational structure is however inadequate for managing projects.

Project management requires resources and competencies that belong to *different* functions (not only Design, but also Marketing, Manufacturing, Purchasing, etc.); hence, when working to achieve certain goals (those of the project) instead of merely following specific procedures (those of the department), the available resources must be reorganized.

The typical organizational structure adopted in Project Management is the so-called *matrix structure*, which combines the conventional division by functions with structures that are specifically set up for every project.

There are therefore two dimensions and two different types of managers: the *line manager* (i.e. the head of the function) and the *project manager* (i.e. the owner of the process). The task of the former manager is to preserve the standards of efficiency/efficacy characterising a given functional unit, as well as managing, preserving and cultivating similar resources and competencies, and making them available for a variety of projects within the firm. The task of the latter manager is to exploit all available resources in the best possible way, allocating them so as to achieve the goals of the project, and manage extra resources brought in if needed.

Inevitably, this *two-dimensional* structure is anything but simple to achieve in practice, since it goes against the Taylor principle of uniqueness of command. In the presence of limited resources and when there are conflicting demands made by the various managers, who has priority and decisional power?

There are two alternatives—two organizational structures that differ according to where authority (i.e. the power to *reward and punish* the resources) resides, either in the function or in the project. The structures are as follows:

- The *lightweight matrix*, where authority is still retained by the line manager, whereas the project manager plays a side role, coordinating, allocating, using and managing the resources for the project (*lightweight* project manager)
- The *heavyweight matrix*, where authority is in the hands of the project manager (who in this case becomes *heavyweight*), and the line managers are in charge of supplying the resources to the project, while preserving a minimum of performances in the line function (productivity, technical updates, etc.)

Clearly, in the latter case, many problems still exist and there is a greater likelihood of conflicts arising from competition among projects: to avoid this risk, the board of directors should appoint a *multi-project manager* to coordinate the various projects and their managers (Multi-Project Management—MPM).

On the other hand, even the first solution has some drawbacks that can be ascribed to the role of the project manager, who is responsible for the resources (used for the project) without having the authority to do so, and therefore risks becoming a mere coordinator/facilitator. For this reason—when deciding to adopt this type of solution—the project manager must have great leadership, competency and experience, since his/her role does not rely on any concrete power.

Figures 11.2 and 11.3 illustrate the two types of matrix structures, and indicate the position held by the project manager. In other cases, when the project is of the utmost strategic importance, the resources are asked to form a *task force* that is independent of the original functions. This task force is at the opposite side of the *steering committee*, in which the percentage of individuals involved in the project is close to zero.

Usually, in order to be effective, matrix management requires certain social and organizational conditions, such as the following:

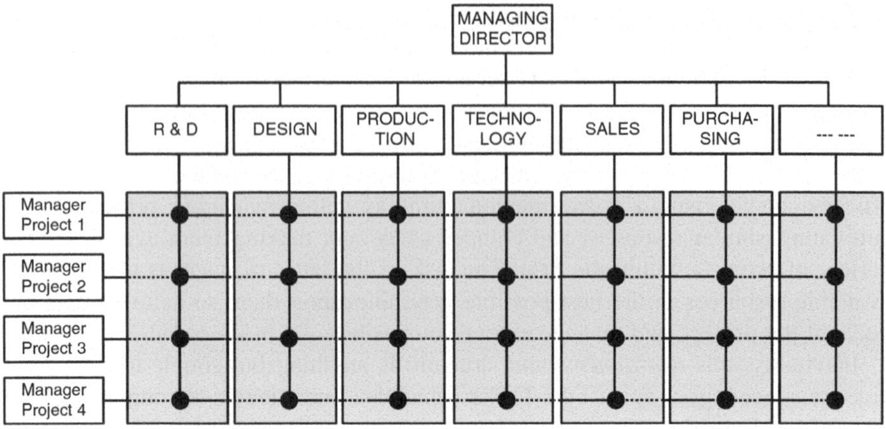

Fig. 11.2 Lightweight organizational matrix of the project (continuous line = hierarchical power; dotted line = coordination only)

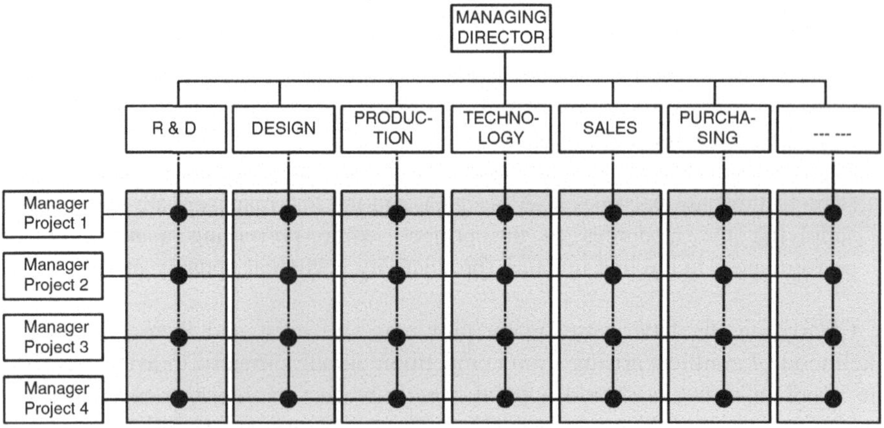

Fig. 11.3 Heavyweight organizational matrix of the project (continuous line = hierarchical power; dotted line = coordination only)

- A high level of communication, because of the notable amount of dependency and the need to establish relationships.
- A predisposition for teamwork, given the presence of individuals with different competencies and backgrounds.
- The ability to operate by objectives, because the teams are working on projects, and consequently an adequate system of reward should be established to encourage the teams to achieve these goals.
- Since projects are not routine activities, delegation of power should be more widespread, so as to ensure significant margins of individual autonomy.

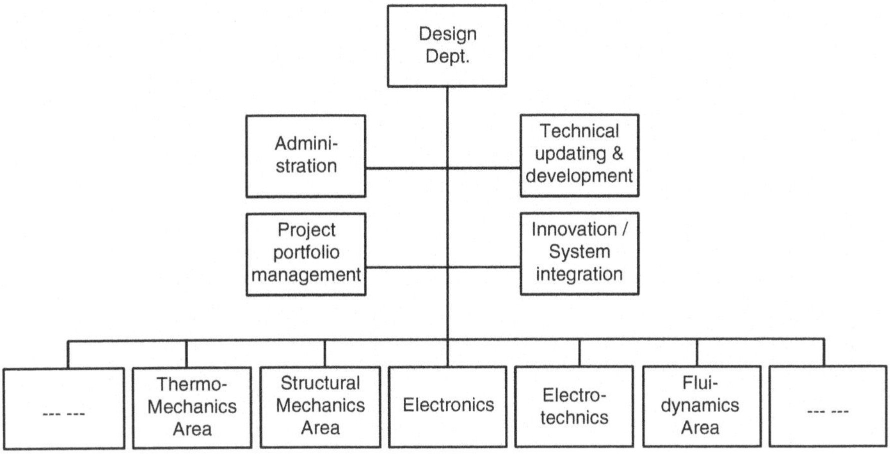

Fig. 11.4 Internal organization of the Design department

– A proactive approach leaning towards innovation and change, with the acceptance of the consequent margins of risk; independent quest for information and availability to listen.

Independently of the matrix structure (and therefore of the existence of the project manager, with or without hierarchical power), the fact that most resources used in a project belong to the Design department/unit implies that the latter must possess a specific internal organization. An example of this is shown in Fig. 11.4: the offices of the technical line (including, for example, designers specialized in the field of thermo-mechanics, structural mechanics, electronics, electro-techniques, aerodynamics, etc.) are supported by the staff in charge of administration/contract management, project portfolio, technical updates, system innovation/integration.

MPM has important consequences at an organizational level.

According to Cusumano and Nobeoka (1998), there are four steps in the evolution of *organizational strategies* that lead towards a more modern organisation-by-projects and that however can co-exist:

1. Separate functional units and sequential development stages
2. Integrated and overlapping functions and development stages, coordinated by a lightweight project manager, who may also become chief director if the events require so (heavyweight project manager)
3. Integrated and parallel functions and projects, coordinated by a multi-project manager, with a possible involvement of chief executives (such as the *platform managers*)
4. Intra- and inter-firm design teams (*joint project management*)

As a consequence, there are different types of organization that Aoshima (1993) classifies according to inter-project learning (which is linked to the transfer of

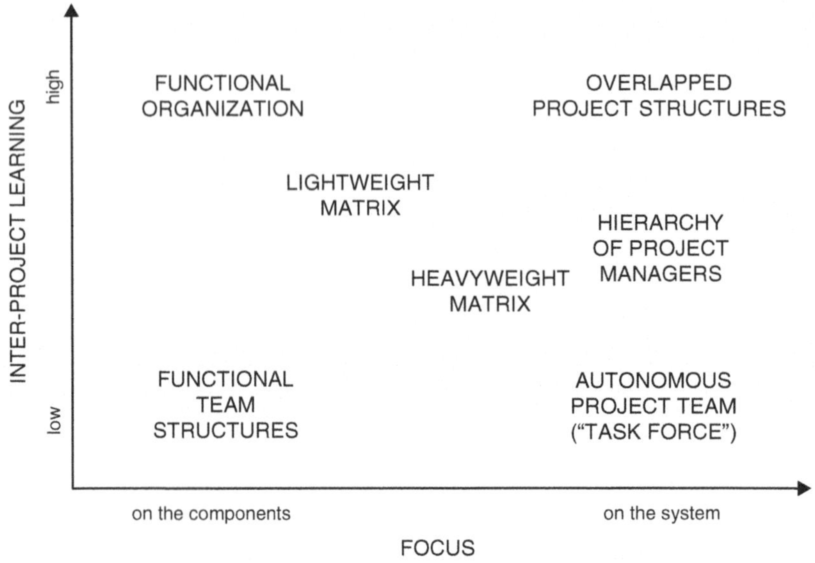

Fig. 11.5 Multi-project organizations (Source: Aoshima 1993)

knowledge among projects) and the project's focus (on the single components or on the system and its connections), as shown in Fig. 11.5.

Nobeoka (1993) observes that it is possible to create *differentiated matrices* (Fig. 11.6), where, according to the case, the line and the project managers are either *heavyweight* or *lightweight* (in Fig. 11.6, the overlapping area indicates an authority over the underlying area), while the functions can serve various projects at different levels of intensity, distributing their resources among them.

The case of product platforms should be analysed separately. As described in Sect. 5.4, these serve to develop a *core sub-system*. Other modular sub-systems are then attached to it, and the combination of different modules gives rise to a family of products. Product platforms require a specific form of organization, which consists of the following (Meyer and Lehnerd 1997):

- A *platform team* (which must not be too numerous—the Renault Mégane platform was managed by 15 persons only), coordinated by a *platform manager*
- *Module teams* in charge of developing the single modules
- *Product teams* in charge of producing the *derivatives*, starting from the platform and its interchangeable modules
- An *informal network* of experts and designers that can include even a hundred persons or more, etc.

The multi-project organizational structures can also differ extensively according to the strategies adopted by the firm. In the automotive sector, for example, Cusumano and Nobeoka (1998) observed the following in the case of Design (not including Research & Development or similar departments, which are usually external):

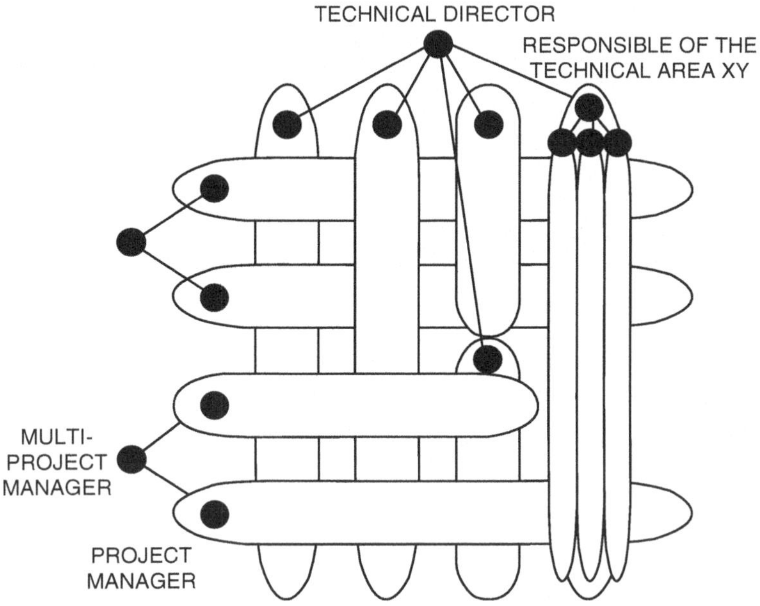

Fig. 11.6 *Differentiated matrix* organizational structure (Source: Nobeoka 1993)

- Rather classical matrix structures, as in BMW, Mercedes and Volvo (i.e. companies that focus more on strategies for product lines than on platforms)
- Matrix structures featuring a platform manager (but without establishing a centre for each platform), as in Fiat and Renault
- An organization totally divided into platform centres, each of which features a platform manager who coordinates the project managers, as in Ford and Toyota
- An organization partially divided into (*semi-centred*) platforms, sharing certain engineering functions, as in the case of General Motors and Nissan.

11.3 The Role of Project Manager

As the name suggests, the *project manager* is in charge of managing the project, identifying, allocating and coordinating the resources to achieve the objectives of quality, costs and time specified by the project itself.

As we have seen, the project manager may or may not have a hierarchical *authority* over the resources, depending on the type of organizational structure (heavyweight or lightweight, respectively) that exists in the company. This person however is *responsible* for the project and the entire Project Organizational Breakdown Structure (POBS).

The project manager must therefore possess the indispensable technical knowledge to understand and carry out the various tasks, and because projects are usually complex and strategically important, also remarkable managerial and relational skills. Often these abilities are more important than the know-how, because the technical aspects of a project can be manifold and different, so it would be impossible to know all of them in depth. It helps however to have a complete and ample knowledge of the process rather than a more detailed one focusing only on certain, specific aspects. Usually in the case of high-profile projects, the persons chosen are those who have gathered great experience over the years in different disciplines, who then consult the specialists in the different sectors and it is these individuals who will update the single technologies.

The most important thing for project managers is to possess great organizational, managerial and relational skills. In particular:

- They must have a global vision of the project, and be capable of planning it and breaking it down into clear *chunks*, defining all the parameters and controlling the progress of the project right to its conclusion.
- They must sense and anticipate both market/customer demands and the possible technical problems linked to the various solutions, identify the persons who can help them make the best decisions and establish a constant dialogue with them.
- They must be capable of coordinating a multitude of heterogeneous persons, creating an environment that will promote constructive work and communication, overcoming the intrinsic obstacles deriving from different *languages*, cultures and work styles.
- Their leadership must be universally recognized, so that they can converse with different persons, motivate them and delegate responsibilities while still maintaining a not too onerous control.
- They must possess the ability to negotiate, but at the same time be able to make decisions and solve the conflicts that naturally arise when dealing with different persons and when there are various, competing projects.

Hence, the various skills of a project manager must also include the ability to manage interpersonal relationships. Various *principles* must be kept in mind for this purpose:

1. *The persons must be treated as individuals, as well as members of a team.* It is not enough to manage a team, since a lack of motivation can arise because an individual's objectives do not match those of the team, i.e. of the project.
2. *Increase and stimulate the efforts of the team members for the project.* One of the keys to success is to involve team members in the decision-making processes, so that they can play a more active role, and not feel that they are mere suppliers of competencies or the mechanical performers of tasks. The responsibility entrusted to them stimulates them to make a greater effort. It is also advisable to supply incentives, not only in terms of money, but also, for example, making them more visible inside the company or allowing them to attend training courses.
3. *Inform the collaborators about all that occurs during the project.* In order to do so, willingness is not enough, because the team members come from various functional units, and may have different backgrounds, objectives, habits—all

elements that hinder communication. It is therefore necessary to establish a common language, which can be consolidated by using standard documents. As well as promoting communication, the project managers must also be able to listen, and so become aware of any conflict brewing.

4. *Build consensus to channel conflicts*. Conflicts are inevitable in any project and a good part of the project manager's task is to manage diverging opinions and behaviour. Conflict however can also become an opportunity to develop new ideas and solutions for the project. Ideally, they should be channelled, so that they may become a tool for promoting creativeness and proactivity.
5. *Share informal authority with the collaborators*. The best project managers accumulate an informal, transversal power inside the organization, which derives from their relational skills, their leadership and charisma, and then share this *power* with their collaborators, letting them perceive that the success of a project is also advantageous for them, because it can boost their informal authority. The team's success therefore becomes the success of its individual members; in this case, what is expected from a project manager is mainly honesty, competence, determination, drive and motivation.
6. *Encourage creativity and propensity to risk*. Being innovative and creative is mostly an innate aptitude, but it can be stimulated and rewarded; a similar definition can also be applied to risk propensity. Even though this subtracts time from management and control, the project manager should devote particular attention to these aspects.

Gilbreath (1986) examines and compares certain operational features that distinguish a project manager from a more *technical* one; the former manager in particular should do the following:

1. Find feasible solutions given the time, budget and resource constraints, rather than seek ideal, perfect ones
2. Aim at delivering the project, even to the detriment of technical perfection
3. Look at the overall results of the project instead of focusing only on the technical output
4. *Despecialize* in order to improve, and find new synergies and nourishment
5. Win together with the team

Project managers are pivotal for the project, because they are the driving force for the flow of information that is established with the following people:

- The board of directors (from whom they receive the indication of the company's policies and to whom they must report the progress of the project)
- The customer/purchaser (from whom they receive indications and/or specifications and to whom they must report the progress of the project)
- The line managers (whom they direct/coordinate and from whom they receive reports)
- The personnel allocated to the project (whom they direct and from whom they receive interim reports)

Project managers therefore play a variety of roles:

- A borderline role between the organization of the project and the external environment (either inside or outside the company).
- They act as interface in the exchange of information, both inside and outside the team.
- They also act as intermediaries between higher hierarchical positions (board of directors) and the staff employed in the project.
- Finally, they are the *agents of change* and promoters of innovation, which is the essence and most typical feature of a project.

11.4 The Profession of Project Manager

The project manager develops over time his professionalism. The project manager not only possesses a technical preparation and experience, but above all a strategic vision, job organization and interpersonal relationship management.

In general, team members must be able to provide the technical expertise needed to develop the project; so the project manager, as the leader of the project, must hold primarily managerial skills. However, such skills often conflict with the potential project manager, who normally has a technical background (unless he is already a management engineer). The propensity of individuals to assume inter-functional management charges contrasts with the vision of stability and continuity of the role of a technical-specialist.

The transition from the role of an experienced technician to a manager does not need to be either instantaneous or automatic, but the result of a gradual training process led by the organization that attributes value and prestige to this role. However, this training process must only begin after an assessment of the individual's predisposition to the management issues. The alternative growth path for a technician is that of senior expert and functional manager.

The growth stages of a project manager typically include:

- participation in training courses
- gradual assumption of responsibility within the projects and subsequent assignment of "small" projects
- continuous feedback and supervision by those who have more experience in the role of project manager.

The education and training of the future project manager plays a key role in its growth path. For this reason, some successful companies have developed, within their organizations, internal Masters (or Corporate Universities), which take advantage of the involvement of academics and professionals at a broader, national and international, level.

In organizations where there is a permanent project management structure [such as the PMO—Project Management Office (Sect. 14.1)], it is easier to plan the career path of future project leaders, through the continuity of project assignments and progression in responsibilities attributed with a standardized monitoring and assessment system.

For example, a potential project manager may start as a "planner manager" (responsible for project's scheduling) for a small project and subsequently act as a "project controller" (responsible for time/cost tracking). If the guy demonstrates managerial skills, he can take on the responsibility of sub-project leader in an important project and then project manager for a small project. The responsibility for managing large projects should only be attributed after a long training and field-proven experience, which also heightens its credibility within the project team to be coordinated and outward.

The profession of PM has been established since many decades, creating some professional organizations. The most important one—as seen in Chap. 2—is the Project Management Institute (PMI), a no-profit organization founded in 1969 and recognized internationally as the most authoritative institution in the field of PM. Besides collecting and maintaining the "knowledge of PM" (through the PMBOK), the other mission of PMI is raising the standard of the project manager profession by certifying:

- the figures of project manager (PMP—Project Management Professional and CAPM—Certified Associate in PM) and similar roles (PgPM, RMP et al.)
- the training courses on Project Management.

Furthermore PMI members have determined that *honesty*, *responsibility*, *respect* and *fairness* are the values that drive ethical conduct for the project management profession. The PMI's "Code of Ethics and Professional Conduct" applies those values to the real-life practice of project management, where the best outcome is the most ethical one.

11.5 Project's Work Loading

Organization also means the management of the work in order to complete the tasks (i.e. activities) described in the Work Breakdown Structures (WBS). The completion of a task implies the production of the deliverable of that task. The *duration* of a task is the time between the start and the end of that task. The *work* (human or equipment, in general "capacity") is the amount of resources necessary to produce the task's deliverable, starting from the task's inputs (these latter are the deliverables of the immediately preceding other tasks).

So it's fundamental do not be confused *duration-time* and *work-time*: the first is represented by the length of the bar in the Gantt chart; the second is the work "inside" the Gantt bar (not represented through the basic Gantt chart, unless to report—below each bar—the resources involved, and along the bar—by numbers—the *work load* of each resource). For this specific representation, see the Annex (Microsoft Project's views: "Task Usage" and "Resource Usage").

The work load also depends on the task's setting: "Fixed units", "Fixed duration" or "Fixed work". "Fixed units" setting engages for a task a precise percentage of the overall resource availability, while "Fixed duration" setting preserves fixed the duration of the task (eventually the resource/s becoming overloaded). For details, see the Annex.

References

Aoshima, Y. (1993). *Inter-project technology transfer and the design of product development organization*. New York: Mimeo.

Cusumano, M. A., & Nobeoka, K. (1998). *Thinking beyond lean: How multi-project management is transforming product development at Toyota and other companies*. New York: Free Press.

Gilbreath, R. D. (1986). *Winning at project management*. New York: Wiley.

Meyer, M. H., & Lehnerd, A. P. (1997). *The power of product platforms*. New York: Free Press.

Nobeoka, K. (1993, September). Multi-project management: Strategy and organization in automobile product development, Ph.D. Dissertation, MIT, Boston, MA.

Project Communication 12

12.1 Communication Management

Communication is largely important to the ultimate success of any project. Project managers spend most of their time communicating with team members and other project's internal and external stakeholders. Most communication's characteristics and skills are common for both general management and project management.

The basic *communication model* includes:

1. *Encoding* by the sender
2. *Transmitting* through a media
3. *Decoding* by the receiver
4. *Acknowledge* (i.e. comprehension of the message, not necessarily agreement)
5. *Feedback/response*.

Communication can be:

- *Verbal*—it is related to the conventional meaning of words
- *Para-verbal*—it is related to tone, volume, rhythm of voice, etc.
- *Non-verbal*—it is the language of the body (look, facial mimicry, hands and gestures, posture, relationship with space, physical contact, clothing, etc.).

Communication can have different dimensions, within which there may be variable communication modes:

- In relation to the timing of upstream to downstream activities (sequential or overlapping/simultaneity)
- In relation to the media (direct i.e. personal, in presence or by phone, through paper documents, in an electronic way e.g. e-mail, etc.)
- In relation to the communication frequency (by blocks or continuous transmission)

S. Tonchia, *Industrial Project Management*, Management for Professionals,
https://doi.org/10.1007/978-3-662-56328-1_12

- In relation to the directionality of communication (one-way, two-way, two-way with feedback)
- In relation to the timing of information flows from upstream to downstream (late delivery of complete information, early release of preliminary information).

12.2 Project Leadership

Leadership—the *art of managing people's talents* and not simply getting things done through people—brings together aspects of technical competence and results orientation ("functional component") and the ability to create a good climate, identification in the mission/goal and sense of challenge to bring it to a successful conclusion ("socio-emotional component").

A "participatory" leadership is very careful to consider people as the primary source of success—'a leader is a man who needs others' (Paul Valery)—and to encourage the conditions for the valorisation of each team member, who becomes a protagonist.

The leader (from the Latin "cum ducere" = pulling together) is the person who exercises most influence in a group. Leadership, as a consequence of this influence, determines a *voluntary consensus* in relation to certain objectives of the group or organisation, becomes *persuasion* of others to achieve a common goal.

Leadership is the ability to convince and motivate others do what they have to do, without having authority over them… and leave them happy.

Leadership is:

1. Guiding. The winning weapon able to highlight the direction that the company must take is the *vision*. The vision predicts what we will be and where we will go. Without *envisioning* to lighten the present path, it is difficult to find any orientation and motivation. Without a vision there can be no leadership, because there can be no people's guiding. For this reason, the future represents an attractive, motivating and seductive force, in that it contains the expected possibility of completion of the work and fulfilment of current needs.
2. Sponsoring. The leader helps his or her collaborators—according to a partnership principle—"to help each other", guaranteeing their potential and focusing their attention on the development of professional identity and personal values that each individual feels they can express. The relationships between leaders and collaborators focus on the individual subject rather than on the role.
3. Teaching. Simple and immediate access to information and knowledge for all individuals allows for autonomy and widespread assumption of responsibility. The leader facilitates the timely *sharing*, allowing the transition from a logic of volume of information to one of value of information, teaching to learn.
4. Coaching. It is expressed through the assignment of clear tasks, challenging objectives, within defined programs and strategies. The "coach" approach, borrowed from the sports model, is to support people by bringing them to the pinnacle of their skills. The leader sets up a system of observations and

suggestions, managing a positive communication, strongly interwoven with feedback.

5. Mentoring. Leadership can only be *exemplary*. Employees should feel comfortable in shaping their behavior to that of the leader. The mentees acquire new meanings and utilities from the exemplary teaching of the mentor.
6. Awakening. The leader is a *motivator*. His desire "to do" translates into energy capacity. He conveys the love of his work with a passionate and enthusiastic testimony. It mobilizes the energies of men, encouraging them to constantly find in their work the hope of a fulfilling self-realization.
7. Caretaking. It is essential to take care of one's own people at a time when their conscience is wavering and their sense of inadequacy towards requests from the outside world perceived as superior to their own possibilities is pronounced. It is about cultivating the growth of people in the same way as the gardener who accompanies with devotion to the most beautiful flowering of his plants. Involvement and care are the most congenial means for getting the best from people, motivating them to act beyond expectations.

Leadership is to develop and enhance people's skills and competencies. *Skills* are the practical or cognitive things one can do; *Competence* is the proven use of skills, knowledge, and abilities to accomplish something. Competencies, therefore, may incorporate a skill, but are *more* than a skill: they include abilities and behaviours, as well as knowledge that is fundamental to the use of a skill.

Competence can be defined as a cluster of related knowledge, attitude, skills, and other personal characteristics that affect a major part of one's job, correlates with performance on the job, can be measured against well-accepted standards, and can be improved via training and development (also see the following section).

The PMI's "Project Manager Competency Development (PMCD) Framework" provides a framework for the definition, assessment and development of Project Manager's competence, based on the premise that competencies have a direct effect on performance; it defines the key dimensions of competence and identifies the competencies that are most likely to impact project manager performance (also see Sect. 12.3).

12.3 Project Teamworking

A *working group* (the Project Team) can be defined as a group of people:

- Interdependent
- Interacting in a given period and space
- Linked to a sense of belonging (more or less formal)
- With values, norms, declared and accepted roles, rules
- Committed to achieving selected and assigned objectives
- Motivated by professional and corporate's interests.

Factors constituting the group may be identified:

- *Objectives* (they must be clear and shared, defined and described in terms of results, through possessed knowledge, comparison and negotiation with and among the members of the group, built on data and facts observable by all, adequate to the resources and potential of the group, which can be pursued and evaluated with coherent tools and criteria)
- *Methods* (should enhance individual skills, creativity but also group synergies, allow the free expression of group members and the resolution of possible disagreements and differences)
- *Roles* (intended as a set of behaviours expected from those who occupy a position within the group, subdivided into functional tasks and functional to relations within the group)
- *Leadership* (balance between "membership" and "groupship", i.e. between belonging to the group and pressure from the group—leadership in a working group should be as a "service", i.e. *for* the group and not *on* the group)
- *Climate* (indicates the quality of the group's relations, internal and external).

A working group can be described according to several *variables*:

- *Composition* (functional, inter-functional, intra-company, inter-company), location (possible need for remote work) and size (number of members)
- Degree of *Formalization* of the group
- *Scope*, *KPIs* (Key Performance Indicators) and *Duration*
- *Procedures* for carrying out the activities, and the internal/external *Coordination* (regulated by precise company's rules and/or defined by the group's members).

The relational part of a group's activities is of fundamental importance. The attitudes in relationships can be distinguished between the following

- Members' attitudes towards their own actions
- Mutual attitudes (need for mutual trust and shared responsibility, with a "generous" approach to finding solutions)
- Attitudes towards team's goals.

In a team the *Competencies* must not only concern what people know or are able to do, but also the behaviour, that is to say, there must be provided for and possessed different "competitive" competencies, linked to:

1. *Knowledge*, i.e. what people *know*, derived from institutional systems and learning developed in organised contexts (they can be divided into thematic areas and relate to: techniques, technologies, methodologies, tools used, product or process, standards and procedures, etc.)

2. *Skills*, i.e. what people are *able of doing* or *how to do*, by their experience and the practical application of knowledge (including: assessment of alternatives, early sighting of problems, problem solving, coverage of roles, etc.)
3. *Behaviours*, i.e. how people *act*, both as individuals and within groups, because they have objectives, values, beliefs and habits (the thematic areas in this case may concern: motivation, assumption of responsibility and delegation, leadership, etc.).

During the first meetings, the team will have to harmonize/standardise its "language", making it common and comprehensible to all the members, exchange basic "cultural" elements that must be shared, assimilate correctly the objectives to be achieved (including the relations between the different objectives or aspects of them), the roles of each component and individual responsibilities.

The teamworking will then be based essentially on good management of the meetings (also see Sect. 12.5), which are necessary for planning/assigning the tasks and activities until the next meeting, and for the examination of work carried out on the basis of the previous meeting/assignments.

In fact, teamworking is not done entirely *in group*, but on the contrary, individual work can still prevail (although it may be done physically contiguous).

12.4 Definition of Responsibilities

The *Responsibility and Relationship Matrix* formalises beforehand the responsibilities and relationships among the various managers (both company directors and managers who are directly involved in the project) during the different development stages of the project (Table 12.1). It specifies the types of responsibility and relationship, which can be summarised as follows:

1. *General responsibility* (concerning the final/overall deliverables)
2. *Operational responsibility* (being delegated to perform groups of tasks)
3. *Specific responsibility* (for single tasks i.e. activities in the WBS and their specific deliverables)
4. *Necessary consultation* (in order to make decisions that are relevant for a particular stage—a consultation is a bi-directional flow of information)
5. *Possible consultation* (not compulsory)
6. *Compulsory communication* (to a specified person)
7. *Necessary approval (by the person involved)*

Responsibility is the authorization to carry out a job aimed at achieving certain objectives, operating the necessary choices and being responsible for the results or *deliverables*. It does not necessarily imply changes in *Authority* (the formal power *to promote or punish*). A *Delegation*, on the other hand, is the formal assignment of

Table 12.1 Responsibility & relationship matrix

Stages	Chief Executive Officer	Managing Director	Project Manager	Project Engineer	Project Controller	Marketing Manager	Contract Manager	Technical Director	Contract Technical Resp.	System Engineer	Technical Area Managers	Procurement Manager	Field Manager	Production Manager	Production Coordinator
Preparation of the offer	F	A	B	C	C	D	D	E	B				E	D	E
Offer approval	A	G	G			D	D								
Contract bidding		A	B	E	E	E	C		E						
Contract signature	B	D	E				D								
Work order planning		A	B	C	C		C		D			E	E	D	E
Design		A	B	C				E	C	C	C				
Checking of specifications		E	A	B			D		D	E				E	
Execution		A	G		D			E	B	D	D	C	C	B	C
Evaluation meetings		A	B	D	D		D		E	E	E		C	D	E
Validation meetings		F	B	C	E		E		E	E			D	E	
Customer relationships		E	B	E	F	D	C						E		

A general responsibility, *B* operational responsibility, *C* specific responsibility, *D* necessary consultation, *E* possible consultation, *F* compulsory communication, *G* necessary approval

one's tasks (deriving from one's position in the organizational chart) to someone placed at a hierarchically lower level.

The *Responsibility and Relationship Matrix* is an evolution of the *RACI Matrix*, which is a common organizational standard. RACI is an acronym for: *Responsible*, *Accountable*, *Consulted*, and *Informed*:

- *Responsible*—Those who do the work to achieve the task; there is at least one role with a participation type of *responsible*, although others can be delegated to assist in the work required
- *Accountable*—The one ultimately answerable for the correct and thorough completion of the deliverable or task, and the one who delegates the work to those *responsible*; in other words, an *accountable* must sign-off (approve) work that *responsible* provides. There must be only one *accountable* specified for each task or deliverable
- *Consulted*—Those whose opinions are sought, typically subject matter experts, and with whom there is a two-way communication
- *Informed*—Those who are kept up-to-date on progress, often only on completion of the task or deliverable; and with whom there is just one-way communication.

12.5 Meeting Management and Project Reporting

The meeting is a fundamental communicative activity in the initial and final phases of the project, but also in central and advanced project phases. It is necessary to maintain the cohesion and motivation of the participants and to monitor the relationship between the objectives set and the results achieved. It is necessary, after the first meetings aimed at organizing the group and planning project activities, to share information on the progress of the works and, above all, to negotiate and agree on intermediate decisions, possible delicate or conflicting steps and eventual corrections.

The operational reasons why project meetings should be called, and which sometimes correspond to formalized moments in the planning, are as follows:

- *Start of the project* ("kick off meeting"), to share objectives, risks, and organization of the project with the members of the team
- *Project status*, to periodically check progress and discuss corrective actions with team members
- *Design review*, to periodically carry out a critical, documented, exhaustive and systematic analysis of what is planned
- *Major problems*, in order to resolve new critical issues if necessary
- *End of the project*, to close the activities and receive acceptance of the results by the customer.

In summary, project meetings are essential for the following reasons:

- *Information sharing*
- *Problem-solving*: brainstorming and generation of ideas, options and alternatives
- *Decision-making*: direction of action and commitment by the team
- *Planning*: what, who, when, how, where e why
- *Evaluation*: monitoring, measurement, revision and re-planning.

The project manager is responsible not only for communication but also for scheduling and managing project's meetings. Meetings cannot be improvised. It is essential that they are prepared with due care so that they are an extremely productive moment.

As it should be a planned practice, even if there were no members of the team, it would be a mistake to postpone the meeting.

The main cause of a meeting's low productivity lies mainly in the improvisation with which it is normally organized and managed.

A meeting is in any case an *activity*, even if of short duration, therefore it must also produce a *deliverable* in the terms of a *meeting report* called "Memory of Meeting" or "Minutes of Meeting" (MoM).

The activity of *reporting* (internal and external), i.e. updating and dissemination of information, is fundamental for a correct management of the project communication. Very often internal communication can take place through the reporting process.

A good reporting process requires that they be defined a priori:

- Moments in the development cycle of the project in which it is to be carried out
- The information to be disseminated and their degree of detail
- The persons to whom information should be passed on according to their degree of responsibility or stake in the project
- Different types of reports for different levels of responsibility or stake.

The information shall be presented in the documents in an essential and concise manner, avoiding insignificant or overly detailed information.

The structure of a good project report should contain the following information:

- Report's title/Project's name/PM's name
- Referring Work Package number
- Number of pages
- Distribution list
- Edition or version
- Authorisation signature and name of the compiler
- First issue date
- Date of update/revision
- Comments.

12.6 Conflict Management

The innovative characteristics of a project, the lack or limited amount of reference standards, the risks and uncertainty concerning decision-making, as well as the involvement of resources from different functional units determine a stronger chance of conflicts rising within the framework of projects, as compared to other routine activities performed in the company.

On one hand, conflict is an intrinsic and essential part of a project, because it ensures a source of confrontation and spurs the resolution of problems; on the other hand, conflicts may indicate (as well personal interests) that the members of the team are fighting for the project's success and to achieve their work tasks. Conflicts arise because different individuals come into contact with each other: where there is no contact and no relationship there is no conflict, and the likelihood of problems increases as the interdependence between the individuals grows.

Conflicts of a technical nature often give rise to non-technical ones that affect business organization and management, as well as interpersonal relationships; these can be mainly ascribed to the following:

- Limited resources, both human and financial, allocated to a single project, due to the contemporaneous presence of and competition with other projects (project priorities)
- Work overload borne by the human and technical resources involved in more than one project (conflicts with the persons in charge of the departments)
- The structure of the POBS and the assignment of sub-project responsibilities
- Underestimated costs or their inadequate partition among the various activities of the WBS
- Badly scheduled WBS activities and blaming each other for the delays
- Problems linked to the clashing personality of the individuals involved in the project.

Successful conflict management results in greater productivity and positive working relationships; when managed properly, differences of opinion can lead to increased creativity and better decision making. If conflict escalates, the project manager should help facilitate a satisfactory resolution; the success of project managers in managing their project teams often depends a great deal on their ability to resolve conflict.

The classic solutions for managing conflicts can be summarised as follows:

1. *Withdraw/avoid*
2. *Smooth/accommodate* (by avoiding the diverging topics and underlining the points of agreement)
3. *Compromise/reconcile* (in the event of standstills, be available to give something in order to obtain something else)
4. *Pressure* (trying to prevail, even if there is the risk of losing)
5. *Collaborate/confrontation* (analysing the conflict in order to reach an agreement).

The polarization of a team following internal conflicts, in other words having the members taking sides, only augments the conflict: polarization must therefore be avoided or restrained. It should also be kept in mind that often conflicts that apparently revolve around content conceal latent, but deeper and more pervasive issues regarding power, personality, etc.

The psychological and social approach to conflicts can be aided by the use of various techniques and tools, such as the following:

- MPM practices, objectively assessing the priorities of the various projects, monitoring in detail the workloads and resources and analysing their commitment over long periods of time
- Predictive tools such as *baseline* and techniques for analysing budget variances
- Anticipated formalisation of the assignment of external responsibilities, linked to the project, and internal ones (of sub-project), monitoring throughout time the workloads on all the resources in relation to all the activities
- Techniques, such as CPM, to determine the margins of flexibility and the causes of the delays
- The *Responsibility & Relationship Matrix* (see Sect. 12.4).

Project Cost Management and Finance 13

13.1 Types of Project Costs

Traditionally, Cost Accounting consists of three stages:

1. Collecting, classifying and recording all elements of cost
2. Allocating costs to cost centres (identified on the grounds of homogeneity, the existence of a person responsible for the costs, and of a certain budget calculated using prospective or standard costs)
3. Summarising and analysing costs to determine the selling price of the products

Costs are classified according to various criteria; the main ones are described in Fig. 13.1. *All* these costs must be considered in a project: the total cost (100) of the project must therefore be analysed according to the nature and origin of the costs, and according to the project size (some costs are fixed; others vary according to the size of the project, whether small, medium or large); it is moreover essential to distinguish between costs that can be directly (=objectively) ascribed to a certain project and those that are indirectly (=using subjective parameters) ascribed "pro quota" to each project. Finally, a classification must also be made on a time basis, distinguishing between prospective, current and final costs.

Particular attention must be given to distinguish between variable and fixed costs on one hand, and direct and indirect costs on the other.

Since variable costs are often easy to ascribe to a product or a project (considering the materials used, the labour and power required by dedicated machinery or single-product lines, although also in the case of multi-product plants, it is possible to calculate them using the routing cycles), sometimes variable costs and direct costs are considered synonymous. It is easy to make this mistake because *direct costing* techniques, aimed at calculating the gross margin of contribution, consider this margin as the difference between profits and variable costs, a margin to cover fixed costs—and, only in this case, *direct* coincides with *variable*.

S. Tonchia, *Industrial Project Management*, Management for Professionals,
https://doi.org/10.1007/978-3-662-56328-1_13

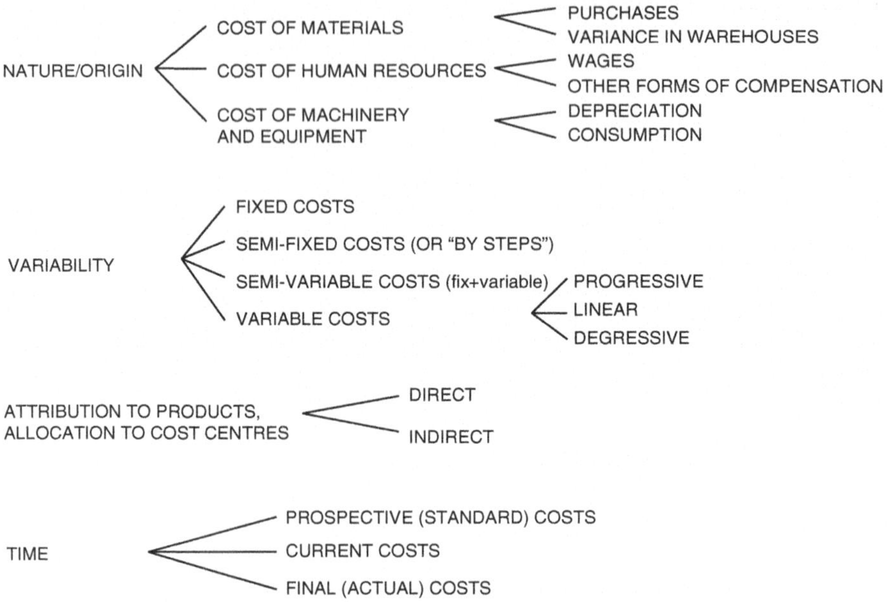

Fig. 13.1 Criteria for cost classification

In practice, there are also indirect variable costs (e.g. those relative to consumption that are not specified by any standard or any bill) and direct fixed costs (e.g. the depreciation cost of a single-product line, which can be directly i.e. objectively ascribed to that product).

13.2 Cost Estimating and Project Budget Calculation

According to the Project Management Institute (PMI), *Cost Management* can be divided into three stages, which will be described in detail in this section and the following two:

1. *Cost Estimating*, the stage during which the cost of the resources exploited for the project is forecasted, so as to estimate the overall cost of the project.
2. *Cost Budgeting*, the process of establishing a budget by summarising the estimated costs of the work packages (according to the Project Budget Breakdown Structure—PBBS—see Sect. 7.2), optimising this budget in relation to the available amount of money and finally defining the *baseline* (i.e. a curve describing expenditure throughout time).
3. *Cost Control*: once the project has started, the progress of expenditure is monitored, analysing the variances from the budget in real time.

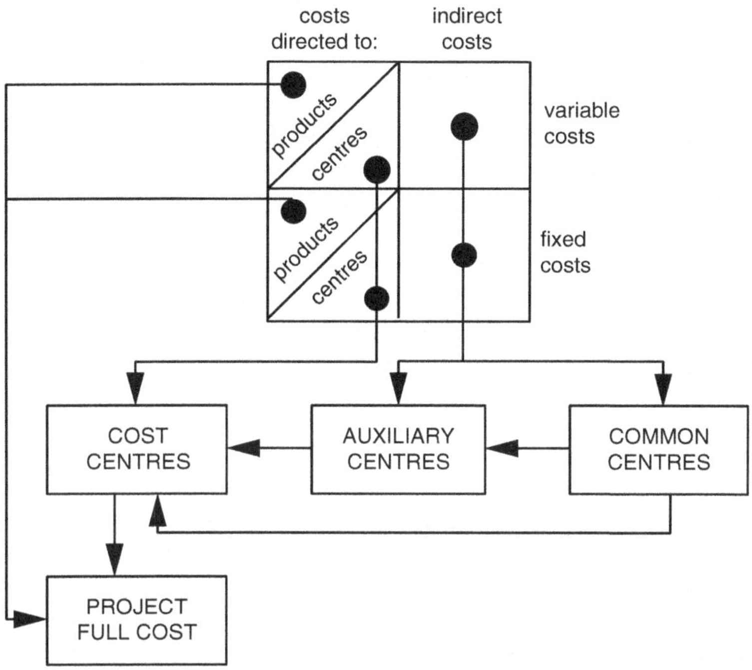

Fig. 13.2 Traditional allocation of indirect costs through cost centres

Hence, Cost Estimating analyses the cost of the resources required for a certain project; the type and amount of necessary resources are identified taking into account the activities of the project, which are described in the Work Breakdown Structure (WBS).

In this section, we see how Cost Estimating in Project Management requires a different approach from that used in Cost Accounting procedures, typical of operations management, procedures based on the allocation of costs—classified as direct or indirect—to cost centres. The innovative approach of Cost Estimating in Project Management follows certain trends that are already present in advanced Cost Accounting, which in turn are based on *Activity-Based Costing* (*ABC*) techniques.

The difficulty of accurately and reliably calculating the *full cost* of a *product/project/work contract* lies in the problems of allocating *indirect* costs, which by definition cannot be immediately ascribed to a certain product/project/work contract. Figure 13.2 illustrates the traditional passages from costs (classified as fixed/variable and direct/indirect) to the cost centres (allocation) and the products (attribution).

ABC was created in response to the traditional technique used for calculating the full cost of a product (or project or work contract) and based on the division of indirect cost on a single base or multiple bases (bases, i.e. criteria, such as direct labour, cost of materials, number of machine-hours per product, etc.). When—like in today's world—variable costs decrease (in particular the percentage of labour), while indirect costs (i.e. those relative to technology, design, trade, etc.) increase (such as, for example, the depreciation costs of highly automated plants), allocating indirect costs on the grounds of direct variable costs (particularly on the traditional amount of direct man-hours or standard times) can cause consistent distortions in the calculation of the full cost.

Back in 1985, Miller and Vollmann claimed in their famous paper published in the *Harvard Business Review* that indirect costs are independent of production volumes but depend instead on various types of *transactions*:

- Logistical (such as those relative to customer orders)
- Balancing (namely, all those activities linked to the availability in space and time of the materials, labour and machinery needed to meet the demand)
- Quality (e.g. the quality control procedures)
- Change (as a consequence of starting new projects or modifying pre-existing ones)

Cooper and Kaplan (1991) believe on the other hand that indirect costs are driven by complexity, whereas direct costs are mainly variable and are therefore driven by volumes; they distinguish between four types of cost-generating activities, each of which has different *drivers*:

- Manufacturing activities (direct labour, materials and power required, the driver being the production volume)
- Activities for running the production plants (such as maintenance)
- Production-supporting activities (such as machine tooling)
- Product-supporting activities (such as product development)

Except for manufacturing activities, in all the other cases there can be a variety of drivers: the number of customer orders, manufacturing orders and supplier orders, the number of lots, of items (common or special) and levels listed in the bill of materials, the amount of handling, tooling, the number of defects, complaints, etc.

ABC is an innovative technique aimed at calculating the full cost of a product (or project or work contract) starting from the cost of the resources used. Assigning the cost of the resources to the single products does not occur through cost centres, as tradition dictates (Fig. 13.2), but occurs via the activities (Fig. 13.3): in other words, it is the activities that exploit the resources, whereas products do not use up resources, but activities.

There are therefore two *stages*:

- During the first, the cost of the resources is assigned to the activities (by means of *first-stage drivers*, also known as *cost drivers* or *resource drivers*).
- During the second, the activities are linked to the products or projects (by means of *second-stage drivers* or *activity drivers*).

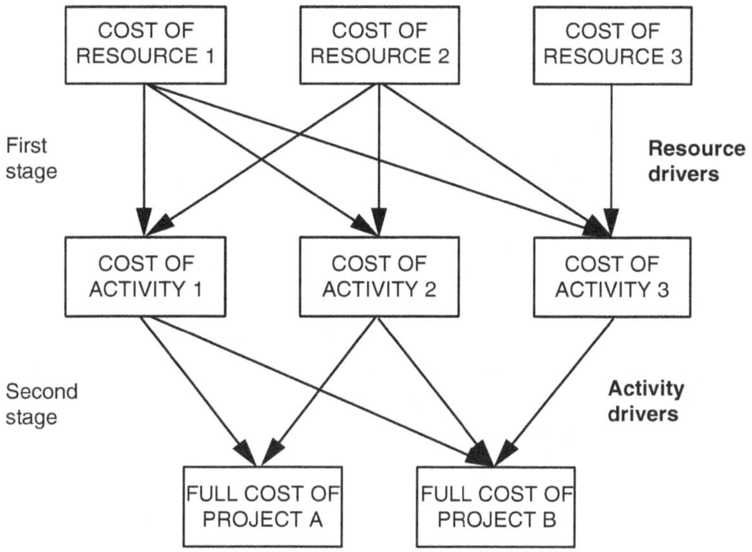

Fig. 13.3 Calculation of full product cost according to Activity-Based Costing (ABC)

Cost Estimating in Project Management adopts the ABC approach to calculate the full cost of a project, but in this case the technique is used to estimate the costs of a project or a work order, and not of a product.

The starting point is represented by the activities needed to accomplish the project, which are described in the WBS.

The second step is to link each activity (also called *task*) to the resources needed, which are organised and described in the POBS (which in turn is connected to the Organizational Breakdown Structure or OBS).

This link is established by criteria that in practice coincide with the *resource drivers* found in ABC: for example, the allocation of the resources for the entire duration of the activity, either in proportion to its duration or as work hours required.

This allocation must be compatible with the calendars and the levels of saturation of the resources, which can be exploited by more than one project.

When estimating the cost of human resources, it is necessary to include both the cost of *calling* or *exploiting* the persons for a certain activity (independently of the time required) and their cost per hour.

Finally, using criteria similar to the *activity drivers* of ABC (e.g. the duration of a certain activity in a project), it is possible on one hand to calculate the overall cost of the project so as verify if there are adequate funds (according to the company's overall budget breakdown structure—BBS) and, if necessary, revise the activities of the WBS, and, on the other, create and develop the PBBS of the project.

13.3 Cost Budgeting: Budget Breakdown Structure, Baseline, Cost Accounts

As described in the previous section, Cost Estimating allows the total budget for a given project to be defined. This budget is part of the overall budget of the company destined to projects (such as those for product development, work orders, improvement plans, etc.). The portion set apart for projects can coincide with the entire budget, in the case of ETO companies or in general, those working to order.

The budget of a project can be considered as a management tool: it is expressed in monetary terms and refers to a project that must be carried out (whereas in management control, it is considered a future accounting period). The budget takes into account all the necessary resources, unfolds in time according to a certain *baseline* (with interim checks, in the case of management control) and is articulated into *cost accounts* (while in management control, there is a profit and loss statement, the assets and liabilities sheet and estimates of the cash flow during the year). In both cases, the issues of leadership, confrontation, coordination and motivation are the same.

The budget of a project is structured according to the PBBS mentioned previously (Fig. 13.4). In the case of work orders, the expected margin of profit (*return*) must be detracted from the available amount (which can be the value of the contract or the starting price of a bid): the remaining amount is the *base budget*; this can include *target costs* and *estimated costs:* the former are those negotiated (e.g. x million euro per kilometre of motorway), whereas the others are not, and can only be predicted, a task that always hides a certain margin of risk (e.g. the cost of setting up a plant, before its turnkey delivery to the buyer).

The base budget is subdivided into two parts: the main one is the *baseline*, whereas a smaller amount is set aside for contingencies (*management reserve*). Management reserve is particularly important in the case of work orders, when the contract includes penalty clauses for late deliveries; considering the inevitable randomness of some variables, a certain budget percentage is necessary to reduce the risk of project delays.

The *baseline* on the other hand is the available sum that can be used to carry out the activities relative to the project. It is so called because its evolution in time can be depicted as a line. It usually has an *S* shape, since during the first stages expenditure is low, and then rises sharply, to finally decrease in the last stages of the project (Figs. 13.6 and 13.7). Therefore, in a cost/time Cartesian representation, the baseline curve starts at the origin of the axis to reach a final point, which indicates both the duration of the project and its overall cost (commonly known as *budget*), which corresponds to the maximum value of the baseline.

This sum includes the so-called *Cost Accounts* (*CA*) that are relative to the *Work Packages* (*WP*) carried out by well-established resources. The integration between costs and work schedule is termed as “Cost/Schedule Control System Criteria—C/S CSC” (or also “C.Spec.”) by the Department of Defense—DoD—or “533 System” by the National Aeronautics and Space Administration—NASA.

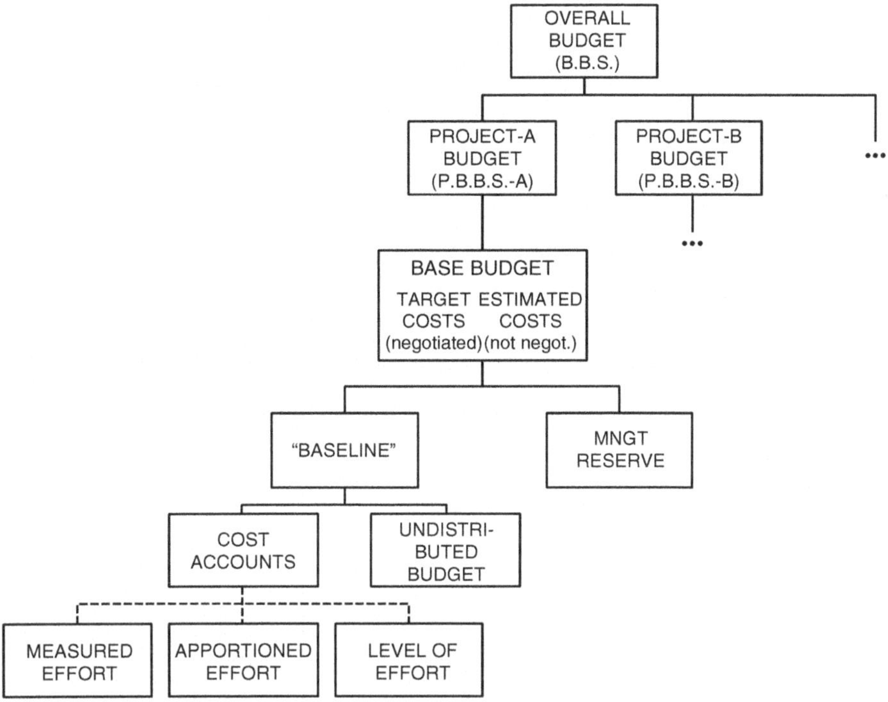

Fig. 13.4 Project budget breakdown structure (PBBS)

In other words, the CA are defined by combining the POBS with the WBS (*who* does *what*, respectively): the tasks described in the WBS are aggregated into work packages; organizational managers are responsible for these WP, and they have a certain CA at their disposal to pay for the necessary resources. Figure 13.5 could also include a third dimension, rising up from the page, which structures the elementary units of the CA, to rebuild the PPBS.

Therefore, the project manager delegates technical and managerial responsibility for the various chunks of the WBS (i.e. the activities) to WP manager, who is also responsible for the CA. The project manager retains a part of the baseline for the direct, overall management of the project: this portion is known as *undistributed budget* (Fig. 13.4).

The criteria used to define each CA must be based on the *efforts* needed to deliver the activities of the project; these efforts can be classified as follows:

- *Measured Effort* (*ME*)—an effort that can be divided into work packages (WP) Consisting of activities (30–40 at the most) of limited duration, and whose outputs (*deliverables*) can be measured directly
- *Apportioned Effort* (*AE*)—an effort that by itself is not readily divisible into short-span work packages but that is directly proportional to the ME, and therefore to

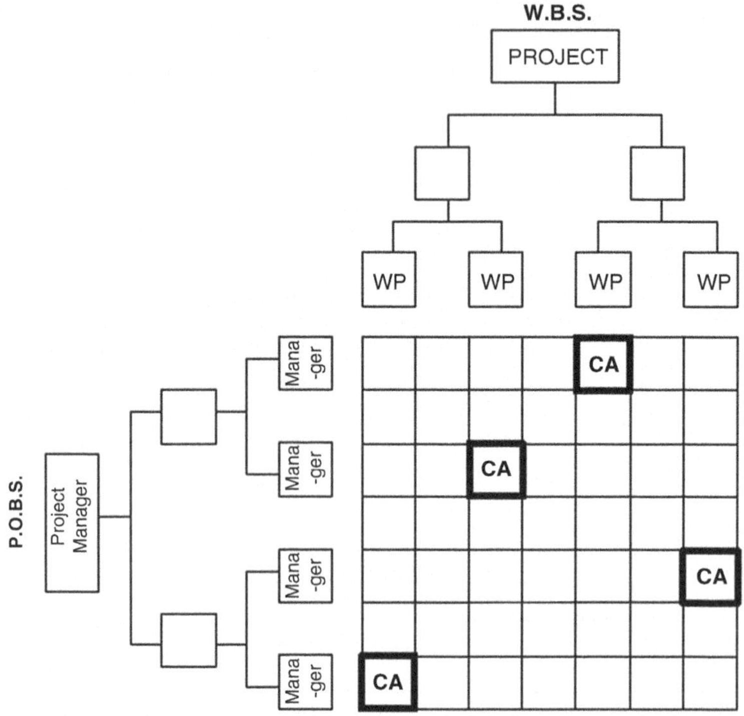

Fig. 13.5 The origin of cost accounts

the WP (these are indirect activities, but the effort they require is linked to the amount of direct work expressed by the WP)

- *Level of Effort* (*LE*)—an effort of a general or supportive nature, which is not proportional to the ME and cannot be ascribed to the WP (fixed activities, independent of the size of the WP)

Hence, in order to carry out the operations described in the WP, it is necessary to have a CA that takes into account the ME, AE and LE; work packages deliver partial outputs known as deliverables, that document/concretise the work carried out, enable assessments to be made (checking whether the technical–qualitative goals have been achieved, and the times and costs conform to those established) and trigger the subsequent activities depicted in the network diagram of the project.

As shown in Fig. 13.6, all decisions concerning or affecting costs—in other words, the *commissioned* costs—are made a long time before paying the actual amounts: the decisions are made when defining the concept of the product and planning the project, therefore before the activities start. These first stages are crucial, and must be carefully managed, not only because of the technical repercussions on the product but also to determine its costs. The fact that it must be calculated so long beforehand is what makes cost management in a project so

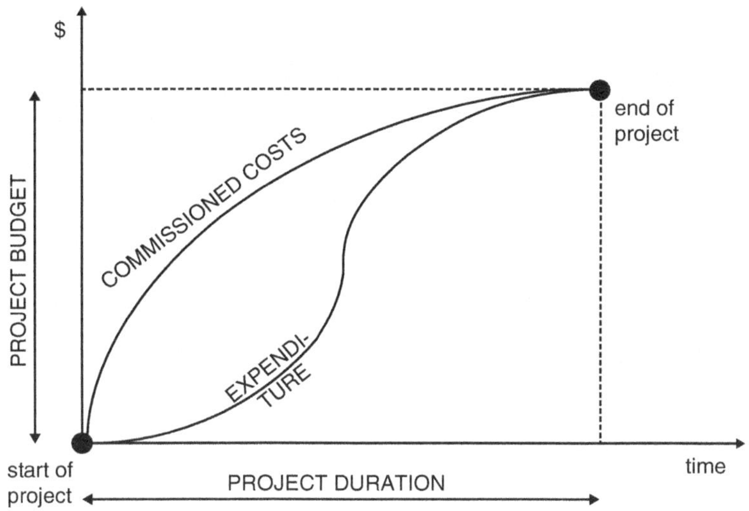

Fig. 13.6 Decisions affecting costs, and payment moments

difficult. The curve of costs incurred in accomplishing work coincides with the planned baseline which in turn—if the prediction is correct—will also coincide with the ACWP (see following section).

It must be underlined that cost-related decisions made during the planning stage not only serve to estimate the cost of the project and therefore the necessary budget, but will also have a profound impact on the overall costs, once production has started. Planning plays a crucial role, even from an economic viewpoint; although it accounts for only 5% of the cost of the product, it can often be responsible for over 75% of the overall manufacturing costs.

13.4 Cost Control and Variance Analysis

As mentioned in the previous section, the baseline curve illustrates how the predicted costs relative to a project unfold throughout the duration of the project. Since it considers both costs and times before they actually occur, the baseline is also known as *Budgeted Costs of Work Scheduled* (*BCWS*).

Once the project has started, it is necessary to *control* its progress, not only in terms of work and technical performances, but also in terms of its cost, so as to intervene in time if the events require it. Cost control is carried out using, as well as the baseline, two other curves with acronyms similar to that used to indicate the baseline:

- The *Actual Costs of Work Performed* (*ACWP*) curve, namely that describing the work actually performed and its actual costs

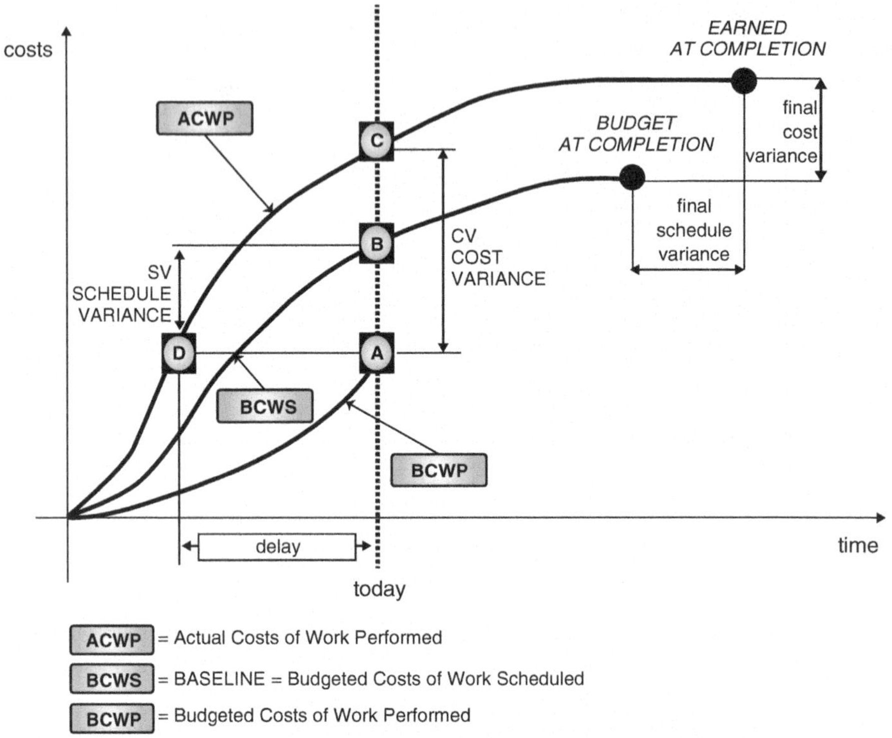

Fig. 13.7 Curves used to monitor project costs with respect to the budget

- The *Budgeted Costs of Work Performed* (*BCWP*) curve, in other words, the curve describing the actual work performed but at the budgeted costs

The ACWP curve, considering both the actual costs and times, is an ex-post curve and is therefore the opposite of the baseline (or BCWS), which is ex ante.

However, the ACWP and BCWS curves alone are insufficient for analysing the budget variances for projects in progress. A third curve is necessary—BCWP—which is a hybrid of the previous two, considering the actual work performed but at the budgeted costs. A lower expenditure than that budgeted can reveal delays in the work in progress.

The need for the three curves to analyse budget variance when controlling progress is clearly shown in Fig. 13.7: at any given time (*monitoring date*), the future trend of the ACWP is unknown; moreover, a simple comparison between ACWP and the baseline would be insufficient, since it is possible that by that date, the exact budgeted amount could have been spent (so ACWP and baseline would have the same value on the *y*-axis), but less work could have been carried out, at higher costs. Thus the need for the BCWP, since this curve shows how much should have been spent at the budgeted costs for the work that was actually carried out: if, for instance, on a given day the BCWP is below the baseline, it is clear that less work was done than that scheduled, since both curves refer to the budgeted costs.

Figure 13.7 shows a typical example of cost variance and a contemporary delay: the ACWP curve is above the baseline (BCWS) and the BCWP is below it. On the monitoring day, the difference on the y-axis between the value of ACWP and BCWP is the *cost variance* (CV), while the difference between baseline and BCWP indicates the *schedule variance* (SV), expressed in terms of costs. It would be a mistake to consider the CV as the sole difference between ACWP and baseline, since they would only coincide if the work progressed in perfect time. The SV gives an estimate of the cost variances during the progress of the project; the amount that has not yet been spent is simply due to the fact that certain expenses have not occurred because the respective activities have not yet started.

It is therefore possible to calculate the percentage of cost and time variances at a given time t during the progress of the work, using the following formulas:

$$\%\mathrm{CV}(t) = \mathrm{CV}(t)/\mathrm{BCWS}(t)$$
$$\%\mathrm{SV}(t) = \mathrm{SV}(t)/\mathrm{BCWS}(t)$$

Only at the end of the project will it be possible to calculate cost and time variances in a different manner, measuring the differences of the final values of the ACWP curve and the baseline both on the y-axis (costs) and the x-axis (time).

The existence of the third curve, namely the BCWP, also gives the name to the method just described: *Earned Value Method*, defined as the value of the actual work according to the budget. Given the total value of the project, or Budget at Completion (BAC), and the percentage of work progress at a given time t ($\%L$), the Earned Value (EV) can be calculated as shown here:

$$\mathrm{EV}(t) = \mathrm{BAC}^{*}\%L$$

When controlling a project and comparing budgeted and actual costs, it is necessary to take into account the efficient use of the various resources.

With reference to Fig. 13.8, for every resource used to complete a work order, it is possible to calculate its predicted cost by multiplying the unitary cost or price of the resource (on the y-axis) by its use rate (on the x-axis), for example, in the case of designers:

$$(\text{€ per hour}) * (\text{hours per corder}) = \text{€ for designing each work order}$$

The total area of actual and final costs is equal to the algebraic sum of the three areas—budgeted cost, variances in the price of the resources (top part of the Fig. 13.8) and changes in operational efficiency (which decreases when moving towards the right hand side of the Fig. 13.8 and improves when moving to the left, because in this latter case less resource units are used for the job). It must be noted that the area corresponding to variances in the price of the resource takes into account both the budgeted costs and the increase (or decrease) in operational efficiency.

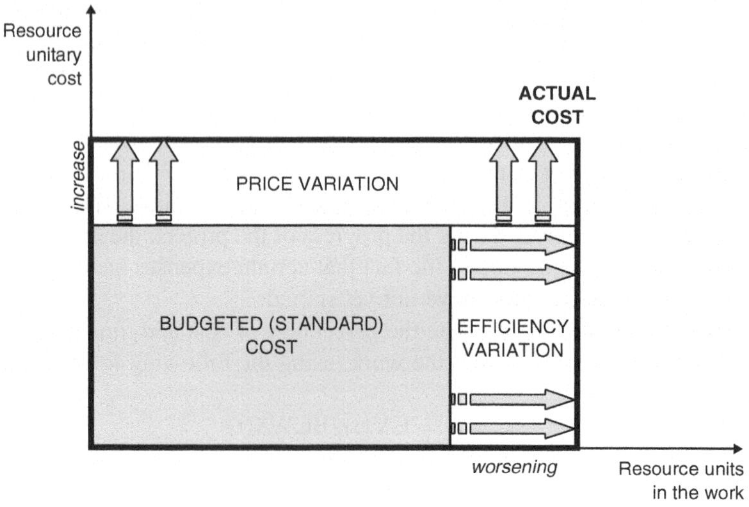

Fig. 13.8 Variance Analysis between budgeted and actual costs due to variations in the price of the resources and in efficiency

13.5 Capital Budgeting: Analysis and Evaluation of an Investment Project

Cost management and its three stages described previously (Estimating, Budgeting and Control) are still insufficient to give a full view of the monetary situation: as well as an *economic analysis*, it is necessary to perform a *financial analysis* of the project, so as to take into account the *time* variable. A project is in fact a form of investment. The capital required for its delivery, and the expected gains, should therefore be allocated and considered throughout time (*Capital Budgeting*).

An *investment*, of any type, can be defined as a sequence of net monetary flows, structured so that at the beginning expenditures (costs) prevail, and only afterwards, the takings (gains or revenues).

Figure 13.9 shows two typical *investment profiles* relative to the launch of a new product, one of which is a *pioneer* (appearing for the first time on the market) and the other a *follower* (appearing later).

These curves are extrapolated from cost and gain histograms relative to discrete periods, so the S-shaped curve of the baseline is the integral (i.e. the area bounded above by the graph of the function and below, by the x-axis) of this extrapolation curve: the value of the baseline on a given date is therefore equal to the sum of predicted costs incurred up to that day, and depicted as upright bars (histograms) of below-zero value (costs).

The moment of maximum slope for the baseline coincides with the bar having the greatest absolute value of cost, which occurs somewhere around the middle of the project's life cycle.

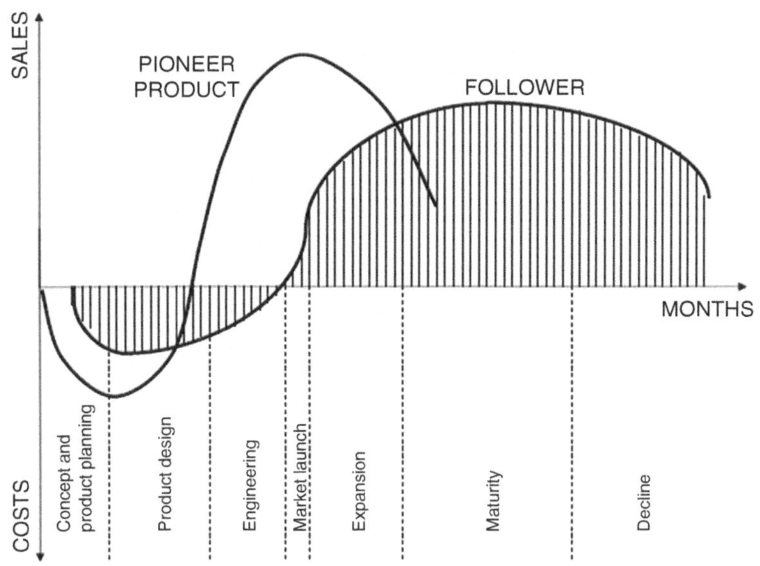

Fig. 13.9 Two typical investment profiles for the launch of new products (stages refer to the follower)

When taking into account the financial aspects of a project, it is necessary to define the *monetary value throughout time* as a cost (periodic percentage) sustained to anticipate a taking (e.g. the cash flow deriving from a loan received) or to delay an expenditure (e.g. when asking for a postponement of payment); in other words, profits are made by accepting to postpone an income or accelerate an expenditure.

The transport of money throughout time requires correction factors: *capitalising* factor if moving from present to future, *actualisation* one if the other way round. As described in greater detail in any financial text, it is sufficient to multiply the amount to be moved as follows:

$$(1+i)^n \, and \, (1+i)^{-n}$$

where i is the interest rate for the period considered, and n is the number of periods through which transport occurs.

Any project, considered as an investment, must be subjected to two phases characterising the analysis and evaluation of a typical investment.

The stage of *investment analysis* includes the following:

- A quantification of the incoming and outbound cash flows that can be ascribed to the project, in other words, its estimated costs and the expected gains (arising from the sale of the product, the delivery of the order, etc.)
- The distribution of these flows in time
- Their monetary value in time, expressed by an interest rate
- The degree of risk, thus the uncertainty of all previous items

The following stage of *investment evaluation* includes the following:

- The identification, selection and application of one or more criteria to evaluate the investment profile calculated during the previous stage
- The definition of the acceptance criteria, which should take into account the company's strategies, its short-, medium- and long-term objectives, technical feasibility, exposure to risk, compatibility with current or future projects, etc.

While the stage of analysis strongly depends on the intrinsic nature of the project, the evaluation stage exploits a set of standard methods to assess the following aspects of a project/investment:

- Net *income* (the difference between revenues and costs)
- *Profitability* (the ratio between all the revenues occurred and the costs sustained)
- *Risk* (intended in this context in an economical, not a technical sense)

The most popular methods for evaluating the financial performances of a project are as follows:

1. Net Present Value (NPV) method for evaluating income
2. Profitability Index (PI), also known as Value Investment Ratio
3. Payback method for limiting risk when estimating cash flows
4. Internal Rate of Return (IRR) method for evaluating the level of risk when estimating the actualisation rate

In the following pages we will briefly describe these four methods, presenting at the end an example of a compared evaluation.

The *Net Present Value* (*NPV*) is the difference between positive (gains) and negative (costs) cash flows, both of which are actualised. This value is the net gain that can be ascribed to the project as if it were already available, in other words, at its present value. It therefore depends, as well as on the amount of forecasted profits and losses, also on the interest rate used to actualise the sum (since the coefficient is always lower than 1, actualisation detracts a certain amount of the future gain to its current value: for this reason, the flows obtained are discounted). A company will choose those projects having a positive NPV, taking however into account the links existing between the various projects (in order to avoid, for instance, product cannibalism), the consequences of path dependency (i.e. constraints on future choices brought about by the decisions made on current investments) and the limited availability of resources.

The *Profitability Index* (*PI*) measures the ratio between positive (gains) and negative (costs) cash flows, both of which are actualised, i.e. discounted; the negative cash flows (i.e. the denominator) often coincide with the initial investment, whereas successive costs are deducted from the gains at the numerator. It must be noted that a good PI does not necessarily imply a good NPV. Since $\text{PI} = R/C = (C + \text{NPV})/C = 1 + \text{NPV}/C$, where R are the revenues and C the costs, there could be a modest value of NPV but an excellent PI if the costs sustained were very limited; on the other hand, a very high value of NPV could be obtained at remarkable costs, and therefore the PI would be low. When applying the PI, attention must be paid to the data considered: the value of the index changes if all the costs and revenues are considered, or the positive and negative periodic flows only (a periodic flow is the difference between revenues and costs in a certain period of time). The acceptance criterion for a project, according to this method, is that the PI value is greater than 1; given a certain sum to invest (in the event of a limited capital), the project chosen is that with the greatest PI.

The *payback* considers the time needed to retrieve the investment made (in other words, the initial stage during which costs prevail)—thanks to the first gains. The fact that this period of time is short—considering that all cash flows are only estimated—reduces the economic risk of the project when income and profitability performances are equal. From a graphical viewpoint the payback is the value of the x-axis (time) when the positive area (indicating the proceeds) is equal to the negative one (the costs): all proceeds from that date on are pure profits. Obviously, both cost and revenue flows must be actualised.

There is another risk factor in the process of actualisation required by these three methods, as well as that concerning the amount and distribution of the flows, namely the value of the interest rate i (mean capital cost, opportunity cost, etc.). To solve this problem, it is necessary to calculate the *Internal Rate of Return* (*IRR*) of the investment project, namely the value of i that causes the NPV to be equal to 0. The NPV is a function of the interest rate i and follows a exponentially declining law: the value of NPV peaks when $i = 0$, then—as i grows—NPV decreases till it reaches negative values. To appeal, and therefore have a good NPV value, an investment must have a real i rate that is much lower than the IRR. With this method however, it is unnecessary to define a precise i value, but only a range of probable values, to be considered at a safe distance from IRR, which is calculated as the rate that would make NPV equal to 0.

Table 13.1 presents an example of a financial evaluation of three different investment projects using the above methods. The table illustrates a simplified estimate of a cash flow (the costs and revenues expected from the project), which is then actualised. It can be observed that, according to the method used, different projects are selected (those in grey) because different parameters are taken into account (income, profitability, risks concerning the estimated cash flows and the interest rate). It is therefore necessary to perform a comparative analysis of the projects, considering the various aspects here described.

13.6 Project Financing

Project Financing is an alternative to the more conventional company financing, from which it differs in one important aspect. While the latter is aimed at evaluating the economic and financial statement of the company and the effects of new investment or new debts on this balance, Project Financing is aimed at evaluating the economic and financial statement of a specific project, linked to a certain

Table 13.1 Evaluation and comparison of three investment projects

		Cash flow					
		Project A		Project B		Project C	
	Actualisation rate ($i = 6\%$)	Not discounted	Discounted	Not discounted	Discounted	Not discounted	Discounted
First year	0.9434	−30,000	−28,302	−60,000	−56,604	−10,000	−9,434
Second year	0.8900	5,000	4,450	−30,000	−26,700	0	0
Third year	0.8396	10,000	8,396	20,000	16,792	10,000	8,396
Fourth year	0.7921	20,000	15,842	50,000	39,605	20,000	15,842
Fifth year	0.7473	15,000	11,210	50,000	37,365	10,000	7,473
Sixth year	0.7049	10,000	7,049	40,000	28,196	0	0
Net value		30,000	–	70,000	–	30,000	–
NPV			18,645		38,654		22,277
PI			1.66		1.46		3.36
Payback			Nearly 4 years		4 years 9 months		3 years 1 month
IRR			25.1%		19.1%		61.8%

investment, and in economic and legal terms it is independent of all other activities made by the company.

In the former case the underwriter, usually a bank, assesses the company requiring the funds and then, on the grounds of this evaluation, finances the enterprise, in exchange for concrete guarantees or bonds. In the latter case, the underwriter finances a single idea or project; the attention is focused on the prospects of the project and its ability to return the investment. If the project is unsuccessful, the underwriters can only make up for the loss with the goods of the project, and it is only possible to claim a limited recoup from its promoters.

Project Financing can be defined as an organisation formed by a plurality of subjects to finance the production and management of an economic good, giving rise to a cash flow that can pay back the initial investment at a market return rate. The aspect characterising Project Financing is the fact that the underwriters accept as a guarantee the cash flows generated by the management of the project, which are made as stable as possible by means of contractual terms defining how the risks are shared.

This definition underlines a typical feature of Project Financing, namely the need to identify the technical and economical risks linked to the project and define risk management procedures. Since usually Project Financing operations require large funds, but the assets of the company promoting the project are in most cases insufficient as a guarantee, the risks relative to the project and its estimated payoff are divided among all the members of the operation, by means of a harmonised system of contracts. It is the management of the financial risk linked to the operation that is the most typical and characteristic feature of Project Financing.

Project Financing is therefore an operation that occurs *outside the budget* of a specific company, which is independently assessed according to its ability to generate profit, and whose cash flows are considered as the main source of the capital invested. This type of logic focuses on the fact that the underwriters' expectations lie on the project's estimated profits and not (only) on the economic reliability of the companies, or their capital.

Consequently, the general goals of Project Financing are the financial support and the delivery of the project under conditions of great autonomy with respect to the underwriters who are its promoters. It is—thanks to its—outside the budget financing logic that Project Financing ensures a financial leverage that is usually difficult to obtain.

When the project concerns technological innovation, Project Financing, rather than supplying resources in the form of medium–long term loans, is aimed at widening the ability to collect risk capital by releasing shares of the company. The technique of Project Financing, as well as responding to the need to finance autonomously the investment projects, can be included among the policies of Risk Management. Obviously, Project Financing is not the only solution to finance research and innovation projects: other opportunities also exist, such as the establishment of joint ventures, pools of enterprises, public research centres and venture capital.

Reference

Cooper, R., & Kaplan, R. S. (1991). *The design of cost management systems: Text, cases and readings*. Englewood Cliffs, NJ: Prentice-Hall.

Part V

From Single to Multi-Project Management

Portfolio Management

14

14.1 Managing Project Portfolio

Multi-Project Management (MPM) is becoming increasingly important, because a good project portfolio management ensures a greater competitive advantage than that achievable by means of good management of single projects. There are at least two reasons for this: (a) the turbulence of the market environment, which makes it necessary to limit the otherwise increased risks by presenting a continuous set of innovations through an advanced project portfolio, (b) the links of interdependency that exist among projects, which are becoming increasingly stronger and restrictive due to the limited amount of available resources.

Project interdependency can refer to the following:

- *Benefits*: market synergies obtained by the one concept idea or family feeling
- *Resources*, which are shared by the different projects and pose certain restraints in their use
- *Contents*

Content interdependencies can in turn refer to the following:

- The *commonality* of certain parts with ones from other projects, either past (*carry-over*), present (*commonality* in a stricter sense) or future (new generations)
- The fact of being part of a *vaster system* (for example, the project for a sub-assembly is part of a global product project)
- The link with previous *platform* or *naked* projects (see Sect. 5.4)

MPM, or Project Portfolio Management, introduces the concept of *families of projects* that are characterized by similar development cycles and the same methods for scheduling activities and distributing workloads.

S. Tonchia, *Industrial Project Management*, Management for Professionals,
https://doi.org/10.1007/978-3-662-56328-1_14

The MPM approach consists of the following steps:

1. Identifying the families of projects
2. Defining the *reference structures* (times, workloads, etc.) required by each family
3. Defining the project portfolio, on the grounds of the company's objectives, and mapping the competencies and financial resources needed
4. Analysing and managing risk
5. Inter-project organization, appointment of a multi-project manager and identification of critical resources/bottlenecks
6. Defining a system for managing the projects, with common routine specifications and integrated planning of the projects
7. Defining the priorities and allocating the available resources, taking into account the interdependencies existing among the projects

Innovation Index and *Ready Plan* methods are particularly useful for determining the *reference structures* for each family. Defining on one hand the complexity of the product structure and on the other the number of families in the project, the Innovation Index method can be effectively used in the presence of high levels of product complexity and limited family extension, whereas Ready Plan is indicated when the families are large, but product complexity is rather limited.

According to the *Innovation Index* method, the product is divided into functional groups, for each of which the index is calculated as the percentage of planning required as compared to that needed to design it de novo; the reference project family is then identified. Each family consists of a common part (the macro-activities) and a variable one (the duration of the activities and workloads on the resources). Hence, the projects—listed in a decreasing order of innovativeness—can deliver a new product, restyle it, give it a facelift or simply customize it, drastically or slightly.

The *Ready Plan* method, on the other hand, classifies interventions into standard, first-level interventions (e.g. a change of push-button panels) requiring a certain family of projects (in the case of push-button panels: functional aesthetics and ergonomics), which in turn determine standard second-level interventions (a change in the shape of the keys, changes in the disposition of electrical contacts, etc.) and relative *ready-use plans*.

Each project of the portfolio should have a *standard reference form*, easy to consult and compare with that of the other projects; the form should list the following:

- Type/class of project
- Client/internal sponsor
- Person in charge/*project manager*
- Allocated budget and resources
- Specifications and scope
- General and detailed technical programme (*project plans*)
- Start, interim—for the milestones—and final technical documentation
- Documents stating formal approvals, the minutes of the meetings and the lists of official external contacts (with suppliers, customers, etc.)

Once a *project management system* is defined, establishing common routines (so as to seek efficiency in the planning process) or sets of repetitive, clearly defined activities, and having made an integrated project plan (scheduling the activities and use of resources), it is necessary to assign the priorities of the projects, fundamental when conflicts arise for the exploitation of the available resources.

Keeping in mind the interdependencies among projects, the *definition of the priorities* can be based on different criteria, such as technical difficulty (so the management can decide to give priority to the most expensive projects, those requiring greater resources and/or those that are considered more risky), commercial opportunities (success rate, outcome in terms of payoff and image, opportunity costs in the event of exclusion or demotion, importance of the customer), time criticalness (compulsory time-to-market stated by contract or established by competition, the penalty due in the event of delays, number of critical activities having no margin of flexibility), etc.

The definition of priorities requires a clear and univocal vision of the projects, their objectives and the technical–managerial aspects: it postulates a clear view of the interdependencies existing among the projects and a precise definition of ownerships, as well as the mechanisms for the periodical revision of priorities. Moreover, it is important to keep in mind all the insurmountable constraints that cannot be levelled out (such as a sufficiently long test period).

14.2 The Project Management Office (PMO)

According to the MPM approach, *organizational interventions* should be carried out, focusing on the following:

- *Organizational mechanisms*, from a simple charismatic control carried out by the head of one department to the creation of rules, procedures and committees coordinating the activities of the project managers
- *Organizational structures*, where each department is in charge of a specific responsibility and there is an office responsible for multi-project management
- *Organizational figures*, introducing one or more levels of multi-project managers, positioned either at a higher hierarchical level than that of project managers or simply in a transversal position acting as "connectors".

The advanced typical solution for MPM is today the PMO, which stands for both Project Management Officer (i.e. the Organizational figure) and Project Management Office (i.e. the Organizational structure).

The PMO reports directly to the Chief Executive Officer (CEO).

The PMO has as its goal the centralized and coordinated management of the projects of an organization. The PMO has the formal authority for coordinating projects and project managers; he decides the priorities in the implementation of projects and assign them resources (both technical, financial and human) in order to achieve projects' performance objectives and *overall* business results.

Within the PMO there may be a dedicated coordination committee, who takes strategic decisions based on project life-cycle, resource availability and integration between projects.

The PMO delegates its authority to the single project managers, to whom provides administrative and experienced support. The project managers and any inter-project teams (if they exist) will refer directly to the PMO.

Another key function of the PMO is to provide support services to project's activities through training, software, standardized rules and procedures.

In summary, the most important activities of the PMO are:

- Manage priority conflicts between projects
- Analyze project's risk in a comparative way between projects
- Name the project manager of each project and sometimes its team
- Evaluate project managers and team members
- Keep the experience of the projects in a "repository"
- Transfer to the new projects the "lesson learned"
- Cultivate, disseminate, promote and improve the *good practices* of PM
- Create and consolidate shared standards to apply to different types of projects
- Assist the PMs, especially during start-up and closing phases
- Support the administrative activity of the projects
- Act as a meeting point and communication link between PMs
- Put the basis for continuous improvement of project performances (quality level, lead time, cost reduction).

14.3 The "3R" Matrix

Projects can be represented as circles in a two-dimensional relevance–risk matrix. The area of the circles is proportional to the use of critical resources, so this matrix is also known as *3R-matrix* (Relevance, Risk, Resources). The main interdependencies between projects are indicated by dotted arrows connecting them, and when one is excluded, all connected projects are affected (as in case D with A, in Fig. 14.1). All the projects that fall into the matrix are considered acceptable; however, on the grounds of the utility of each project for the company, and considering the restraints upon the resources, it is possible to define a triangular area representing the selected *project portfolio* (the shaded area depicted in Fig. 14.1).

According to this type of representation, projects can be classified as follows:

- Class I projects (having great importance and a high level of risk), for which no delays are admitted; the aim is to reduce their development times, and in the event of conflicts, they have priority for the use of critical resources.
- Class II projects (having less importance and a lower level of risk).

When scheduling activities, care must be taken to limit the number of contemporaneous class I projects.

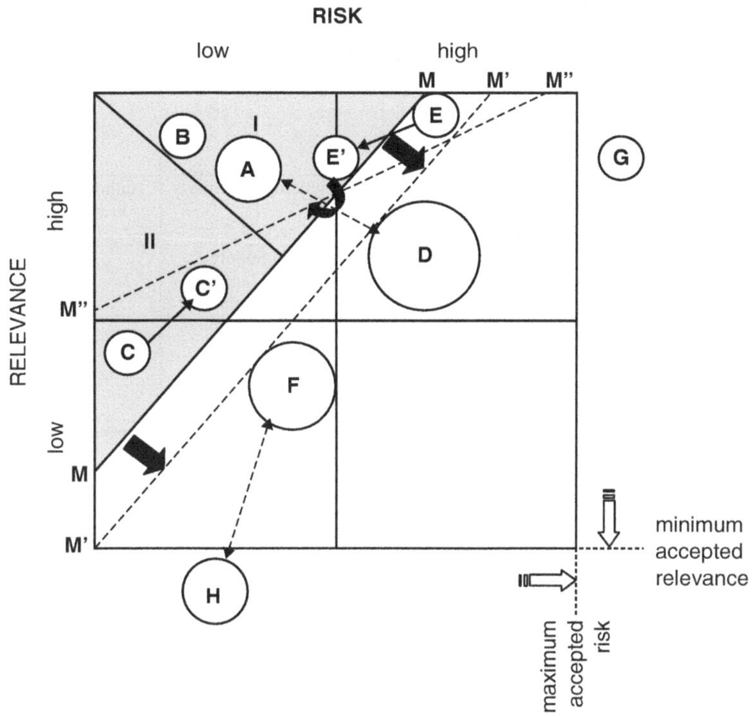

Fig. 14.1 The "3R" matrix (Relevance—Risk—Resources)

It is also possible to depict multi-project management (MPM) improvement interventions, which can lead to the following:

- An increase in the availability of critical resources, or alternatively, an increase in their productivity (the project portfolio expands, passing from M–M to M′–M′, incorporating project E but excluding project C)
- An increase in the ability to manage risk (the project portfolio changes, passing from M–M to M″–M″, thus incorporating project E)
- An increase in relevance, hence accepting a greater level of risk (for instance, the project is placed in position C′)
- A reduction in risk, accepting a reduction in its relevance (for example, project E is allocated in position E′).

14.4 PMI Standard for Portfolio Management

According to PMI, 'a *Portfolio* is a collection of projects or programs and other work that are grouped together to facilitate effective management of that work to meet strategic business objectives'. The portfolio components may not necessarily be

	Defining	Aligning	Authorizing & Controlling
Strategic Mngt	Developing Portfolio Strategic Plan + Developing Portfolio Chapter + Developing Portfolio Roadmap	Managing Strategic Changes	
Governance	Developing Portfolio Management Plan + Defining Portfolio	Optimizing Portfolio	Authorizing Portfolio + Providing Portfolio Oversight
Performance Mngt	Developing Portfolio Performance Plan	Managing Supply & Demand + Managing Portfolio Value	
Communication	Developing Portfolio Communication Management Plan	Managing Portfolio Information	
Risk Management	Developing Portfolio Risk Management Plan	Managing Portfolio Risks	

Fig. 14.2 The 16 processes for Portfolio Management (elaboration from PMI "Standard for Portfolio Management")

interdependent or directly related. A portfolio may consist of a set of past, current and future portfolio components.

Portfolio Management includes identification, categorization, evaluation, selection of components to best accomplish organizational strategies. Portfolio management balances conflicting demands between programs and projects by allocating resources (e.g. people, funding) based on organizational priorities and capacity.

Program and project managers focus on "doing work right". Portfolio managers focus on "doing the right work".

The PMI "Standard for Portfolio Management" (3rd ed., 2012; incoming edition: end of 2017) listed 16 processes, classified according to two dimensions (Fig. 14.2):

1. "groups of processes" (Defining, Aligning, Authorizing and Controlling)
2. "knowledge areas" (Strategic, Governance, Performance, Communication, and Risk Management).

In particular, "Defining" concerns how the organizational strategy will be implemented in the portfolio (includes the "strategic plan", the "portfolio charter" and the "roadmap"), while "Aligning" manages and optimizes the portfolio. The "portfolio charter" is a document which describes the high-level scope, high-level timeline, success criteria, and main stakeholders; the "portfolio roadmap" is the components' key milestones and dependencies in a chronological sequence.

Strategic Management ensures alignment with the organizational strategy; Governance Management includes portfolio oversight and high-level decision-making activities; Performance Management manages resource allocation and tracks portfolio performance and value realization.

15 Program Management

15.1 Definition of Program Management

This chapter addresses ***Program Management***, a fairly new development in the Project Management world. It covers definitions, the recent evolution of Program Management, relationships between Project/Program Management and organizational strategic planning, Program Management operating environments and methodologies, program offices, and the future of Program Management in the current race of technology environment.

Definitions of ***Program Management*** abound and sometimes create confusion for Project Management practitioners. For example, Business Dictionary (www.businessdictionary.com) defines Program Management as the "process of managing multiple related projects at once. Where Project Management is often used to describe one project, Program Management involves multiple projects that are all related and working toward the same goal or result."

The Association for Project Management (https://www.apm.org.uk) defines *Program Management* in a much simpler fashion: "the coordinated management of projects and change management activities to achieve beneficial change."

Local practices frequently determine the use of Program Management definitions. Other current working definitions of *Program Management* that organizations use adhere to or derive from the Project Management Institute (PMI) treatment, a "group of related projects, subprograms and program activities that are managed in a coordinated way to obtain benefits not available from managing them individually" (PMI Standard for Program Management—3rd Ed., 2013).

PMI is especially helpful for managers as it defines and clarifies Program Management within the context of the closely related management fields of *Project and Portfolio Management* (PMI Standard for Program Management, 2013). This

This chapter has been written by **Gary LaGassey**, USAF Colonel (Ret.), MBA from University of Maryland. He currently serves as Southern Region International Affairs Advisor for the United States Air Forces in Europe. From 1998 to 2004, he managed "Aviano 2000", at the time the largest base reconstruction program ($530 million, 264 projects) in the U.S. Department of Defense.

S. Tonchia, *Industrial Project Management*, Management for Professionals,
https://doi.org/10.1007/978-3-662-56328-1_15

chapter uses the PMI definition for two reasons. First, PMI's leadership in defining this important management area provides new and experienced project managers a solid framework to follow, as it has with the comprehensive Project Management framework contained in its *Project Management Body of Knowledge* or *PMBOK*. When followed, the baseline references and management structure contribute to project and program success. In short, one is not alone in trying to figure out how to manage projects, programs or portfolios. It's all there in the *PMBOK* and related standards. Secondly, PMI's recognition by the American National Standards Institute (ANSI) and International Organization for Standardization (ISO) as the standard for Project Management application assures managers they are and will be on track with current worldwide standards. As of this writing, 779,577 Project, Program and Portfolio managers have earned PMI certifications (PMI Today, July 2017). And, there are many more who, while not officially certified, follow Project Management techniques as prescribed by PMI.

15.2 Evolution of Program Management

As a standardized discipline, Program Management has come a long way since this author's first exposure to it in 1998, almost 20 years ago. At the time, although there had been large Program Management organizations for decades in such fields as aviation production, research and development, and large government acquisition programs, there was little practical information to be found in the literature about principles and techniques of Program Management, its relationship with Project Management or business practices, or developing program or Program Management offices. A manager wanting to manage mid- to large-scale programs short of sending men to the moon, had to extrapolate from a wide array of Project Management information, adapt PM techniques to senior level management practices, or combine personal management experience with basic PM skills, and attempt somehow to convert project management principles to the program level.

It was a great deal of trial and error. I frequently recount the story of my good fortune of finding Graham and Englund's *Creating an Environment for Successful Projects—the Quest to Manage Project Management* (Jossey-Bass, San Francisco, 1997), as a basic guide in figuring out how to set up a program office for our more than half a billion-dollar base reconstruction program.

Today, the situation is much different. One can find plenty of information about *Program Management* and its sister disciplines, *Project* and *Portfolio Management*. The trick is to not get lost in all of it. That said, theory development and published guidance is still very much a work in progress and a lot of smart people are working on defining and improving Program Management principles and techniques.

The leader in this initiative, as indicated above, has been PMI. In just the past 11 years, it established the initial *Standard for Program Management* and refined it twice. When combined with the *PMBOK*, this standard provides both fledgling and experienced program managers a highly effective compass for success. Interestingly, the evolution of the standard, done through a complex review process which included the contributions of hundreds of Project and Program Management practitioners, produced three editions in the first 7 years. A fourth edition has been published at the end of 2017.

Close examination of the first three editions underscores the volatility of the Program Management debate during this evolutionary process. The first edition, published in June 2006, was 109 pages long. The second (2008) weighed in at 311 pages of text and appendices. The third (2013) was reduced back down to 169 pages. These varied editions attempted to capture a rapidly evolving field of management and strike a balance for pertinent content. The new standard will help refine the subject even more.

Defining, solidifying and standardizing Project and Program Management was a great boost for managers in the field, providing them with solid incentives to implement proven practices. On the other hand, it has often been an uphill battle to convince business owners and public entity leaders to establish Program Management as a means to increase profitability and productivity. It took aggressive and persistent advocacy to bring about change.

A major breakthrough for application of Program Management took place on December 14, 2016, when the United States Senate passed the *Program Management Improvement and Accountability Act (PMIAA) (S.1550)*. This law established guidelines for use of Program Management throughout the U.S. Government through adoption and oversight of implementation of government-wide standards, policies, and guidelines for Program and Project Management for executive agencies. This was a major step forward in standardizing government program and project management and to encourage standardized practices among those dealing with government offices. If one considers the government construction and services contracts alone that will be affected by this legislation, the numbers run into the trillions of dollars. American and foreign firms bidding on projects that require adherence to Program and Project Management standards will be affected, virtually assuring the application and proliferation of Project and Program Management techniques. Among the key provisions of the law is the requirement to establish standards for training and certification of program and project managers. This will assure such training organizations as PMI and its partners will continue to prosper.

15.3 Modern Program Management Relationships

During the past 20 years the business and private sectors have seen many new ideas to improve productivity and profitability. One of the most significant has been the attention given to *strategic planning*, described by the Balanced Scorecard Institute (http://www.balancedscorecard.org) as "an organizational management activity that is used to set priorities, focus energy and resources, strengthen operations, ensure that employees and other stakeholders are working toward common goals, establish agreement around intended outcomes/results, and assess and adjust the organization's direction in response to a changing environment."

Public and private organizations are generally heavily project oriented, that is, in large part their operations center on projects whether in production, manufacturing, services, or support. These organizations must depend on innovation to improve competitiveness. Recent emphasis on innovation as a means of enhancing productivity through streamlined processes leads management right to the doorstep of Project Management. If an organization is focused on multiple projects that support strategic planning objectives, it needs Program Management in order to advance its efforts to maximize benefits. Thus, it is imperative that Program Management be imbedded in the fabric of organizational management and planning approaches.

PMI recognizes the synergy of this strategic planning-Program Management relationship. In the recent past the two seemed to run in parallel. Now they are flowing together. Organizational grand planning has been around for a long time as a methodology. Recent technology advancements, especially in organizational communications, production, fabrication, marketing, and distribution have improved planning by contributing to strategic planners' capabilities. They can now develop plans informed by multiple empirical data streams that are timely, pertinent and accurate. Equally, technological advances have given project and program managers advantages to improve Project and Program Management approaches.

For years, project and program managers have tried to align themselves with organizational planning functions. It has been an uphill battle and many a good PM has gone down in flames trying to convince sponsors of the importance of strategic alignment of organizational planning and execution with Project/Program Management. One example is the efforts by Project and Program Management advocates to convince sponsors to establish Project/Program Offices. Now, as it becomes more evident that strategic project costs take up ever larger shares of organizational budgets, the importance of linking planning and Project/Program Management is on the rise.

The PMI relationship structure recognizes the synergy among Program Management, operations management and organizational strategy (PMI Standard for Program Management, 2013). On the top line, it recognizes that an organization's mission, strategy, strategic alternatives, and portfolios contribute to achieving objectives. Portfolios are described by the PMI standard as "a collection of projects, programs, subportfolios, and operations grouped together in order to facilitate the effective management of that work to meet strategic business objectives" (Standard for Program Management, 2013). Thus, when portfolios exist in an organization, programs and projects are subordinate to them.

Strategic Planning and Program Management. Modern high-performance organizations have embraced strategic planning as a key element of success. A solid strategic plan is a fundamental document to define where the organization wants to be in the future. These plans often include a vision statement, mission(s), goals and objectives, and performance criteria and indicators for all sub-organizations and personnel to adhere to.

Project based organizations, those that produce goods or services, need the guidance provided in strategic plans. The guidance should be viewed as a project's (or program's) north star. Projects further the goals and objectives of the sponsoring

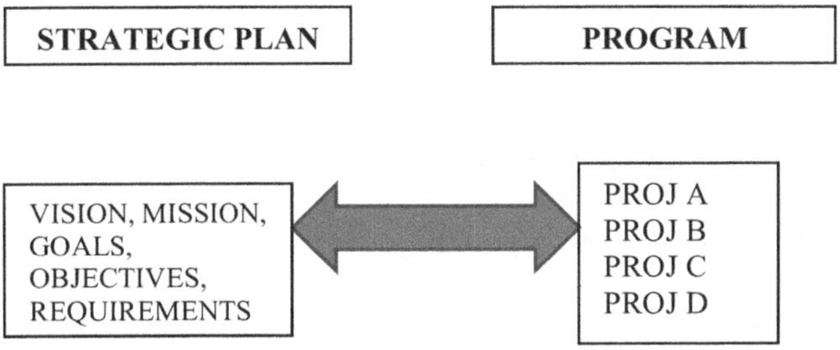

Fig. 15.1 Synergies and communications between strategic plans and program offices

organization. Sponsors don't decide to do projects to benefit a project team or office, they must fit the vision and mission of the parent organization, which in business is to make a profit or in public organizations to provide a service, for example, defense departments for the security of a nation. Gone are the days when project or program offices operated simply to do a handful of projects. They must now be totally aligned with the larger organization for strategic benefit. Similarly, program offices, which are by definition designed to manage multiple projects or subprograms, must be aligned with the overall strategic planning effort. This reciprocal relationship requires synergy and communication (Fig. 15.1).

The Program Management Operating Environment and Elements of Success. Business and public organizations that embrace Projects, Programs and Portfolios for strategic success must provide support to in multiple ways. That is the task of Program Management!

In its most simplistic view, program support elements that contribute to projects might include those shown in Fig. 15.2.

To make the Fig. 15.2 example function properly, program managers can approach their task in a similar fashion to what project managers do with individual projects, by running a Work Breakdown Structure (WBS) scenario to determine the specific work or support elements that might be required. In the case of Program Management, where we need to execute at the program level, the WBS elements will generally be heavy on support tasks. The objective of a good program is to make individual project manager's work easier by reducing the drag of administration. Essentially, program managers and the program support team take administrative burdens on, allowing PMs to focus on their projects. This is done mostly by creating program level processes for such elements as change control, risk assessment, training, finances, and so forth. Every element of commonality of processes should be analyzed to determine if elements of cross Project Management can be consolidated at the Program Management level. Those elements then feed the program WBS, which then becomes the baseline for developing a program methodology which the program team can then follow to success.

Fig. 15.2 Support elements that contribute to Projects in a Program

15.4 The Aviano 2000 Experience

One example of a successful program is the one developed by the Program Management Office (PMO) for the Aviano 2000 base reconstruction program, which ran from 1993 to 2008.

Aviano 2000 was a combined NATO and U.S. Department of Defense program to upgrade the infrastructure at Aviano Air Base, Italy, an Italian Air Force base hosting a U.S. F-16 fighter wing. The impetus for the upgrade was the 1994 bed down of fighter aircraft on a base that had not had permanently assigned operational units before. All facilities, operational, support and personnel quality of life, needed improvement. NATO's decision to deploy the aircraft and personnel quickly meant renovations had to be fast-tracked. At the outset, operating and maintenance facility conditions were poor and personnel and family living conditions far below the standard of the U.S. Air Force.

Looking back, the strategic imperative was to get quality facilities up and running to an Air Force standard as quickly as possible, contributing to the fighter wing's overall mission capability. The strategic alignment message was clear to the early project managers. The early challenge was lack of a cohesive management plan to meet the task. NATO had approved an initial tranche of almost $100 million for approximately 30 design and construction projects, but execution was slow due to the fact that individual projects were the focus, not the overall program. By 1998, the program was in trouble. Something clearly had to be done.

The senior Air Force operational commander on the scene determined a Program Management approach was needed. In December 1998, he directed startup of the Aviano 2000 PMO, assigned a new leader, and directed the fledgling team to move out on aligning its goals and objectives with the strategic goal of having a world class facility for the fighter wing.

The methodology (see Fig. 15.2) for the multi-agency, 100-person program team comprised of engineers, architects, financial, administrative, acquisition and supply type managers became the baseline to execute the total $500+ million program.

The overall approach was to provide project managers a "fully integrated project management process," "integrated PM tools," and all program level support needed to complete their individual projects. Common support elements, for example, quality control, change/configuration management, risk assessment, and safety inspections/mitigation, was managed at the program level and PMs took advantage of common practices and procedures. With many construction projects going on concurrently, PMO leadership realized individual PMs, who were handling six to seven projects each, did not have the capability to handle the white space between the projects. Such items as tying facilities together with communications and utilities was getting dropped. The fix, instituted by the program manager, was to add "sector" managers to focus on the space between projects. These sector managers, who were more senior engineering PMs, also served as mentors to junior PMs. This decision, identified by program reviews, resolved a lot of problems that were festering.

We have touched on five of the eight slices of the Program Management pie shown in Fig. 15.3, project management, quality, change, risk and safety management. These five are the process oriented elements, those that require standardization across the organization. The remaining three tasks (PM tools, personnel development, and communications management) are also critical Program Management elements. These can only be addressed at the program team level as they invariably involve financing and allocation of resources.

PM Tools The Program Management level must provide individual PMs and their teams with all the necessary tools to do the job. In the Aviano 2000 program these were in the form of a standard PM schedule (GANTT), a project data base, necessary operating funding, a project web page, and standard reviews. PM effectiveness depends on all of the tools. More importantly, the overall program team depends on functional and effective tools that everyone understands and uses. For example, many programs use off-the-shelf PM scheduling tools and force the PMs to use them, which ultimately can create challenges. The Aviano 2000 methodology (and philosophy) was to create tools that got the job done but that had buy in from the individual PMs. The dog has to wag the tail, not the other way around.

Communications Management Effective communications among leadership and team members of large program teams is key. The Aviano 2000 program methodology recognized that by providing effective technology solutions the team members

Fig. 15.3 Aviano 2000 Program Management methodology

used. When something was ineffective, we scrapped it. Here again, the organization has to adapt to the people doing the job.

Personnel Development Strategic success of the Aviano 2000 program depended on quality personnel doing quality work. That took great attention to detail, to their training and development. Senior Program Management must commit to personnel improvement. The rewards are great as employees feel that commitment and produce accordingly. This applies not just to the PMs but to every team member.

Note this methodology encompasses only part of the support elements shown in the Fig. 15.2. The learning point here is Program Management methodologies must be tailored to the situation and operating environment. Organizational Program Management methodologies will be different and each will contain the support elements that are driven by the program WBS, which ideally is informed by the parent organization's strategic plan. If the WBS elements contribute to making PMs jobs easier and the projects contribute to the organization's strategic objectives you've got a winner. One final takeaway for this part—once you have developed the program methodology, it must be communicated! It does no good to develop it and put it on a shelf. Each new member of the team must be trained on it and it must be reviewed periodically for currency. In that sense, it is similar to the strategic plan—it should be a living document.

The linkage between strategic plans and program management is a positive one. Each informs the other, each contributes to the organization's short- and long-term success.

15.5 Program Office

As the field of Program Management evolved, project offices of every sort began to spring up in organizations. The impetus of growth was the realization multiple project initiatives had to somehow be corralled and supported if they were going to be successful, not only as individual projects but as contributions to the organization's objectives. In their infancy, project offices frequently consisted of two or three project managers thrust together without a great deal of executive or material support. Often, they did more marketing of PM concepts within the organization than actually managing projects. First line managers and supervisors of production units were skeptical of this thing called Project Management and were loath to committing to change and "managing by project" because it might reduce their importance, authority or resources. PMs who could not obtain solid sponsorship from top line leadership were doomed.

In the early years project offices were invariably heavily matrixed organizations, with PMs fighting for manpower resources with front line managers. Many were nothing more than allocated office space where PMs, still owned by their primary functional managers, met in a semi-collaborative project oriented environment. This began the breakthrough as the dialogue about project delivery that took place eventually turned to the professional aspects. PMI's founding fathers recognized the movement and through its efforts provided frameworks and focus for widely dispersed PMs who then became the voice for project management and project offices within their organizations. Then, during the 1980s, the tidal wave force of computers and technology took over. Computerized PM management techniques improved GANTT and PERT methodologies, supervision, record keeping, and data analysis that lent to new, faster and more accurate ways of managing projects from "cradle to grave."

This evolution led to significantly more interest in PM fields and, because projects could now be analyzed with greater accuracy, senior managers grew more confident in their new decision-making capabilities that led to the bottom line. Project managers took on new stature and new respect within their organizations and enlightened senior leaders became project advocates. This led to ever increasing importance of project offices and more definition of their roles and responsibilities. It also impacted organizational methodologies related to the business of managing projects and programs, mostly aligned with the type of parent organization. We now hear the terms project office, project management office and project coordinating office related to specific projects. Program office, Program Management office, strategic program office, enterprise program office, etc., are frequently used terms organizations use to identify their specific program office. Again, these are related to the overall type of organization, the specific mission of the office, the tie to the principal organization's strategic planning process or, on occasion, to the personal whim of some senior manager. In reality, it probably doesn't matter what one calls the project or program office, only that it is consistent within the organization and that people know what its function, roles and responsibilities are. Often, people waste too much time on the importance of title.

It is much more important to determine what the program office is supposed to do. Depending on the complexity of the mission, a program office generally organizes and aligns the contribution of projects (or sub-programs) to the main organization's overall strategic plan. For example, in the automobile business, an automobile manufacturer's program office might provide organization and support to the design effort for future car models. The individual project offices might be organized by models (Fiesta, Focus, Fusion, C-Max, Mustang, Taurus) or product types (passenger, crossovers, trucks) or by individual design studios by brand. Ford's overall strategic plan might be to increase sales of individual models or by product type by 10% by 2020. The program office would organize the marketing and design effort to reach that strategic goal. For illustration, each model, brand or type project team might have a project team plugged into the overall car development/ marketing program team, which provides day to day guidance and alignment with Ford's strategic objective.

The key takeaways from this section the concept that Program Offices, no matter what they are called, link the Project and Program work with organizational strategic goals by providing an organizational framework, coordination, control and support for the actual work.

15.6 The Future of Program Management

Technology will continue to drive the work of program managers and they must stay attuned to advancements which are coming at ever increasing speeds. Product development and related marketing efforts depend on meeting defined and critical organizational goals. For example, the speed with which specialized companies are moving toward self-driving car capabilities demands Program Management efforts designed to capture and take advantage of technologies before they are fully defined. Recent investments of $1 billion each by Ford and GM in technology innovation are just two examples of how important Program Management will be for those companies. This is not a game of chance, organizations must be ready.

Ford and GM both seek innovation in self driving technologies as part of their strategic plans. They have elected to invest large sums of money in research and development. R&D requires well-conceived and managed projects to meet fiscal and time constraints. The stakes are high in order to be first to market new and innovative products. These high-level programs require managers who are well versed in the project and program aspects as well as technology advancements. The fast pace of technology development will require program offices that are tuned in and poised to react so as to take advantage of new technologies. The winners will be those first to innovate and first to market new products.

Program management is the key element for success.

Project Risk Management 16

16.1 Managing the Risk

Risk affecting organizations can have consequences in terms of economic performance and professional reputation, as well as environmental, safety and societal outcomes. Therefore, managing risk effectively helps organizations to perform well in an environment full of uncertainty.

According to PMI, 'Project risk is an uncertain event or condition that, if it occurs, has a positive or negative effect on one or more project objectives such as scope, schedule, cost, and quality. A risk may have one or more causes and, if it occurs, it may have one or more impacts'.

In a project, the *sources of uncertainty* can arise from the following:

- Identification of the customer's needs
- Knowledge and management of technology
- Behaviour of the competitors
- Type and availability of the resources

These sources of uncertainty determine the so-called *risk areas*, which can be defined as those factors that are not fully dependent on project choices, in other words, that cannot be completely controlled or predicted, but that may have a great impact on the project and its performances, namely the following:

- Time—namely the duration of each activity and therefore, that of the entire project
- Costs—meaning both the cost of the resources and their exploitation rate
- Quality of the project's outputs (or deliverables)

Risk is an intrinsic element of a project, and the more the latter is innovative, the greater the risk. *A project without a risk factor is not a project.* Moreover, since its

S. Tonchia, *Industrial Project Management*, Management for Professionals,
https://doi.org/10.1007/978-3-662-56328-1_16

success depends greatly on its degree of innovation, the ability to manage risk is one of the most qualifying skills required when working on projects.

Three strategies typically deal with risks: avoid, transfer, and mitigate. Risk Management techniques deal with *management* in a stricter sense ("avoid", or—at least—"mitigate"). The approach to risk *management* can be either reactive or proactive. A *reactive approach* is used to (1) manage risk, (2) correct mistakes and (3) compensate the negative effects that have occurred; a *proactive approach* on the other hand will attempt to (1) identify and manage risks, so as to avoid the onset of problems, (2) eliminate the causes of risk at the root, acting on the risk factors.

Since risk factors are an intrinsic feature of the projects, a proactive approach will attempt to prevent, rather than simply eliminate, risk. Five stages can be defined as follows:

1. Risk Identification
2. Risk Analysis
3. Risk Quotation
4. Risk Response
5. Risk Monitoring

In the stage of *Risk Identification*, an attempt is made to identify the potential risks inherent to the project in group discussions, by examining previous, similar experiences, formalising project restraints and opportunities. A risk can in fact be considered both as a restraint and as an opportunity, although characterized by a certain degree of uncertainty. Possible risks can be mainly ascribed to the aspects and features of the project's output (its innovative technical and technological content, excessive expectations regarding performances, low level of detail in the specifications, integration with other projects) and the management of the project (underestimated costs and times, overestimating the supplier's capacity, unskilled labour, wrong assessment of the workloads borne by the available resources). A checklist of the risks is thus prepared, together with supporting documentation, which is signed by the project manager after discussing the matter with the persons in charge of the project and the heads of the technical offices and the functional units.

Having been identified, each risk is subjected to *Risk Analysis*, determining the likelihood of its occurrence and the severity of its effects on the performance of the project. It should then be possible to translate risks (even those affecting quality, times and workloads) into costs that—multiplied by their probability—express a quantitative valuation of risk.

On the basis of the previous analysis, *Risk Quotation* is carried out to measure and rank the risks and consequently, the interventions; a maximum level of acceptable risk is also set ("cut-off line"), so that all risks having a higher value must be brought down to a level lower than the threshold using the appropriate actions.

Ratings are assigned to risks based on their assessed probability (also called "occurrence") and impact (also called "magnitudo" or "severity"); evaluation of *each* risk's importance and priority for attention is typically conducted using a "Probability–Impact Matrix" (Fig. 16.1—the points inside the matrix are the

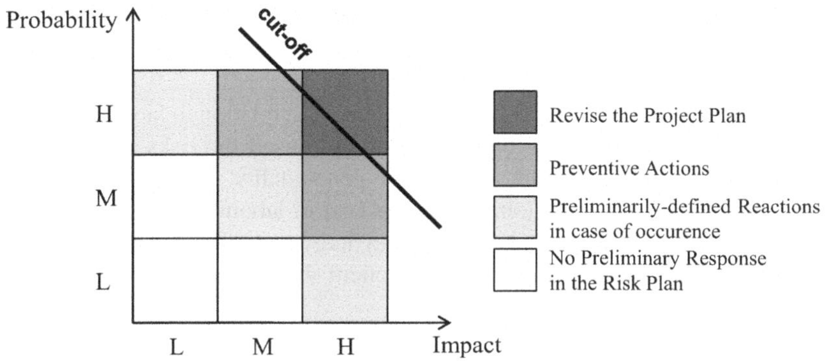

Fig. 16.1 Probability–Impact matrix

identified risks). Descriptive terms (L—low, M—medium, H—high) or numeric values can be used depending on the organization's choices.

The first three stages (Risk Identification, Analysis and Quotation) are also known as *Risk Assessment*.

Risk Response includes all those actions that are necessary to bring the risks below the previously mentioned threshold levels. The interventions may be synergic and must follow the scale of priorities, since the efforts are made by a limited number of resources and must be addressed univocally. There must be a person in charge of organizing the actions, and once these are carried out, risks must be calculated once again (*residual risk*). The actions can be classified as follows:

(a) *Context actions* (defining specific internal programmes, operational agreements with customers and suppliers, revising the contract and excluding certain parts or price increases, getting insured, etc.)
(b) *Internal actions* (collecting further information, carrying out trials and experiments in advance, involving experts, divulging the issue of risk and organizing *problem-solving* meetings, eliminating activities and design solution characterised by a high level of risk, allocating the best resources and/or greater ones to the more risky activities as well as paying more attention to them, changing partners and resources or involving them more, moving the activities that are considered more risky out of the critical path, protecting the *core output* of the project, seeking guarantees for a minimum result, developing alternative solutions for the parts considered more risky and trying to modularise them, preparing contingency plans, etc.)

Risk Monitoring finally helps keep risks under control and reduce risk areas. Often paying a little more attention to this issue can reduce the risk itself—as in a basketball game, a time-out can work wonders.

If the project is linked to the execution of a contract job, the risk factor also depends on the type of contract issued:

- In a *fixed-price contract*, the price is defined, hence the supplier is responsible for all the risks, and in the event of greater costs occurring, the latter's margin of profit will be eroded.
- In *time and expense contracts*, the price is defined in relation to labour costs plus a certain margin, hence the customer takes upon himself the risks, even accepting longer times and hence greater costs to privilege quality.
- *Maximum price contracts* are based on the cost of labour plus a certain margin; these can reach a certain threshold level beyond which all costs are taken by the supplier (in this case the supplier and the client share the risk).

16.2 RM Standards and Certifications

A number of standards have been developed worldwide to help organisations implement risk management systematically and effectively. These standards seek to establish a common view on frameworks, processes and practice, and are generally set by recognised international standards bodies or by industry groups.

The different standards reflect the different motivations and technical focus of their developers, and are appropriate for different organisations and situations. Standards are normally voluntary, although adherence to a standard may be required by regulators or by contract. Commonly used standards/certifications include:

- *ISO 31000:2009 (UNI 2010)—Risk Management Principles and Guidelines*
- *ISO/IEC 31010:2009—Risk Assessment Techniques*
- *Document 4245.7 M "Transition from Development to Production" issued by the US Department of Defence*
- *Risk Management Standard (published by the Institute of Risk Management—IRM)*
- *Certified Risk Management Professional (by the Risk Management Society—RIMS)*
- *Professional Risk Manager Handbook edit. 2015 (by the Professional Risk Managers' International Association—PRMIA)*
- *"Governance—Risk—Compliance" GRC Certification (by the non-profit organization OCEG).*

ISO 31000 *Risk Management—Principles and Guidelines* provides principles, framework and a process for managing risk. It can be used by any organization regardless of its size, activity or sector.

Using ISO 31000 can help organizations increase the likelihood of achieving objectives, improve the identification of opportunities and threats and effectively allocate and use resources for risk treatment.

ISO 31000 cannot be used for certification purposes, but does provide guidance for internal or external audit programmes. Organizations using it can compare their risk management practices with an internationally recognised benchmark, providing sound principles for effective management and corporate governance.

ISO/IEC 31010 reports 31 "risk assessment techniques", including: brainstorming, structured or semi-structured interviews, Delphi method, checklist, structured what-if analysis, scenario analysis, business impact analysis, root-cause analysis, Failure Mode and Effect Analysis (FMEA), fault-tree analysis, event-tree analysis, and others, not all applicable to all the sectors or the risk types.

16.3 PMI Risk Management Professional (RMP)

PMI Risk Management Professional (PMI-RMP®) certifies the individual ability to identify and assess project risks, mitigate threats and capitalize on opportunities. The relative exam is based on PMI's book *Practice Standard for Project Risk Management*, and consists of 170 multiple-choice questions (to answer in 3.5 h); the prerequisites are 3000 h of project risk management experience and 30 h of project risk management education (for those who hold a 4-year bachelor's degree).

The competence is articulated in five domains: (1) Risk Strategy and Planning, (2) Stakeholder Engagement, (3) Risk Process Facilitation, (4) Risk Monitoring and Reporting, and (5) Specialized Risk Analyses.

The *Practice Standard for Project Risk Management* can be used as a general, not specific, check for an organization's processes; in fact, different projects, organizations, and situations require different approaches. In other terms, risk management is more effective if it is "tailored to the project and congruent with the organizational culture, processes and assets".

Project Risk Management (PRM) aims to identify and prioritize risks in advance of their occurrence, and provide action-oriented information to project managers. It is "not an optional activity: it is essential to successful project management. It should be applied to all projects and hence be included in the project plan and operational documents". It addresses the uncertainty in project estimates and assumptions, and it is based upon and extends other project management issues (e.g. time, cost, quality, resources, supplying etc.).

In the early stages of a project, the level of risk exposure is maximum but information on the project risks is minimum; anyway, risk management is to be applied throughout the project's life cycle. The PMI's Standard uplights some "critical success factors" for PRM the *risk manager* (as a specified figure) or otherwise the project manager should consider and pursue:

- Recognize the value of PRM
- Individual commitment and responsibility
- Open and honest communication
- Organizational commitment
- Risk effort scaled to project
- Integration with Project Management as a whole.

In particular, PRM should be consistent with the value of the project for the organization and other organizational factors (such as objectives, policies, practices and constraints), and the RM cost should be appropriate to project's risk level.

A RM expert, certified as RMP®, must also know the *tools and techniques* to perform both qualitative and quantitative risk analysis (The Appendix D of the *Practice Standard for Project Risk Management*—to which reference is made—presents an overview of the tools, techniques and templates more targetted on projects than the ISO/IEC 31010; for example: Probability-Impact Matrix, FMEA, Critical Chain, Earned Value Analysis etc.).

Part VI

International Contracting

17 From Bidding to Project Launching

17.1 The Commercial Phase

Danieli & C Officine Meccaniche S.p.A. has been in business for more than 90 years and is one of the three foremost companies in the world for the supply of machines and plants in the iron and steelmaking industry, and is expanding his OEM offer in other industrial sectors like the Aluminum.

Danieli & C is a leader in the design and construction of minimills, of which it is a forerunner, and comprises an integrated network of companies specialised in the design and construction of plants for the entire steelmaking process.

It also possesses the necessary flexibility and versatility to meet the needs of the steelmaking market by, in fact, supplying a wide range of diversified products and services from single machines to complete *turnkey* plants, thereby giving its customers the benefit of the high-level integration and coherence among the various departments. Figure 17.1 gives the base network of Danieli Group.

Today the minimill is the ideal solution for the needs of a primary steelmaking industry, because of the flexibility and limited investment required to purchase and run it, compared with full-cycle plants.

The constant demand for evolution and high-tech products is also applicable to markets characterised by continuous growth such as Asian countries and Central and South American countries. This is particularly true in *emerging* countries—especially in India and China—which, thanks to the support of their governments' economic policies, are in the process of acquiring the necessary primary industries required to sustain their future economic development.

These countries are extremely sensitive to cost and time variables, which in terms of company strategy represent a competitive lever of fundamental importance.

Danieli's motto *INNOVACTION* also refers to a strategy that comprises various and continuous actions of a technological, economic and managerial nature.

This chapter has been written by **Flavio Cozzi**, actually Member of the Board of MME Gmbh, Director Projects, Sales and Marketing Manager—former Danieli's Vice President Projects, on behalf and with permission of Danieli Group.

S. Tonchia, *Industrial Project Management*, Management for Professionals,
https://doi.org/10.1007/978-3-662-56328-1_17

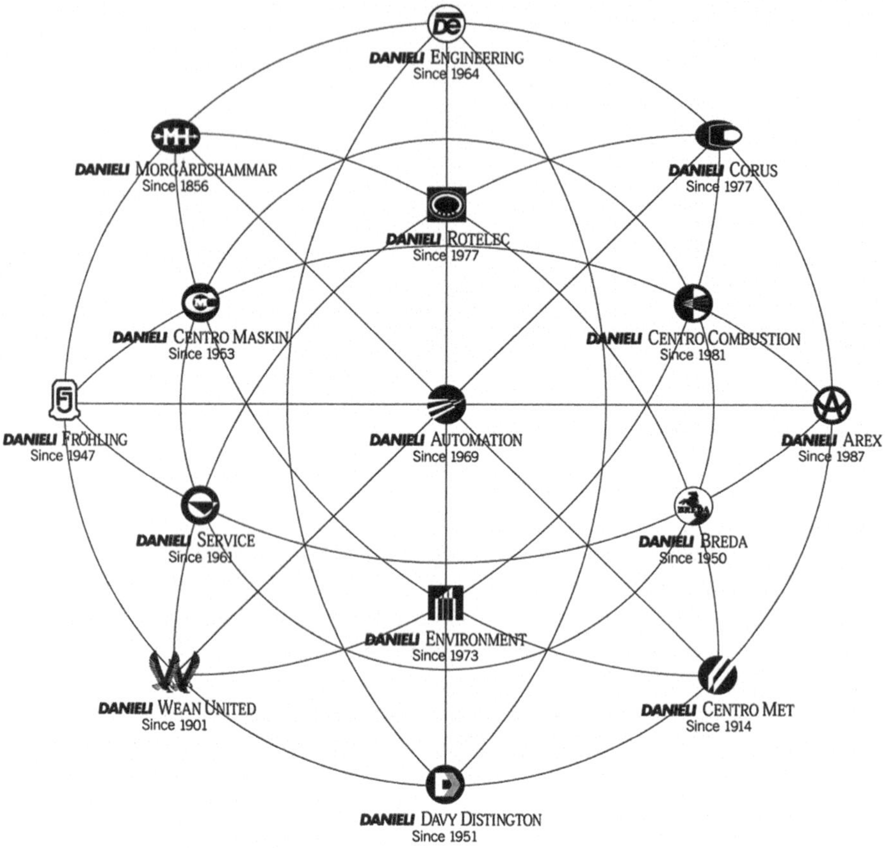

Fig. 17.1 Danieli group

It is a strategy aimed at rationalising all activities and minimising those without added value, developing more efficient management and organisational tools and associated Information Technology media.

Danieli's goal is to create global purchasing strategies in *low cost* countries to reduce material purchase costs.

It has decentralised some of its manufacturing businesses that are not critical to product quality and performance while maintaining its in-house machining facilities with their consolidated *know-how* and exclusive technology.

Plant output, reliability and quality continue to increase with a view to constant improvement backed and made possible by investments in Research and Development.

Danieli is a matrix-oriented organisational structure.

Added to the vertical structure, divided into product lines (long products, flat products, BREDA, Construction and Service and the subsidiaries as Danieli Centro Combustion (DCC), Danieli Automation, Danieli Construction International and

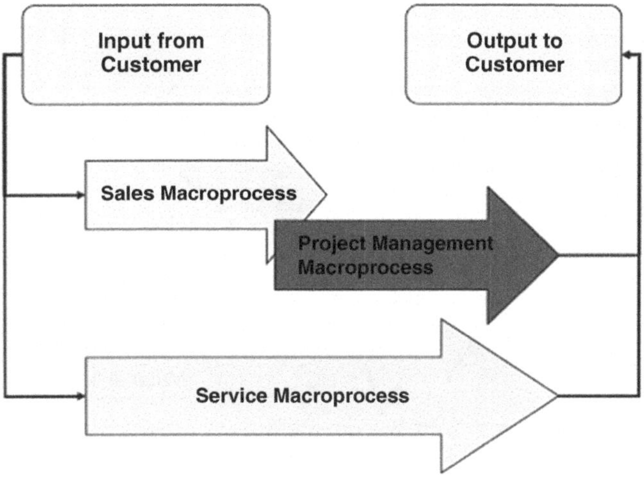

Fig. 17.2 The business macro-process

others), where each division is backed by its own functional organisation (technical and sales management, production, logistics, purchasing, human resources, administration, management control, quality, research), is the horizontal structure of the three business macroprocesses: Sales—from the first contact with the customer to Project Economic Analysis; Project Management—from Re-Examination of Offer (REO) to Contractual Guarantee Period; Service—from the first contact with the Customer to delivery of spare parts and services, analysis of customer satisfaction allowing us to follow all the stages of a plant order from technical–economic negotiations to the installation of the plant, startup and after-sales service. Figure 17.2 shows the business process flow during the project execution.

Given here is a possible representation of a Project Manager's duties in a complex company that designs, makes, installs and starts up its own products, and not only single machines but a group of machines making up complete plants that can be extremely complex (Fig. 17.3).

The cost $C1$ for contracted deliverables $D1$ is calculated internally; the actual delivery of the project is $D2$ with actual cost $C2$. The flow is shown here.

The project manager's task is to minimise $\Delta\ (C2,C1)$ and $\Delta\ (D2,D1)$ although $C2$ is always greater than $C1$. $C1$ here is the budget given to the PM after being discussed internally and is unrelated to contract negotiation.

The reasons for these cost and deliverable variations can be explained as follows:

The sales team tends to bid a low price for a project with the risk to underestimates the deliverables (scope of supply, cost for contingencies, project life cycle control) $\rightarrow D1 < D2$.

To gain a price advantage in a tender, the profit $\Pi = \mathrm{P} - C1$ is forced to be squeezed to a minimum while the client has higher expectations $D1' \gg D1$ when signing the contract, and since D1 is estimated by the sales team, it is often difficult for the PM to keep the actual delivery under $D1$.

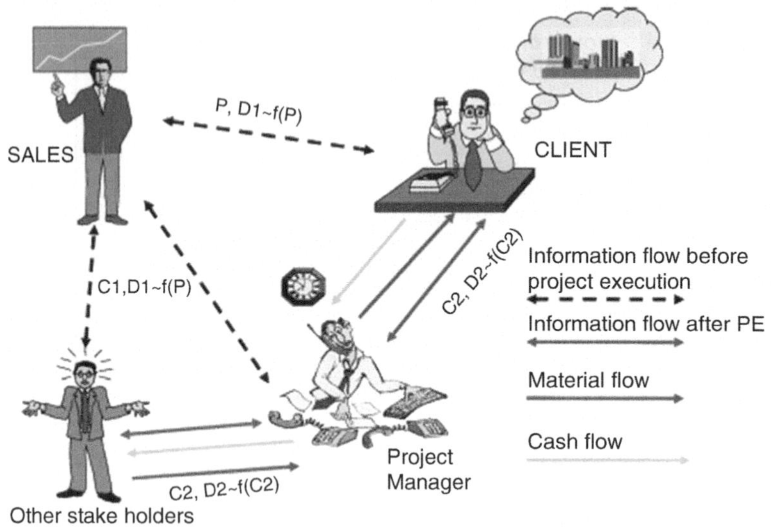

Fig. 17.3 The Project Manager at work

It is required to monitor the progress of different projects and plan strategies, so that in case of overload, delivery D2 is greater than D1 and results in an actual cost of $C2 > C1$.

Therefore in critical situations we have to consider how to implement corrective actions at the various stages of the corporate organisational process.

The commercial phase of the Project Manager activity includes the section relevant to commercial proposal in which the commercial conditions are discussed; at this stage, the offer is reviewed with the PM and the contract is signed.

A job order must first of all be assessed as a whole in a time diagram with various stages that can overlap.

In its most complete form, the contract time schedule includes all the deadlines and stages of the job order process, as shown in Fig. 17.4. Let us now take a look at the first stage, which goes from REO to Contract Signing.

The sales stage is not the responsibility of the Project Manager, who is only involved in macroplanning together with the other company departments and other business units.

In addition to the customer, the Sales Macro Process (SMP) involves the following company parties and departments: Proposals Department (UTC), Economic and Commercial Management (GEC) as a business line, Proposals Management (DTC) and Commercial Office as a support function. Naturally, the proposals preparation stage involves the participation of the technical and technological departments, in addition to the Financial and Legal departments.

In project management, the starting point has to be the review of the proposal in the final stages of negotiation, when, before the first meeting with the Customer, the technical and commercial issues were discussed and all the fundamental aspects of

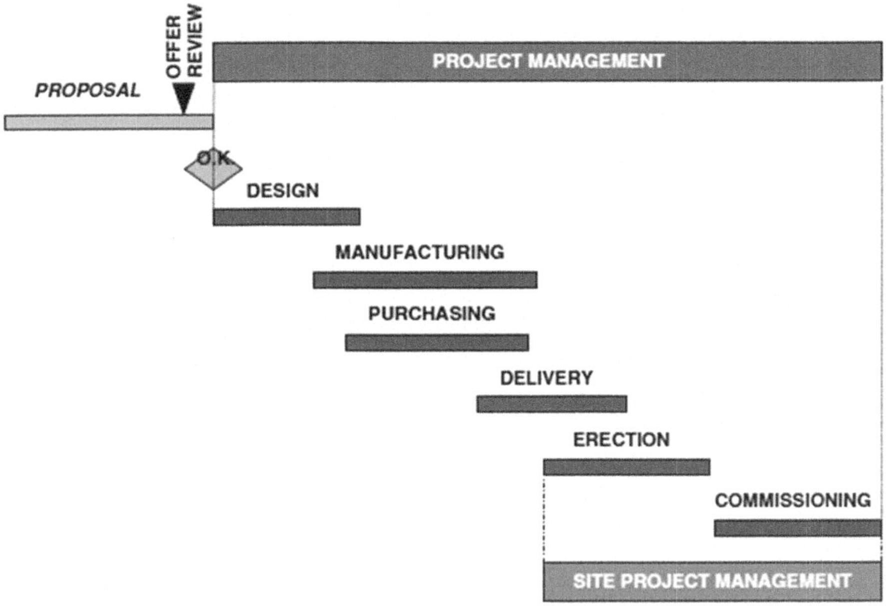

Fig. 17.4 The project schedule

the contract were defined, such as its supply and associated activities, the technological process and naturally the economic and financial terms.

The Project Manager's duty during the proposal stage is essentially concentrated on the event referred to as REO involving macroplanning throughout the entire process until the Contract is awarded and signed by the Customer and Danieli.

Furthermore, the Project Manager, by taking part in the final stages of the negotiations, together with the Proposals Manager ensures the transfer of information to the operational stage and creates a concrete outline of job order progress, thus consolidating the exchange of information between Sales and Project Management.

It is during this stage that the Project Manager uses his/her experience to assist the UTC Proposals Manager, primarily in contractual and planning matters, by specifying the departments involved and the main purchasing strategies on which the Economic Proposal is based, thus ensuring continuity in the implementation stage.

The contractual aspects having the greatest impact on the implementation stage are generally the project schedule, payment milestones, special clauses, etc.

Together with the Proposals Manager and the Operations staff, particular attention must be focused on strategies for the purchase of critical *external* products that may feature a particular technology, or special market products influencing the time schedule.

For most contracts, UTC prepares the documents that will regulate the contract.

The technical specifications are essential for the acquisition of technical knowledge specific to the supply of the order. These are made up of various sections such as the main features of the proposed plant, the design concept and the process (depending, obviously, on the product line in question). Specifically, the detailed contents are as follows:

- Mechanical equipment: detailed description of the machine being supplied, its technical–structural features
- Automation and electricals: a detailed description of the components being supplied such as Programmable Logic Control boards, electrical boards, etc.
- Motor and sensor list: one of the documents linking the functional mechanical part to the electrical part
- Auxiliary plants: for example the compressed air production plant, water treatment plant, which the Customer can also purchase directly and for which the specifications are needed
- Civil works and buildings: can also be purchased directly by the Customer and Danieli for *turnkey* plants
- On-site services: a detailed description of the duties and activities to be carried out on site by Danieli supervisory personnel for erection, startup and production
- Training: complex plants where Danieli is the supplier of the training courses. The theoretical courses are held in classrooms while the practical courses are held in similar plants
- Operational parts and spare parts: in most contracts, the recommended running and startup spare parts are defined right in the first phase
- Engineering and technical documentation: this is the engineering required to make the equipment, install the plant systems and make the various auxiliary plants
- Tests and Guarantees: where agreed on with the final Customer, they provide a description of the contents and the operating procedures, all the tests to be done on equipment once it is installed on site and all the performance tests to be done in order to prove the contractual capabilities of the supplied machines or plants
- Time schedule: a diagram showing the project due dates defined together with the final Customer. It contains the deadlines for all the main process phases and the time required to carry them out as well as the milestones; this is a fundamental aspect of the contract
- Definition of scope of supply: a description of the contractual items according to Work Breakdown Structure (WBS), indicating if they are to be supplied by Danieli or the Customer and the required engineering
- Alternatives and options: possibilities of future expansion
- Drawings: the drawings included in the offer
- Documentation: specifies the type and format of documentation, i.e. hard copy, electronic medium, language, logos, etc.

The offer is analysed according to the documents issued during the negotiation phase.

The project manager will focus his/her attention on the engineering attachments (trying, wherever possible, to standardise the documents or personalise them for loyal customers), and share the information with the Proposals Manager and the Sales Manager, to study and set up a strategy for manufacturing, procurement and purchase.

He/she will put all his experience on the table.

Organisation and general coordination in this phase are the responsibility of UTC, which coordinates the various work groups and decides when to schedule the REO if there is a chance of getting the order. Before this there are macroplanning activities to be done, not mentioned here, that are very important to establish an operating and market strategy.

To proceed with the evaluation a meeting is called to present the job, which must be attended by the level 2 Managers of the Organisation Breakdown Structure (OBS) of the product line (LdP) or their delegates representing the various business units involved in the project in accordance with the characteristics of the job in question. In any case a senior project manager must be present to convey his/her experience and carry out the necessary analysis.

The people called to the meeting have to establish the milestones and job order strategies looking through the conditions and the availability of the market.

In the REO the Project Manager assists the Sales Manager in checking and analysing the following aspects:

- Contractual project schedule and payment schedule.
- Lead-time clauses.
- Coverage required in terms of bank guarantees and invoicing procedures.
- Means of transportation and any provisions as per Incoterms 2000.
- Conditions required to obtain the Provisional Acceptance Certificate (PAC) and the Final Acceptance Certificate (FAC).
- The vendor list, which is a specific detailed list containing the names of Danieli's sub-suppliers for products featuring technologies that do not belong to Danieli & C., such as for example motors, hydraulic pumps, etc.
- Any particular clauses referred to in the contract.
- Scope of supply: it is checked to make sure that it is consistent with the technical attachments.
- Purchasing strategies: analysed and defined together with the departments concerned are the purchasing strategies for critical *E* products (civil works, buildings, erection, long delivery items, etc.) in order to guarantee continuity during implementation.

In the spirit of *Velocity* and prompt reliable organisation, the REO Meeting is run in a *concentrated* manner, where forces and experience combine to draw up a checklist that highlights critical points.

The technical aspects refer first and foremost to feasibility including time, new equipment, technical performances, knowledge of the project through experience, time required to make the equipment, procurement and implementation. All these elements are combined to define the job process. In addition to the technical issues, also the commercial aspects such as risks on payments, country, financial limits and rules are obligatorily identified.

This establishes, once the order is obtained, what in the future will become the implementation and job management strategy.

This meeting has two fundamental goals:

- To apprise the sales manager of any corrections and to point out the elements that should be discussed with the customer
- To provide the company with enough information to update its macroplanning model, which includes adding the job order to the workloads of the technical and production departments so that it is always in line and ready before the job is awarded

An example is the checklist as per Table 17.1, that is prepared before the meeting is called and then updated during the meeting.

The checklist contains the general contract references: the customer and the country; the type of contract, i.e. turnkey, joint venture, consortium, etc.; it includes a description of the main equipment supplied, the process requested by the customer and the subject of the contract.

It comprises a list of all the participants so that they can be contacted to provide their assistance and experience.

The participants in each meeting can vary depending on the type of contract. However, following are the departments and business units that may be involved: UTC (Proposals Department); UT (Technical Department); PM (Project Manager); UAC (Purchasing Department); MAC (Erection, Startup and Commissioning); DA (Danieli Automation); DE (Danieli Engineering); DCC; UP (Production Department); UTECH (Technological Department); CC (Cost Controlling); UTI (Plant Systems Department).

The essential documents to be submitted at the offer review meeting are the layout, the scope of supply, the time schedule and the price table.

The REO document contains the Project Manager's notes on the plant and the machines together with the comments of the various people that took part in the meeting. The report states the outcome of the review and the critical points encountered so that, once the order is obtained, everyone is informed and can verify the status of the critical points in the signed contract.

The offer review process is not limited to the REO meeting but continues in parallel in the technical departments, where the specific contents of the technical specifications are examined, and in the production department, where the features of the ordered equipment are studied.

The technical analysis refers to the following: type of product, feasibility of making the plants according to contractual data and requirements, whether the supply contains repeated machines, whether new drawings are required to make certain machines, whether a plant is new or existing and performance evaluations of machines or plants.

The commercial/economic/legal review is done according to the same procedure used for the technical part using a similar checklist that must include the following items:

- Vendor List (if there is one it has to be attached for the purchasing and technical departments): a list of possible Danieli sub-suppliers or the customer may prefer or impose sub-suppliers for off-the-shelf equipment or equipment not made with Danieli technology such as, for example, motors, power transformers and hydraulic components
- Whether it is an onshore contract or includes the local supply of equipment in the customer's country
- Whether payment terms are standard with down payment, engineering phases, shipping phases with pro rata payments, hot test phase, acceptance and performance phases of plants and machines
- Confirmation of any Financial Coverage specifying any foreign amounts to be paid depending on the type of loan
- Monetisation and application of events that involve contractual penalties
- Delivery conditions and shipping procedures according to the limits and rules required to carry them out properly
- Laws and regulations applicable to the contract
- Analysis of critical contractual points with an indication of possible risks
- Any other economic and financial responsibilities of Danieli & C. in consortiums
- Manufacturing strategies and application of cost category to prepare estimates
- The possible existence of restrictions in the manufacturing or shipping countries that could affect goods with certificates of origin
- The possibility of expediting by the customer, which could entail economic impacts, according to the equipment shipping timetable and the corresponding pro rata payments
- Then, in the same manner as the technical part, the result of the economic/commercial/legal review is recorded together with the critical points

Therefore, if we consider the commercial aspects of the contract and their connection to the payments and drawdown periods, we can establish a possible drawdown curve and then determine job cash flow.

It is important for a Project Manager to have detailed knowledge of how invoicing will take place and the events triggering the issue of the invoice.

The down payment, which is the first payment that initiates job activities, generally ranges from 10 to 30% to cover the initial organisational expenses and the engineering and, as far as possible, the machines to be made. We have to state and know which documents need to be processed to set this procedure in motion, as in certain countries transactions such as this one could take a long time because they are subject to approval from state institutions, financial institutions with lengthy bureaucratic procedures, which can therefore impact the general job schedule especially in determining the starting point, which is the coming into force of the contract (CIF—Coming into Force).

More and more contracts require that payments be made in various instalments, tying them to the delivery of the working drawings and therefore to the progress report. In this case coordination with the technical departments is important, which means that macroplanning takes on considerable importance in the strategy and management of the various product lines.

As for the payment of the equipment and therefore the shipments, it is important in drawing up the documents that are shipped together with the goods according to INCOTERMS 2000. Also, special attention must be paid to particular import regulations in the various countries where the goods are shipped.

Supervision to startup and commissioning of plants and issue of commissioning certificates represent the final stages of the contract, which must be coordinated with MAC and which are followed by acceptance of the plant by the final customer. This is of considerable importance as it triggers payment of the final contract instalment.

Financial conditions for collection are as follows:

- The letter of credit and its transcription procedures, its contents to be defined with the customer and the banks
- Telegraphic cash transfer to a bank, which has received instructions directly from the customer
- Promissory notes and all related information such as release date, expiry date, etc.
- Bank guarantees issued by Danieli linked to financial events.
- Contractually specified taxes and those levied in the various countries
- Penalties tied to job events and milestones agreed with the customer

The Contract signing event is the milestone where the participation and presence of the PM during contract signing is generally required for special and large projects to implement the synergy company actions.

This is because the contents of these contracts are more extensive and the Sales Manager may require the assistance of the Project Manager for the final confirmation of certain parts of the contract that will be managed directly by the PM after the signing of the contract.

The Project Manager's presence is usually required during final negotiations for the reasons explained here.

Modifications to the contract that the customer requests at the last minute during final negotiations to avoid in-depth discussions (these modifications are generally penalising for the supplier who is *forced* to accept in order to avoid starting up the commercial discussions again or having to deal with any competitors). In addition to providing support in making the final decisions, in these cases the PM can *filter* the customer's requests and explain the impact that these last-minute modifications can have on the agreed overall job order process. The Project Manager's negotiating experience in this stage is important in getting to know the customer and laying the groundwork for a relationship that must be or become *friendly*, to gain the customer's trust by ensuring the final result *at any cost*. Modifications to the contract attachments may also be requested if they are related to or affect the contract time schedule. They could, for example, refer to the delivery of engineering and equipment. At the last minute the customer tends to *wring* an earlier delivery date out of the sales manager for the engineering or the equipment. In this case the PM, having done the macroplanning, should be able to realise the importance of the modifications requested by the customer within the general framework of activities and try to turn the situation to his/her own advantage since these are last-minute changes that would involve a contractual compromise and which could weigh on or be significant in terms of project economy. He will therefore have to provide clarifications on locally made or co-manufactured equipment included in the scope of supply.

During commercial negotiations, the scope of supply, engineering included, is split up. In this stage it would be useful, especially if requested by the Customer, to clear up the general aspects related to locally made equipment. This means, for example, that Danieli's engineering may contain particular control devices that are not easy to find on the local market and that have to be imported by the customer. These details are often not dealt with during commercial negotiations but during execution, which becomes a supply problem within a budget that does not take these devices into account. In this case, the PM, for example, can come up with an operating procedure for a technical–economic solution or the Sales Manager can intervene simply by changing the supply.

In co-manufacturing we provide a general explanation of the contents of Danieli's supply. For example, when drawings are converted (from European to Chinese) Danieli will make an effort to find materials that are equivalent to those found locally, on the specific request of the customer, but without intervening directly or independently in the conversion of the drawings to the local standards of the country.

In the final negotiation stages the PM must provide support to the sales manager and instil trust in the Customer by showing operational synergy.

17.2 Offshore Technical Phase

The offshore technical phase begins on the date of coming into force of the job and ends once all the shipments have been made for a typical supply of machines and plants, but it also provides support to the site activities coordinated by MAC through a Project Manager of the business unit, which has to answer to the job Project Manager (Fig. 17.4).

The PM immediately focuses on the first phases of the project and constantly interacts with the Sales Manager to maintain the relationship that was established with the customer during the commercial stage.

The coming phase of the job is called CIF and at this stage the Project Manager, together with the Financial Department, will instruct the departments concerned to issue a downpayment invoice and any bank guarantees.

He/she will also follow up to ensure that the customer makes the downpayment allowing the job to come into force.

This follow-up becomes important in countries that are known for complex procedures for obtaining authorisations and carrying out financial transactions. The

PM works towards coordinating the various departments and monitoring this first important contractual step, which, among other things, should already be specified in the preconditions.

For example, the down payment invoice will have to be submitted to the Customer's bank together with an Advanced Payment Bond Bank Counter Guarantee agreed on with the customer and added to the contract's commercial conditions. Therefore, we have to be clear on who does what in order to monitor the situation properly.

Another case that could seriously complicate the situation is a contract with a Ukrainian customer, requiring that we produce the documents to obtain a building permit from the Ukrainian National Bank, according to current Ukrainian legislation, a list of documents to start the project as well as statutory documents of Danieli & C, justification of the equipment supply times, technical description of the supplied equipment, detailed commercial specifications and overall layout drawing of the supplied equipment with itemisation as per detailed commercial specifications.

The lead time for this transaction obviously has to be as short as possible.

As regards the activities to be coordinated in the job order process, the offshore technical phase includes the following:

- Detail design and workshop drawings
- Fabrication and manufacture of equipment
- Purchasing of the various components and materials required for the project and its completion, including both equipment and plants
- Planning and shipping of materials and equipment to their place of destination

The Project Manager gives oneself up completely to analyse the document of the acquired contract and start its job organisation and resource coordination.

This includes all the typical operations that follow internal information flows, data processing, data implementation, data exchange and information for the preparation of engineering, purchasing, manufacturing and shipping.

Therefore, in order to have a clear vision of the project, the PM examines the contract documents that are now definite and final, looking into the Technical Specifications and the Commercial Part.

Figure 17.5 represents a flow chart that can drive the PM on its steps giving the overview of the first stage of the project activity.

The first study is basically done by carrying out the same analysis that was done during the commercial negotiations and the offer review, only more thoroughly.

All the aspects in the contract are examined in order to come up with a realistic planning schedule taking into account the contract goals without exceeding the budget.

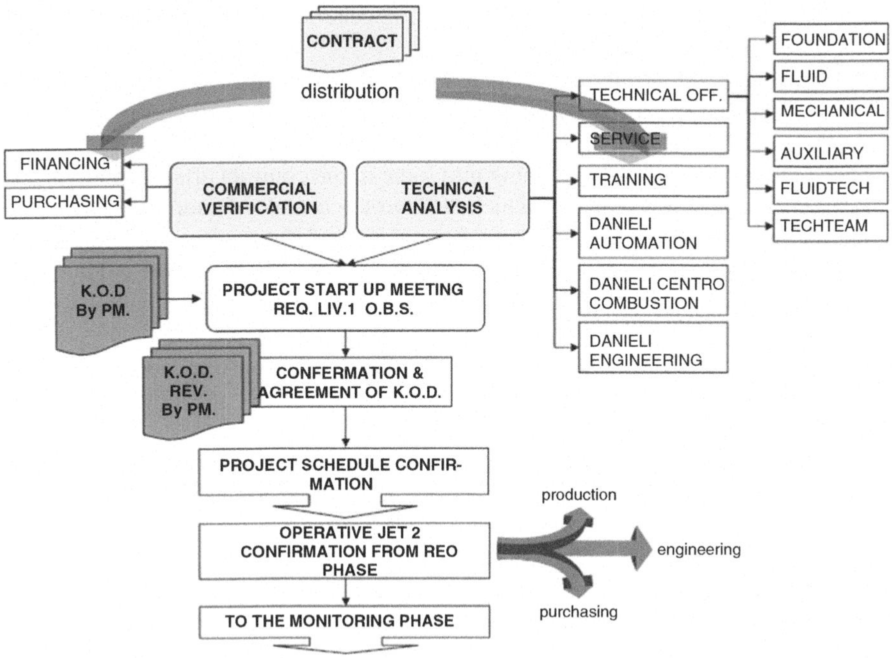

Fig. 17.5 Flow chart activities for job startup

To launch the job the PM must have a flow plan to assist him/her in the operating sequence, which could be as follows:

- Analysis: analysis of Commercial Offer, analysis of Technical Specifications
- Planning: checking the extended JET to make sure it complies with the contract, drawing up an initial macroplan, preparing for the internal job launch through Kickoff Document (KOM), calling the Esame Economico Commessa (EEC), agreement on job order planning and macroplanning of job order cash flow
- Offshore activities: checking the progress of engineering and any invoicing, checking the progress of purchasing, checking the progress of manufacturing and checking the progress of shipping and invoicing

17.3 Analysis of Commercial Offer

We should first of all take a look at the commercial offer because the economic conditions of the contract allow us to carry out actions to launch the project; they allow us to make assessments, do some preliminary planning and prepare a preliminary curve of expected drawdowns; they allow us to take into consideration particular or critical conditions related to drawing conditions and guarantees in order to make an initial assessment of possible risks.

Following are some important events representing the project's milestones on which we will base the job order planning goals:

- Down payment: triggers the coming into force of the contract after signing, and it is the first payment instalment, which leads to the actual job launch.
- Goods payment: the payment of goods is payment for the contractual supply. It can be divided into payments for the development and delivery of the engineering and payments for the actual delivery of the equipment and goods making up the plants.
- Last payment: the last contract payment is connected with final acceptance and therefore to the completion of all aspects related to machine and plant performance.
- Performance Bond: these are bank guarantees that must be given to the customer in the various contract stages to guarantee contract performance.

These are contract milestones and the Project Manager will have to coordinate the resources and departments involved in carrying out financial transactions.

For the down payment, the Project Manager must ensure coordination of activities such as issue of down payment invoice, obtention of the necessary authorisations for collection such as, for example, export licences, as well as letters of guarantee.

In this phase, the financial department assists the Project Manager in dealing with the customer's financial department and Danieli's bank as well as the customer's bank. The PM's knowledge of the financial aspects of the project must include knowledge of the main financial instruments used, such as Letters of Credit, Bank Guarantees and Promissory Notes, as he/she must be able to confirm that the contents of the various documents are in compliance with the contract.

This is when the PM enters into contact with the customer and tries to establish a relationship based on trust that will grow during the course of the project.

Let us now examine goods payment.

As we have already mentioned, there could be payment plans based on work progress (SAL), engineering and, in some cases, manufacturing, which, in any case, come to an end when the contractual equipment is supplied.

Some commonly used international terms are Basic Information (BI), Basic Engineering (BE) and Detailed Engineering (DD).

For a company like Danieli, which can and does prepare all the various types of documents, we have defined three main elements that can be offered, which are then included in the contract depending on the needs of our customers.

- Basic Information: all general data and specifications, which are fundamental for Basic Design and for the purchase of equipment and machinery.
- Basic Design: based on Basic Information, this is the designing work that determines the specifications and the basic drawings (including design of layout,

outline drawings, schematic diagrams and calculation sheets). Basic Design contains all the Basic Information.

– Detailed Design: based on Basic Design, this is the designing work required to prepare the general arrangement drawings, detail drawings and bills of materials, which will provide all the necessary information for manufacturing, purchase, transportation and installation of equipment and machinery.

These documents allow us to make the necessary economic assessments, cost and time evaluation for the delivery of the engineering—prepared and agreed on with the final customer—according to the general contract time schedule.

We have to be aware of the invoicing conditions for engineering, specified in the commercial contract, such as the delivery of hard copies and computer files of documents, their approval by the customer, particular contents and data. We also have to set the corresponding payment milestones and then enter them in the project macroplanning schedule together with the known lead times.

Another work progress report that could be found in a payment breakdown is the issue of all the purchase orders for rough materials, together with the documentation required to prove their existence. This is not very common in Europe where the tendency of our customers is generally to have the entire purchase *financed* until the machines and plants go into production. It occurs in emerging countries such as China and India for orders placed with the customer's own suppliers.

Some of the fundamental checks to be done on the commercial contract are invoicing of goods and preconditions for payment collection.

It is therefore important to study this in detail since we will have to make sure that the Letter of Credit has been issued by the customer with all the bank regulations concerning the shipping documents and the revised deadlines to be included in the initial macroplanning schedule. Other particular conditions such as packing characteristics, customs regulations, etc. have to be checked to determine the time required to procure these documents and avoid delays in delivering all the documentation required by the bank for collection.

At this point we have collected the greater part of the contract price but in general there are two other payment events with a final target, which is the final acceptance of the plant.

A commercial offer may contain the payment of on-site activities carried out by Danieli supervisory personnel, where a lump sum payment is specified in the contract for supervision to erection and startup of the equipment and plants up to the final acceptance of the plant. This involves the issue of documents that certify the presence on site of the personnel, and obviously need to be countersigned by the final customer for acceptance. These are generally time sheets that are attached to the invoice for services. It is important to determine if there are personal or corporate taxes to pay for this payment group that could entail the need to issue or obtain documents to be attached to the invoice. The utmost attention must naturally be paid to the procedures and time required to obtain this documentation, and any problems must be solved by providing facilities and associated costs, which should already

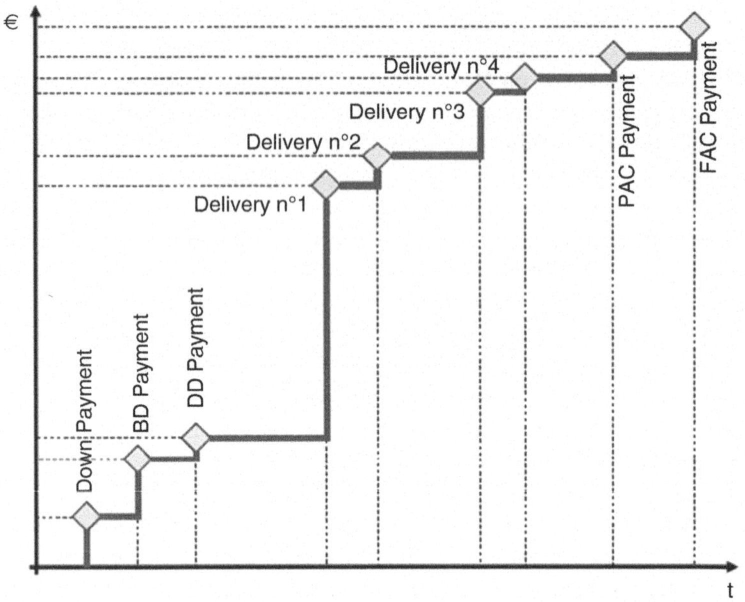

Fig. 17.6 Incoming value

have been entered in the commercial evaluation but which at times may not be clearly identified in the job budget.

The final acceptance of the plant is generally the last contract milestone tied to the collection of a payment (at times it ends at the end of the warranty period). It requires the customer's approval and confirmation after the efficiency of the equipment has been proven by means of testing on machines and plants. These tests are generally referred to as PAC sometimes replaced by a Takeover Certificate (TOC) or FAC. Depending on the equipment and plants supplied or the agreed contractual conditions, there may be other preconditions. However, it is fundamental to identify the documents for the commercial part of the contract needed for Danieli to issue the bank guarantees covering the opening of the L/C by the customer and triggering collection of payment. After having done this the milestones are then associated to the collected payments in a macroplanning schedule (Fig. 17.6).

The subject of possible penalties indicated in the commercial contract needs to be addressed separately. There are generally two types of penalties, i.e. those tied to time (penalty for time—PFT) and those tied to the technical result (penalty for performance—PFP).

The commercial contract specifies these penalties and quantifies them economically, indicating the object subject to penalty, which for Danieli will be considered as an element of risk in the contract.

The penalty tied to time refers to the contract schedule drawn up with the customers and sets forth the time-related milestones to complete the engineering, BD or DD, and time periods connected with shipment of equipment and goods and possibly their quantities if transportation is to be paid by the customer. For turnkey contracts, the time-related penalties can also be tied to events such as plant startup based on a milestone such as the start of Cold Tests (CT), Hot tests (HT) or load tests, or events connected with APC production tests. The value assigned to a delay-related penalty is variable depending on the length of the delay, and its maximum value is calculated and specified in the contract after being agreed on with the client during the negotiations.

The penalty tied to the technical results and specified in the contract generally refers to the technical specifications to explain how it is calculated, because, since it is tied to the performance of machines or plants, the details are necessarily contained in the technical specifications, also because the tests could be of a different type and, in the case of plants, the number of tests is considerable, which, consequently, makes for difficult calculations. Therefore, it is useless to include in the commercial part the details that are already specified in the Contract Technical Specifications. In conclusion, the maximum reachable value determined during the negotiations is generally indicated.

The sum of time-related penalties and performance-related penalties gives the maximum contractual liability value to be considered in the risk analysis.

$$\text{Total Liability Value (TLV)} = \text{PFT} + \text{PFP}.$$

17.4 Technical Specification Analysis

The detailed analysis shall be carried out, of course, by the competent technical department (UT), which will then check and develop its activities in compliance with the contract. However, the PM must have cognisance and be technically aware of the supply, since he shall have to discuss about it with the various bodies involved in the project.

Considering that today specialisation is more and more required also from a project management point of view, in a first phase the technical specification analysis shall aim at checking its compliance with the Scope of supply, with the contractual schedule and therefore with the delivery deadlines of engineering and equipment, and if necessary also of erection, startup and commissioning. The PM shall analyse in detail the contents of the JET by comparing it with the JET REO.

As an analysis objective, all the factors taken into account for the job planning must be checked. Therefore, the PM shall reconsider all the observations made during REO (project review) stage and adopt them in his analysis, in order to ensure that all the items considered are checked.

Let us start therefore from the type of product, by analysing its features. For instance, in case of long product making plants, first it must be checked whether it is a rolling mill for the production of wire rod, or sections, or special steels, etc. This is an introductory activity, since it allows to identify the resources to be involved in the creation of the job OBD. It further focuses attention on the consolidated lead times, if any, by referring to the previous lesson learned. When we refer to the plant development on the basis of contractual data and requirements, we have to be aware that there are constraints dictated by the contract, which have been demanded and agreed by the customer. As a consequence, we must always bear in mind the contractual objectives leading to customer's satisfaction. On such a basis, the PM must check if it is necessary to make any adjustments, which do not involve changes at contractual level, and prepare himself for the project review meeting with suitable clarification questions.

Repeated machines means the equipment included in the product standardisation, and which has therefore already been fully developed at design level, besides having being repeatedly tested. They are therefore consolidated products, which do not require special attention at preliminary stage. We must also think of asking, if necessary, to study a way of utilising standardisation as much as possible, in order to reduce delivery lead times and allow a quick construction of the plants on the customer's part. In this case, lead times are known and can be easily inserted in the initial macroplanning, which gives the advantage of ready-made design, prompt forwarding to production, and therefore possibility of optimising procurements, as well as delivery lead times.

New design machines means not only the equipment that has to be completely designed anew to achieve the desired technical result, but also those machines that, though having a standard basis, must be adapted in dimensional and technical terms owing to layout and plant requirements. Of course, these two conditions imply different lead times and considerations, but nonetheless they must be weighed carefully because of their possible unforeseen developments. At EEC (project economical review) stage, we shall therefore ask for confirmation and for verification, at basic design level, of the machines making part of the contract. We shall also ask for and compare the collected past experiences, consider and request a new engineering lead time of UT, as well as an evaluation on the manufacturing lead times, besides ensuring a Quality Control Plan (QCP). All these elements must be considered as critical conditions for a possible risk analysis, since they could have repercussions on lead times, costs and performances.

On the basis of the requested machines, we shall then pass on to verify the JET 0 *exploded*, i.e. the formalisation of the JET REO plus some changes, but preserving its agreed basis. By means of such document, which is present in the One World IT system, we shall proceed to verify the manufacturing features that the equipment should have if the REO results are confirmed at manufacturing strategy level.

We shall therefore define which, how many and what kind of machines are to be manufactured internally, i.e. in the Danieli workshops; which ones are always to be manufactured internally, but in the Danieli offshore production units; which machines are to be purchased externally on the basis of the Danieli detail or basic design, depending on the product to be bought and manufactured; and which machines are available on the external market and must therefore be purchased, since their technology is not produced by Danieli, such as, for instance, electric motors, fans, hydraulic pumps, special components like load cells, transformers, etc.; and finally which machines can be manufactured directly by the customer according to Danieli design, in which case again it may be either BD or DD, depending on the product and on the customer's specific demands. Remind that in the latter case there could be the need to make a conversion of the Danieli design, as it happens for instance in China and in India, which is then executed by the customer and where Danieli shall give its support by confirming the comparison of its own standardisation with specific customer's demands regarding the materials requested for equipment construction, such as steel grades, sections, heat treatments, etc.

We shall also have to evaluate the contractual responsibility related to products designed by Danieli but manufactured by the final customer, since there may be cases requiring expediting by Danieli to give confirmation of their execution according to the constructive dictates of the Danieli design. We shall therefore check also the budget for the expediting and customer-service activities. The *lower* limit of the supply could consist even in the supply of the sole design, without responsibilities on manufacture but only on design itself. This involves some risks that must be generically pointed out at REO stage or otherwise must be considered in the project budget and held up to view in case of dispute with the customer at project execution stage.

Let us always determine whether the plant is new or existing, since this will imply a completely different approach. When we refer to a new plant, we generally start from a greenfield situation, and therefore from the construction of new buildings, new civil works and auxiliary facilities to be used for the new plant. With new plants, we have the possibility to transfer the Danieli experiences and therefore guide the customer towards the most suitable solution according to its needs, by analysing and, if necessary, supporting the latter in the cost cutting process. With the *turnkey* products, the whole has instead a more complex internal development involving several business units and therefore requiring the creation of a dedicated project team. As a matter of fact the turnkey projects are managed by a special department called Lump Sum TurnKey (LSTK), which, through a project director, coordinates all the involved product lines and their single structures. On such conditions, the initial planning may refer to standardised patterns based on past experience, which can therefore be consolidated.

If we start from an existing plant, we shall provide ourselves with a preliminary check list to identify the possible criticalities to be then solved at REO stage, such as the drawings of the existing building, machines or plants, and the need to make a survey on the possible TOPs, if our supply will be part of the expansion of an existing plant, in which case it might require both *off-line* and *in line* activities. If otherwise it will be part of an existing plant revamping project, it will involve only in line activities. These are all elements that had been already analysed at REO stage, and are now considered again as part of the contract, with all the possible relevant negative and, hopefully, positive sides. In this situation, the initial planning shall be made more carefully, since the installation activities on existing plants generally imply the customer's plant shutdown for either a long or a short period, depending on the intervention to be carried out (in line activities). Therefore, a shutdown, besides being onerous for the customer in economic terms, is risky for Danieli,

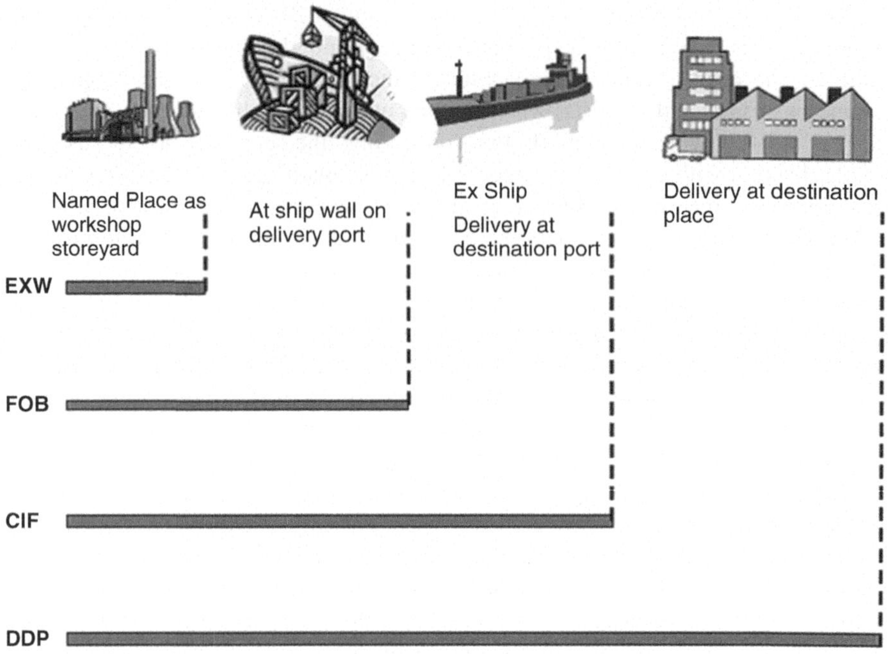

Fig. 17.7 The INCOTERMS

which is bound to strictly observe all the project milestones, and to pay special attention to the equipment dispatch and installation.

The machines and plant performance evaluation to be made by the PM will be efficacious, if the PM has a certain seniority and therefore a fairly good knowledge of the product and its results, besides having the possibility to transfer his direct experience. What the PM must generally evaluate are the contents of the tests requested by the contract (remind that they had been dealt with at the REO meeting, but then what were the contractual provisions in this regard?), their apparent difficulty and the contractual schedule imposed for their execution. The PM must bear in mind these contractual requirements to have confirmation of the achieved targets at project review stage. Moreover, remember that all the formulas for the verification of the achievement of performances to be used also for the technical computation of the contractual penalties will be given at this stage.

When dealing with the shipment, we must never forget the applications of the INCOTERMS 2000, which are included in almost all contracts as international transport regulations. Let us refer to the most frequent cases, i.e. shipments of type EXW (Ex-works—named place), FOB (Free on Board—named port of shipment), CIF (Cost Insurance Freights—named port of destination), DDU (Delivery Duty Paid) (Fig. 17.7 and Table 17.2).

Let us then consider the aforementioned cases, by taking into account the various aspects governing the relevant activities, as well as the commercial contract

Table 17.2 INCOTERMS 2000 (details)

	Loading on truck	Export customs formalities	Carriage to port of departure	Unloading from truck at port of departure	Loading charges port of departure	Carriage to port of destination	Unloading charges at port of destination	Loading on truck at port of destination	Carriage to final destination	Insurance	Import customs formalities	Import taxes and duties
EXW	No	No	No	No	No	No	No	No	No	No	No	No
FCA	Yes	Yes	Yes	No	No	No	No	No	No	No	No	No
FAS	Yes	Yes	Yes	Yes	No	No	No	No	No	No	No	No
FOB	Yes	Yes	Yes	Yes	Yes	No	No	No	No	No	No	No
CFR	Yes	Yes	Yes	Yes	Yes	Yes	No	No	No	No	No	No
CIF	Yes	Yes	Yes	Yes	Yes	Yes	No	No	No	Yes	No	No
CPT	Yes	Yes	Yes	Yes	Yes	Yes	No	No	No	No	No	No
CIP	Yes	Yes	Yes	Yes	Yes	Yes	No	No	No	Yes	No	No
DAF	Yes	Yes	Yes	Yes	Yes	Yes	No	No	No	No	No	No
DES	Yes	Yes	Yes	Yes	Yes	Yes	No	No	No	No	No	No
DEQ	Yes	Yes	Yes	Yes	Yes	Yes	Yes	No	No	No	No	No
DDU	Yes	Yes	Yes	Yes	Yes	Yes	Yes	Yes	Yes	Yes	No	No
DDP	Yes	Yes	Yes	Yes	Yes	Yes	Yes	Yes	Yes	Yes	Yes	Yes

EXW ex-works, *FOB* free on board *FCA* free carrier, *FAS* free alongside ship, *CFR* cost and freight, *CIF* cost, insurance and freight, *CPT* carriage paid to, *DAF* delivered at frontier, *DES* delivered ex ship, *DEQ* delivered ex quay, *DDP* delivered duty paid, *DDU* delivered duty unpaid

provisions surely connected with the invoicing. We shall determine whether a shipment is by land, therefore using trucks or trains; we shall consider the need of using means of transport for wide or long loads and/or exceptionally large goods vehicles, therefore requiring special consignments with police escort and permits; we shall determine an approximately total weight of the equipment subdivided into mechanical and electrical part, in order to better evaluate the necessary volume distribution; we shall consider carriage by sea; whether maritime or transoceanic packaging is necessary or not; we shall consider packaging in wooden crates where fumigation could be necessary, like for China; and packaging in containers of various types, like e.g. open top, high cube, etc.

Of course at this stage we shall not go into detail, but the cases put forward will allow us to make an initial planning of the possible time schedule required for shipment and transport, by comparing our results with the contractual provisions.

These features then depend on the various countries and on the greater easiness, which could derive to the customers or Danieli in connection with the economical aspects that are of course significant in the field of the steelmaking plants, considering the volumes and weights involved. We shall then check, on the basis of the commercial contract and the country of destination, the possibility of direct shipments to the customer also from the offshore Danieli works, without having to pass through the headquarters. At the same time, the shipment documentation to be necessarily attached to the invoices for payment collection must be checked, in order to ask for a confirmation during the EEC (project economical review).

Moreover, in some cases we shall have to consider the possibility that, according to contract, the customer will carry out the inspection of the goods before packaging and shipment, and therefore needs a confirmation of costs and lead time required for these activities.

All this information and any problems that might have arisen must be pointed out and collected in a document to be used as a basis for the project lesson learned, which must reflect both the Commercial and the Technical analysis contract (Tables 17.3 and 17.4).

17.5 Project Planning

After having made all the necessary checks and verifications, the Project Manager shall start to coordinate the resources and to plan the project activities. For this purpose, internal IT tools are used, which of course allow a fast execution of the actions. In any case, this is the activity that needs the greatest engagement on the PM's part and marks the opening of the job.

Let us consider all the characterising elements of planning, such as the interfunctional and the intersectorial consistencies and the agreed objectives.

We have already a clear picture of the involved activities, such as defining the products to be manufactured and the activities to be carried out, identifying the operational responsibilities, assigning the necessary resources to each activity and placing in time the activities and the relevant resources.

Table 17.3 Technical comment summary

Contractual reference	Contractual description	PM comments	Action by	Status
Sect. 3 item 1.3.1	Descaler	The scope of supply foreseen two ramps. To define for which billet sections (working by 120 × 120, 160 × 160 and future 140 × 140)	UT 053	O
Sect. 3 item 1.3.2	VAR outlet furnace	Are foreseen near three hoods for heat protection. Are they needed considering the steel grade produced?	UT 053 technical team	O
Sect. 3	Rolling mill	To define a proper configuration of the stand (i.e. position of the universal stands)	UT 053 UT 041	O
Sect. 3 item 1.5.5	Shear CV1	To define if it is necessary a dog house. Also for the other shears	UT 053	O
Sect. 3 item 2.1.1	QTB	To supply completed with bypass: to verify if it is possible to replace the VAR with funnels or to reduce the number of driven rolls	UT 048	O
Sect. 3 item 2.2.1	VEC	Grease plant is needed centralised as per VRC?	UT 047	O
Sect. 3 item 2.2.3	Cooling bed	Increase the width of the cooling bed: foreseen in the contract 10 m, now are required 14 m	UTC	O

Planning a project implies differently structured phase activities, which require an orderly and systematic approach to ensure the definition of all the activities to be carried out, trying not to omit any elements, and pointing out the interrelations existing between the activities.

Moreover, we know that organizing a project means proceeding to its segmentation into a series of sub-projects, which, on their turn, are subdivided into minor elements that in our case are managed by the single business units responsible for the execution of the detail functions. The latter report and answer to the Business Unit Management as regards the operating and company efficiency, but above all to the PM as regards the technical–economical results linked to the project, as well as the contractual execution vis-à-vis the final Customer.

Having referred the detail phase to the single project business units, we can confirm that the PM will end the project breaking-down process when he has ascertained that the product is definite and measurable, and when he has made sure that the work involved will develop without a break.

Finally let us say, for completeness' sake, that the sum of the elements of each level equals the total for the level.

On such bases, therefore the project frame is starting to be outlined.

Now the PM must start to organise his staff: therefore, he shall start to think who will be involved in the project by working out an OBS for the job. He shall prepare an initial planning resulting from the check of the contractual WBS and from the JET 0 exploded with the time schedule and the lead times as defined at REO stage. He

Table 17.4 Commercial comment summary

Contractual reference	Contractual description	PM comments	Action by	Status
1.10 page 5	Definitions: Contract equipment	Also are listed the *Key Components for Joint Manufactured Equipment* as the goods to be supplied from. At the moment, it is not foreseen—the joint manufacturing but it is not specified if, in case of co-manufacturing, those goods are already included in the contract price	PM in KOM	
2.7 page 7	Local manufacturing advisory service	To foresee the advisory on local manufacturing, including inspections for the not defined *major items* (to define) and equipment of co-manufacturing		
2.11 page 8	Spare parts	List of the spare called *left* (which ones?) to define during the detail design without changing in the value of the contract		
3.4 page 9	Delivery	FOB Italy, to propose FOB Europe in LC for the SUND presence	PE	
4.2.3 page 11	Amount for documentation	The difference between BD and DD is not clear, if it is considered as described in SOS or as it is detailed in Sect. 5, in that case it could risk to delay the cash inflow of BD (50%). As per items 7.6, the BD has to be completed with the meeting in Italy at 4.5 months from EV. To propose a subdivision more convenient	PM in KOM	
4.2.4 page 12	Training amount	A personal tax recipe is required to clarify. It could delay the cash inflow	Dir. Fin.	
4.2.5 page 12	Service amount	A personal tax recipe is required to clarify. It could delay the cash inflow	Dir. Fin.	
5.1 page 13	Delivery terms	Delivery FOB from 8 to 8.5 months from CIF in five lots. Delivery Port: North Italian. To change in LC with European	PE	

shall also prepare a first indicative cash flow curve based on the elements in his hands.

This information must be included in what we call Kick Off Meeting (KOM) Document, which sums up the technical–economical aspects like the milestones linked to contract payments, the supply of the machines and the requested contractual deadlines, with considerations on a proposed planning based on the lead times of the equipment declared as existing in the company. This document will be then used as a guideline for the job-launching meeting to be convened for job planning confirmation.

Its contents shall therefore include the following: scope of works, OBS, WBS, master plant list, risk and opportunity analysis, forecast profit-and-loss account and the cash flow curve.

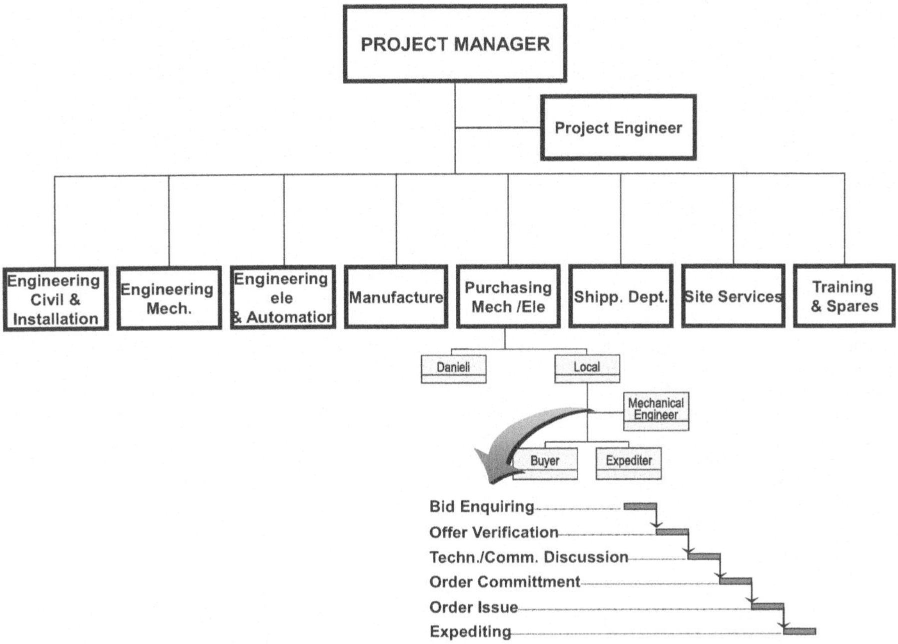

Fig. 17.8 Project OBS

The OBS worked out to appoint and confirm the project staff could result as shown in the following Fig. 17.8.

So, the document which accompanies the summoning of the meeting for the Job Economical Review (EEC) shall contain all the elements described for the Contract definition, the macroplanning of the activities, the scope of supply, the risk and opportunity analysis, and finally the forecast profit-and-loss account and cash flow curve. And also, with reference to the contract, it shall specify the following: the effective date, the contractual payments and currency, the necessary bank guarantees, the contractual deliveries with dates referred to the contractual time schedule, the penalties for late delivery and performance, the requested performance figures, the guarantee limits, miscellaneous such as laws, insurance, standards; criticalities, such as e.g. the forecast difficulty in respecting the contractual time schedule owing to the technical departments' overload, the high penalty figures and their very unusual application conditions, or quite demanding performance tests.

On the basis of the provisional dates set at REO stage, we have then drawn up an activity macroplanning document defining the main interchange dates to be shared by the parties to the meeting and to be finally confirmed during detail verification, but without deviating from the strategic objectives agreed during the meeting itself.

An exemplifying table is given later (Table 17.5).

Table 17.5 Preliminary exchange document plan

	Technical department		Production		Delivery	
Description	Mech.	GRC balance	Months from Mech.	Date up CESPED	DDP customer	Paym. and/or penalty
Coming into force: 14.05.03						
Bar finishing (047–049)						
VCO—PW—VUA modifications	30.09.03		4,93	28.02.04	14.04.04	X
RDM—CMP	15.07.03	15.12.03	7,43	28.02.04	14.04.04	X
VUS—layer preparing—pockets with roller table	30.07.03	15.12.03	6,93	28.02.04	14.04.04	X
Roller table—transfers	30.08.03	15.12.03	5,93	28.02.04	14.04.04	X
Simple single machines	30.09.03	15.12.03	4,93	28.02.04	14.04.04	X
Consumables	30.09.03		4,93	28.02.04	14.04.04	X
Tying machines (SIGNODE)	30.08.03		6,50	15.03.04	14.04.04	X
Lubr. central units	30.09.03		4,93	28.02.04	14.04.04	X
Hydraul. central units	30.09.03		4,93	28.02.04	14.04.04	X
Electric part (744)						
Part (759)	30.07.03		6,93	28.02.04	14.04.04	X
Requisition form for motors	30.08.03					
Order for motors	15.10.03		4,43	28.02.04	14.04.04	X
Motor drawings from supplier	15.11.03					
Motors must not be fitted on the machines						
Mechanical engineering (047–049)						
Preliminary layout	15.07.03					
Motor data sheets	30.07.03					
EDM	30.07.03					
Junction box	30.09.03					
Tying machines specifications	15.08.03					
Fluid consumption	30.08.03					
Functional cycles	30.10.03					
Civil-plant engineering (049–043)						
Final layout	30.11.03					

(continued)

Table 17.5 (continued)

Description	Technical department		Production		Delivery	
	Mech.	GRC balance	Months from Mech.	Date up CESPED	DDP customer	Paym. and/or penalty
Similar foundations (PM)	15.07.03					
Foundations	30.10.03					X
PFD—P&ID air—water	30.10.03		(Preliminary for DA at Sept. 30th)			
Central units layout—Diagr. Distribution	30.11.03					
PFD hydraulic—Lubrication	30.10.03					
Electric devices layout	30.11.03					
Electric engineering	Prelim.	Final				
Motor sensor list	15.09.03	15.11.03				
Functional specifications	30.12.03					
D.A. spare parts offer	30.12.03					
Data at CESPED for transport	15.12.03					

The scope of supply shall be considered according to the contractual one plus any clarifications and requests for clarifications to be discussed. Here follows a practical example of the scope of supply to be written in the report:

1. Modification of roller table with aprons, cooling bed and lining-up rollers
2. Installation of cooling sprayers
3. Modification of roller table at cooling bed outlet and addition of two-layer preparing systems: one north side towards the existing round bar finishing station, and the other south side towards the new CLS and stacker
4. A CM with cutting-to-length beam north side
5. CLS with multi-strand straightener and mechanical shear south side
6. Roller table with trolleys at stacker inlet
7. Stacker with 24-m single magnetic heads
8. Steel strapping stations
9. Stack transfer and storage
10. Lubrication and hydraulic central units
11. Electric part and automation

Design data, such as requested productivity and maximum speed parameters, starting materials and their characteristics, product mix and requirements.

The contents of the document to be discussed and updated at the KOM shall report the specific clarification requests, if any, to be made to the technical, technological, purchasing and production departments.

Then all the delegates of the involved departments and business units are summoned. In this phase, the information flow between the various departments is defined and the first actual dates of which everybody shall take charge are set. During these meetings and the subsequent ones, that will be single and specific, what is important is to fix the correct lead times for the transfer of the information between the different types of engineering and the different business units, with the aim of giving the production and purchasing sectors the information on time for the respect of the contractual schedule.

Today, thanks to the REO phase and to macroplanning, it is relatively easier to agree on the execution time schedule, considering that it is entered officially in advance in the company planning. However, in this phase, to go into details, problems may arise in connection with the new workloads, with the supply time schedule updates, as well as with the updates linked to the external market situation. In this case, it is absolutely necessary to agree on actions and strategies such as to ensure the regular development of the job process with actions sometimes aiming at risking more for the respect of the Schedule.

The important PM action is to highlight and give evidence to itemised list of the possible risks resulting from the technical and economical analysis of the documents, in order to be able to define at a later stage, if necessary, a specific analysis quantifying their importance in economical terms, as well as the possible impact on the contractual time schedule and on the various project execution phases.

If we take for instance a contract analysis, we shall point out a few items as follows:

- Significant percentage of local supplies under Danieli responsibility, which can have an impact on Costs, Quality and Delivery lead times.
- From the schedule analysis, the machine setup times for internal production result to be narrow, with binding shipment dates.
- *Excessively elastic* down payment collection conditions allowing a relatively elastic payment, which has an impact on the job cash flow and affects the issuing of orders to the sub-suppliers, with risks in terms of delivery deadlines.
- With a contract providing for a FOB delivery, there may be late shipments owing to the lack of means meant as either monetary means or actual carriers, which are at the Customer's charge.
- Euro – Dollar exchange rate over the contractual time period reaching a time limit of 18, 20 months.
- Very restrictive Quality Tests requested by the contract, involving possible objections owing to the difficulty of measurement.
- From the analysis of the opportunities we could, for instance, define the possibility to delegate the dedicated structure of the offshore department in charge of the assigned commercial area (Beijing Office) to the issuing of orders.

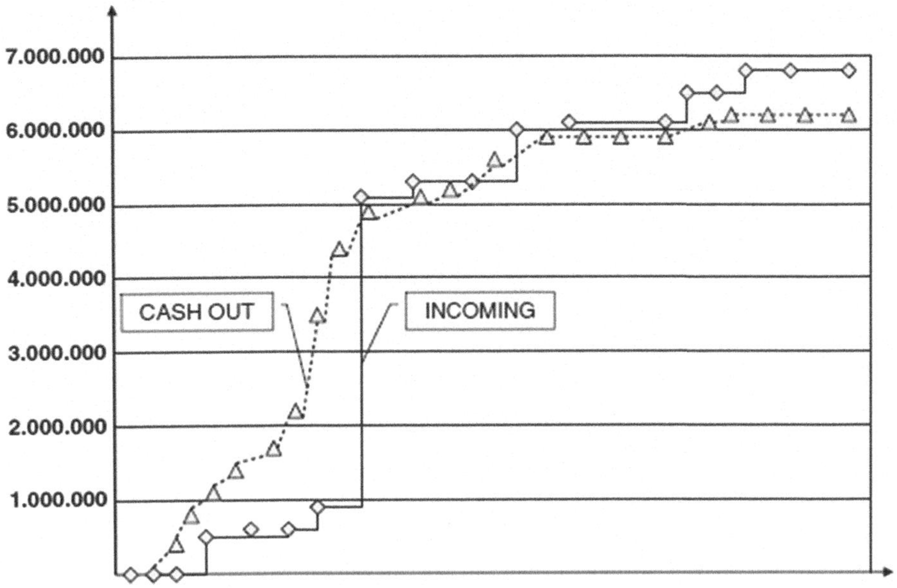

Fig. 17.9 Cash flow forecast

- Customer's direct responsibilities which could favour claims safeguarding its own activities, such as e.g. possible late erection/machine malfunctioning/auxiliary plants to be supplied by the Customer that could *cover* possible delays of supply on the one hand, without affecting payment conditions on the other.
- Absence of final tests, should there be the possibility to enter rapidly into production, therefore reducing the final acceptance lead times.
- If the Customer is satisfied, there could be a further development of relations leading to more direct contacts with the customer in view of future expansions or investments.
- Presentation and study of a risk and opportunity analysis.

As regards the financial part, the analysis of the forecast profit and loss account, and cash flow for the project have the importance to start the follow-up of the cash flow. Figure 17.9 is the first representation of the job result.

On the basis of such document resulting from the updated KOM document agreed during the EEC, the PM shall draw up the updated KOM document to be sent as a reference document to all the participants to the meeting, and as copy to the various levels 2 of the product line OBS.

The document must contain what has been discussed during the KOM, with reference to the single items of equipment and to the necessary clarifications, if any, to be given during work execution.

As a practical example, the Table 17.6 later shows how all this must be described in a short and itemised way.

Table 17.6 Highlight table of critical items

EEC item	Contractual description	Report status	Action to be performed	Action by	Status
1a	Roller table with aprons	The existing roller table must be modified to increase the apron width	Confirmed by UT, weight increase	UT 053	O
1b	Roller table with aprons	The apron angle must be checked because it seems too low (25°) for discharging flats and sections (it was originally designed for rounds)	To be checked within _____ by UT and Technical team	UT 053 Technical team	O
1c	Roller table with aprons	Discussion on the working problems for both *tail* and *head* discharging onto the existing cooling bed. This area is controlled by the mill automation system at GE charge. It is therefore necessary to discuss this subject with the customer and clear matters up officially. Check whether it is worth preparing a short specification	Dedicated meeting of PM with UT and MAC to check details and who, what, by, when	UT 053 UT 041 PM 040 MAC 091	O
Miscel. 1	Engineering	It is necessary to send by July 10 a preliminary layout drawing, which must be final as regards the positioning of machines, and the basic foundation drawings of an identical stacker. The customer needs it to calculate the position of the building columns. (a) UT (047) shall		UT 047 PM040	O

(continued)

Table 17.6 (continued)

EEC item	Contractual description	Report status	Action to be performed	Action by	Status
		prepare the preliminary layout drawing (b) PM shall send the basic foundation drawings of 1594 as typical/similar			
Miscel. 2	Management	Prepare a detailed project schedule		PM040 UT DIR PROD. MAC091	O

At the end the project schedule must come out, which shall include all the phases of the project itself. The project schedule as a database is the JET2. Its possible graphic representation can be developed through the application of the PRIMAVERA SW by withdrawing data from the OW database.

To develop a planning for complex projects involving a series of machines, we must look for a simplification to spare the PM the trouble of a detail drawing up through manual updating. To this purpose, a master plant list must be produced which subdivides equipment according to Plant Unit and allows an optimal management of the project phases. It shall therefore become an interface document for the customer, and a reporting document for the bodies assigned to erection and subsequent startup.

The MPL replaces to all intents and purposes what is commonly known as project WBS, which aims at making the project breakdown systematic: it helps not to forget any operating phases during initial planning; it can be taken to the desired detail degree; it allows a discerning allocation of resources; favours the identification of the responsibilities; with the support of a suitable coding, it allows analyses of selective and aggregate, or just aggregate type.

We have therefore determined that plants and areas can be codified according to a standardisation based on the functions of the plant units and their technical characteristics.

If we take for example a rolling mill, its area subdivision will be as follows in Table 17.7.

With reference to the Fig. 17.10, this type of graphic representation, by use of PRIMAVERA tool, is common to several types of project management. Of course, it allows handling all the connected activities, such as the evidence of deadlines and critical paths, the percentage work progress, etc., and finally the extraction of the project's *s* curve, which describes the development in time of the contract.

It therefore gives a view in time of the project development pointing out the project completion date; it allows checking the interrelation between Work Packages (WP), or between significant external events, according to the actual project

Table 17.7 Master plan list table configuration

Area (plant unit)	Pos.	Description	Pos. of project	Contract item	Detailed description	Weight (kg)
JC22	1	Billet transfer and charging device	0110		Inlet furnace area	0
			0112	3.1.1.1	Hydraulic pawl billet transfer	31,104
			0121	3.1.1.1	Billet charging device	8748
	2	Roller table and stoppers	0131	3.1.1.2–3.1.1.3	Roller table in front of transfer	10,044
			0132	3.1.1.2–3.1.1.3	Intermediate roller table	21,600
			0133	3.1.1.4	Fixed roller table with weighing bridge	8100
			0134	3.1.1.4	Emergency discharging bench	8748
			0141	3.1.1.2	Fixed stopper	108
			0142	3.1.1.2–3.1.1.4	Movable stopper	2376
	3	Weighing cells	0145	3.1.1.4	Weighing system	108
JC02	4	Reheating furnace	0201	3.1.2.1	Reheating furnace	216,0000
JD12	5	Emergency equipment for furnace	0151	3.1.3.1	Roller table in front of transfers	11,124
			0152	3.1.3.1	Emergency discharging bench	8748
			0153	3.1.3.1	Fixed stopper	108

execution conditions; it allows considering its optimisation on the basis of the internal and external constraints, as well as the effectiveness of the balancing between requirements and needs.

The advantage of a subdivision according to areas and its code reference as indicated in the first column of the Table 17.7 (Master Plan List Configuration) is that it allows to streamline project management, as well as to organise the project not only on the basis of one's own internal needs, but also to create an objective towards the maximum result at shipment and installation stage, where—thanks to this system—the site organisation can be decidedly improved by reducing the general standby time.

Table 17.6 reports only a few of the items to be monitored. For each single item, we shall have all the relevant JET data present in One World—the Danieli Corporate ERP system, then the design development and execution dates, everything concerning production, orders to sub-suppliers, weights, etc.

An important consideration must be made at initial planning stage with regard to shipment: right from the start the general situation of the equipment should be

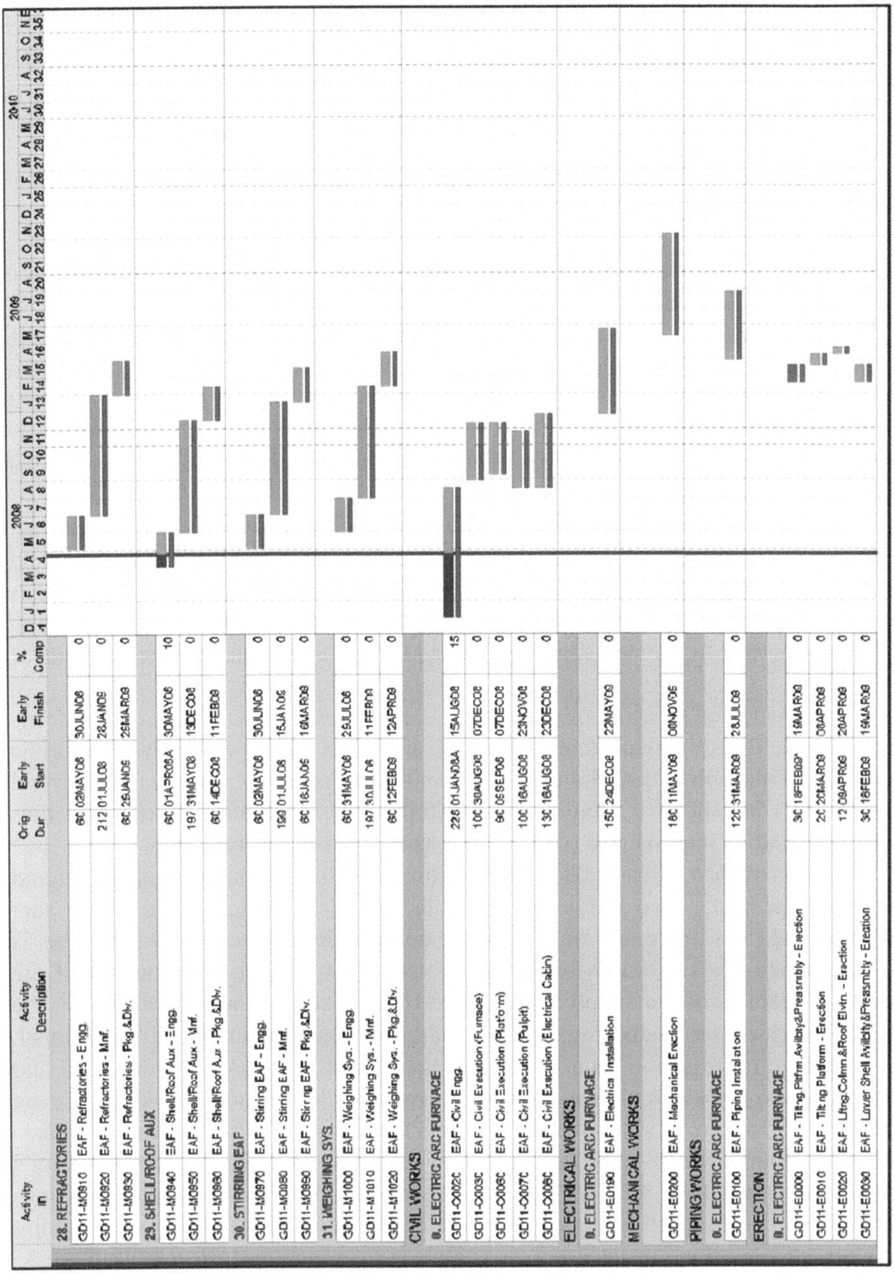

Fig. 17.10 Work-In-Progress

outlined clearly by establishing the possible requirements for exceptionally large goods vehicles and by agreeing on the necessary total time for shipment by land, sea or air.

17.6 Risk Process

The risk and the possibility of suffering damage arising from circumstances that are more or less foreseeable.

The profit and loss account is one of the fundamental elements that determine the result of the job order; this means that the PM needs to know the details to calculate it and, during the job order process, must be able to foresee any corrective actions that may be needed to stay within the budget or improve the result itself. This is not really management of the profit and loss account but of the activities and phases that determine its result.

Consequently, the PM must be aware of the economic impact on the product line, caused by correct management of the job order; therefore, meetings are held periodically to define the general progress of the product line and the various business units to be able to set priorities and take general action.

It is important to provide some definitions to help us understand the contents of analysis.

The presence of an economic job order element, considered as a cost, is determined in the JET tool, which contains all the analytical data to correctly monitor the progress of the job order. Splitting the contractual scope of supply into several components in order for production to determine the operating flows, identifies each item and its internal cost, which is the target cost of the project.

The offer review already takes into account the contractual critical points and therefore the first phase is considered to be complete, which is the phase that analyses the risks that could impact on the commercial and technical part. This is not a detailed analysis but is based on first-phase experience and analysis of the technical features of the supply and is done by the people who participated in the REO. Today, risk assessment needs to be more specific, structured and professional.

In a large organisation such as Danieli, the human factor is very important and since a person in the company is aware of the risks, his/her attitude may influence the accuracy of risk perception as well as how he/she responds. An attempt must be made to create a homogeneous approach to risk management adapting it to the company's organisation based on relationships of trust where communication and decision-making are *open* and truthful since the answers to the risks are their own, showing how the organisation creates a balance between assuming and preventing the risk. The organisation must therefore be actively present for the entire stage of the project.

The risk process is a flow of activities that go from management planning and, through identification, quantitative and qualitative analysis, determine response planning by means of checking and monitoring.

Useful in analysis is a risk breakdown structure (RSB) of a project and its components such as organisational and technical aspects, external factors and finally Project Management.

We feel that financial risks are clearly highlighted and analysed by the financial department, which consequently dictates the guidelines for optimum management based on cash flow forecasts. The analysis of these risks can therefore be included in the general risk package of the job order that the PM will create or request.

For the technical part, the risks reach high levels of specificity because they generally concern equipment or plant performance, and therefore must be studied by the process experts and engineering firms possessing the specific know-how required to quantify and highlight them. These risks can have an economic impact on the job order's budget, and therefore the PM includes them in the contractual analysis.

Technical risks can be numerous. The most obvious is when, for example, the acceptance of certain contractual conditions is not standard but practically a target, which, although it has never been pursued in the past, we are certain we can reach. This is particularly applicable to machines with new engineering where technical design analysis combines theory and practical experience, producing a result that fulfils the customer's requirements.

Technical risk analysis is inevitable, clearly, for prototypes or improvements, which are necessary to the customer and which may occur.

The making of tailor-made plants can increase the risk factor because they incorporate components that have not been fully proven.

The risks described earlier are to be considered internal risks and as such *controllable* by the PM.

External factors are an important variable because they cover a wide range used by various company departments that have to deal with risks; these are *non-controllable* risks because they are affected by external factors such as politics and the market economy.

As for Project Management, we have to consider that the PM acts as coordinator and link within the company, carrying out functions such as estimates, planning and control.

Therefore, we have to target risk verification and analysis more systematically due to the various problems that arise in the management of complex projects in difficult contexts and then analyse the various possible management methods that could be implemented depending on whether it is an LSTK (lump sum turnkey) contract or an EPC Contract.

Today, for example, the market also provides the possibility of analysing the production of large projects of customers, with longstanding industrial tradition in various countries of the world and where various contractual forms are used. This condition requires and necessarily involves excellent Project Management and risk management capabilities. Developing the capacity for analysis—first of all within the company—then allows you to be a contributor to the positive result of the Customer's investment.

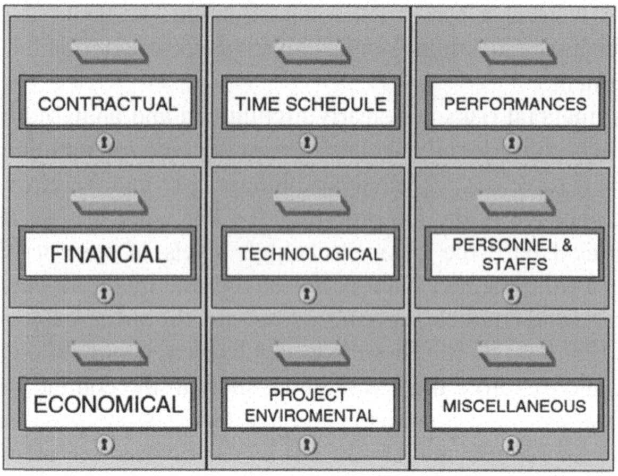

Fig. 17.11 The *risks* chest of drawers

When we start with analysis, we have to set a clear goal that is visible to everyone during the execution and production of the project. Our goal is to classify the types of risks that can arise during the execution of a project in order to systematically analyse risks and consolidate the feedback of past projects in an organised manner.

The field in which the Company operates can definitely propose different routes and specific analyses depending on the sector. In any case, if we begin with what typically constitutes a contract, we can try to first of all determine the possible risk areas to see if they need to be considered and classified immediately to be used in future valuations.

Let us take as an example the standard areas: Execution Risk, Project Environmental Risk, Technological Risks, Performance Risks, Personnel and Plant Integrity Risk, Time schedule Risks, Financial Risks, Economical risks.

Putting all the possible risk areas on the table does not necessarily mean that each one of them has to be weighted in the analysis and thus have a specific valuation. A lot depends on how we decide to consider the contract and its possible impact on the company's profit and loss account. An analysis must definitely be done for projects that the company considers large and important for its products and image.

Therefore, the areas we are going to examine in the risk analysis need to be shared/combined.

In the chest of drawers as per Fig. 17.11, each drawer contains the elements required for our analysis.

Once we have determined and identified the risk sources, we will do another more thorough analysis based on significant risk sources whose occurrence is highly probable, that do not produce secondary/side effects, that have precise reference data, that have other contrasts and facilities, which all together have to constitute a sensibility analysis (Fig. 17.12).

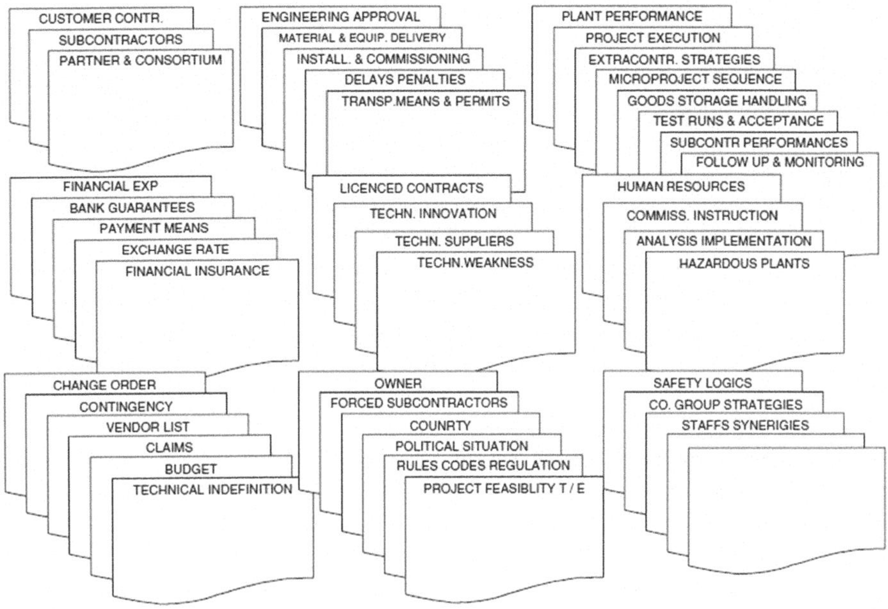

Fig 17.12 Risk detail folders

Considering the type of company activity, we will create two main risk classes and groups, i.e. *repetitive* and *non-repetitive*.

The former can be identified during the offer phase and in the initial project phases and can be run by establishing repetitive strategies according to the type of risk.

Non-repetitive risks are clearly more difficult to identify and manage since they can occur in each executive phase of the project and the signs indicating their presence are often weak. In this case, it is the people involved in the project, who, with their management experience, can ensure early identification and effective management.

It is therefore up to Project Management to assume the main responsibilities for the management of these risks. It is therefore important to be a Project Manager who is more and more sensitive and able to properly identify and manage project risks, as well as applying effective control methods with the participation of company structures in those project events that require solving of risks.

17.7 Risk Evaluation

To analyse risk management we have to necessarily monitor project risks through actions that are important in the first phase of the analysis and then in all the other project phases, identifying possible upsetting events, analysing the extent of their consequences, evaluate and apply the most suitable procedures.

This is all done for the purpose of minimising the results arising from negative events and maximising the results arising from positive events.

After having established a technical risk monitoring process we can use a flow chart to guide us in the various initial planning stages, where, as already mentioned, we will have to identify and describe possible risk events, assess and determine the importance and extent of possible risk events, plan and define actions and prepare a risk plan.

During work progress, we will have to set up actions and update the risk plan.

The risk assessment stage aims to calculate the extent of possible consequences and will therefore have two parts: quality-based to classify the significance of the risk and quantity-based to determine the extent of the risk itself.

The goal is to create a risk exposure matrix through actions focusing on so-called *key* risks, so that the matrix is able to support the definition of more effective intervention strategies, and to define the classification criteria of the *significance* of risk-related consequences.

The following Table 17.8 gives a base representation of how we could summarise and group the possible risk items.

Table 17.8 Risk definition

Conditions to define impact scale on principal target of project (negative impact)					
Project target	Very low	Low	Medium	High	Very high
Financial cash flow	Financed	LC (letter of credit)	Mix LC/TT (letter of credit, telegr. transfer)	PN (promissory notes)	TT (Telegr. transfer cash)
Costs	Not significant cost increase	Cost increase CI < 10%	Cost increase 10 < CI < 15	Cost increase 15 < CI < 30	Cost increase CI > 30%
Time	Not significant cost increase	Time increase TI < 5%	Time increase 5 < TI < 10	Time increase 10 < TI < 15	Time increase TI > 15%
Innovation technology	Typical product	Common ready	Common tailor-made	Research department. New project	Unknown technology to acquire
Quality	DWS 100%	DWS 50% DFE 25% DME 25%	DWS 20% DFE 20% DME 20% EXT 40%	EXT 80%	Unknown Supplier EXT > 80%

Table 17.9 Matrix of probability

Matrix of probability and impact										
Probability	Risks					Opportunity				
90	5	10	20	40	80	80	40	20	10	5
70	4	8	16	32	64	64	32	16	8	4
50	3	6	12	24	48	48	24	12	6	3
30	2	4	8	16	32	32	16	8	4	2
10	1	1	8	4	8	8	4	3	1	1
	0.5	10	20	40	80	80	40	20	10	0.5

Impact (ratio scale) on a goal (i.e. cost, time, quality)

Table 17.10 Risk impact scale condition

Conditions to define impact scale on principal target of project (negative impact)					
	Project 1	Project 2	Project 3	Project 4	Project 5
	Target	Target	Target	Target	Target
Financial	4	32	20	10	24
Costs	12	4	64	16	48
Time	40	12	16	16	24
IT	2	8	16	4	6
Quality	2	2	2	8	16
Total mark	60	58	118	52	128

This will all be entered in a matrix, which will have the characteristics shown in Table 17.9.

The area shaded in dark grey represents the highest risk, and that in light grey with the low numbers represents the lowest risk. We can therefore assign a numerical value to each goal.

Numerical scale representation can be done in a linear fashion such as 1, 3, 5, 7, 9, or a non-linear fashion such as 0.5, 1, 2, 4, 8. By means of a non-linear scale, we aim to avoid high-impact risks to try and take advantage of opportunities with high impact even with low probability of occurrence. The more complex use of non-linear scales requires more detailed knowledge of the meaning of each numerical value and the reciprocal relations (Table 17.10).

In the same way, we can linearise the values of the elements to be analysed in the project and, while maintaining the critical perspective of the table, we can have a graphical representation such as the one in the figure where the major critical points are shown, which need to be *migrated* to more manageable areas. The quality check is shown on a matrix-based graph represented in Fig. 17.13.

In making decisions, various solutions are applied, such as the decisional criterion of probability, which associates to each system status the corresponding probability of occurrence.

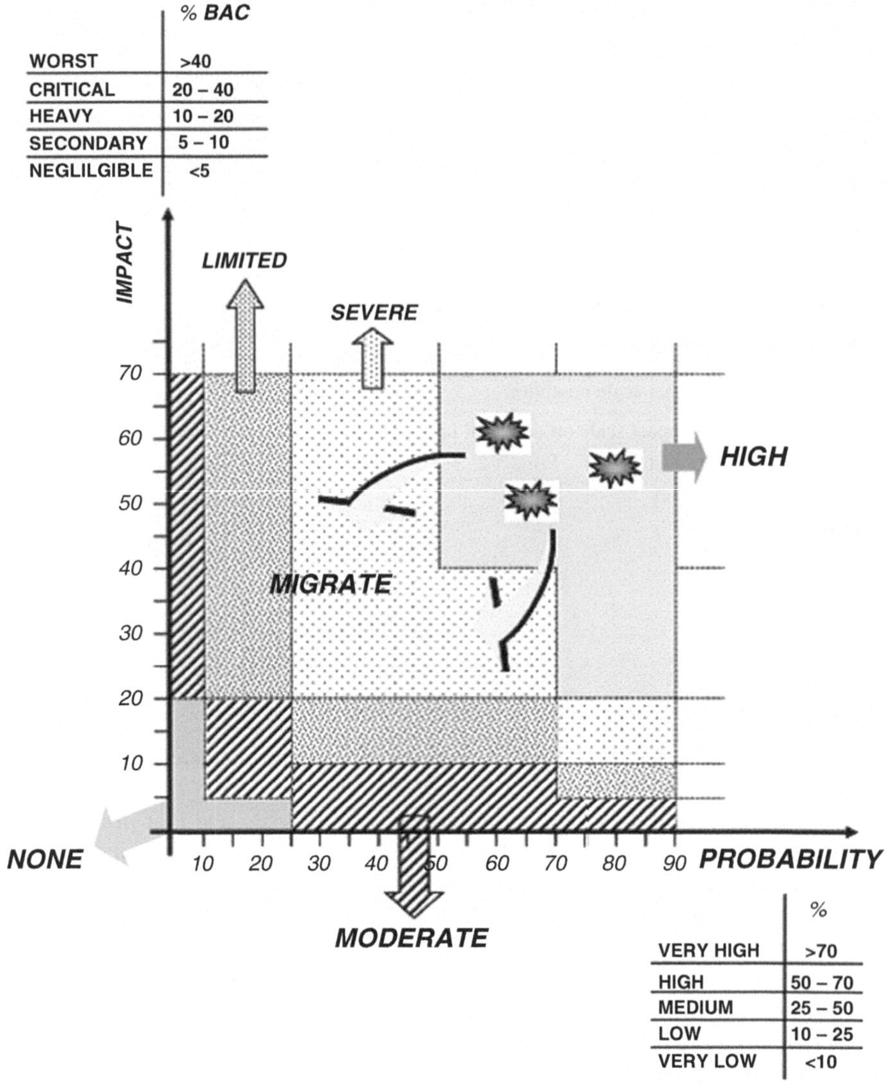

Fig. 17.13 Matrix of risks

We will then have to introduce an executive procedure to assist us in making decisions through a diagram that defines a decision tree showing circumstances, decisions, events, probabilities, terminal nodes—all connected to definable routes.

In creating a structured risk analysis we will try to provide the company with simulation techniques to properly quantify the risks with a risk assessment tied to a time variable and by checking the effects on the entire duration of the project (or intermediate phases) of alternative work cases applied to the project grid.

Table 17.11 Risks: definition of variables

Definition of components and variables	
Components	Variables
Financial	• Down payment
	• Progress payments by event
	• Payment at goods readiness
	• Payment at delivery DDP
	• Payment at hot startup
	• Payment at FAC
	• Exchange rate
	• Payment and performance bond
	• Penalty for late delivery DDP
	• Penalties for performances
Time	• Delivery engineering
	• Receiving rough material first lot
	• Preliminary foundation engineering
	• Final foundation engineering
	• Receiving main rough material second lot
	• Equipment ready for shipment
	• Delivery equipment DDP
	• Hot Startup
	• Finale Acceptance
IT	• Production round, angle, channel
	• Performance on angle, channel, flat, round, rebar
	• Productivity CLS of stacker
	• Cut shear tolerance
	• Straight tolerance
	• Bundles head alignment

Therefore, in applying alternative hypotheses we could take various routes that hook up to the grid resolution in PERT if it was used for planning. We also have to prepare a questionnaire (for each type of project) containing the main risk sources, components of each source, the variables that characterise each component and the dimensioning metric of each variable. We could, for example, consider the total risk assessment of a project, consisting in the supply of a finishing line for a bar rolling mill supplied by one of the companies of the group (Table 17.11).

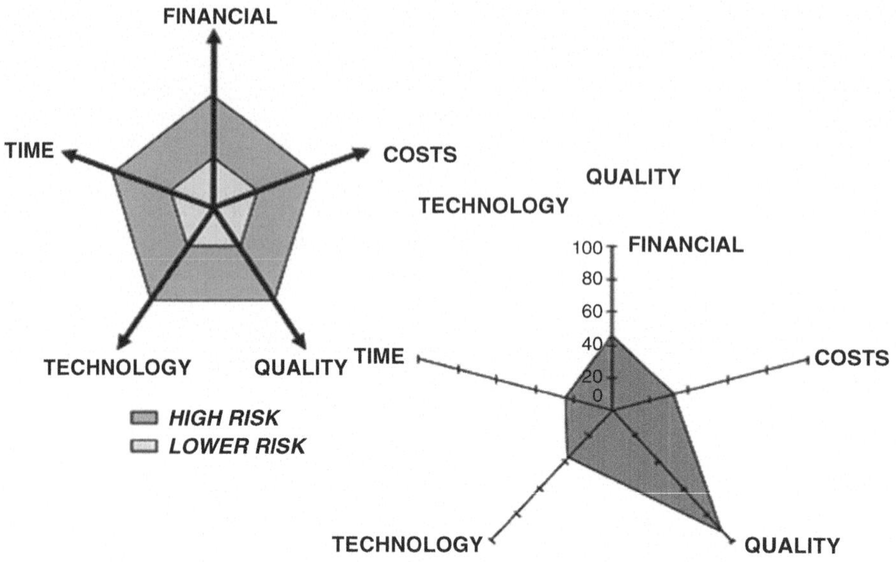

Fig. 17.14 Radar risk configuration

This is also applicable to all the other components in order to obtain the project risk RADAR program configuration as the sample in Fig. 17.14.

18 Project Execution

18.1 Project Manager Role

While everybody theorizes the need of continuous improvement, the high-performance companies are the only ones that fully implement it. Because they consider the ability to change as a key skill to achieve excellence.

At Danieli, the earlier sentence is a boost for development and a continuous challenge for everybody.

To keep constantly in line with market demands and remain always active in the change management by espousing the sense of velocity, Danieli is continuously updating its quality standards and procedures by means of its quality manual on management.

When we present our Project Manager activity outside the Danieli Company, we have to start from the concept of work and engagement *according to job*, by putting forward as an example the execution of a job within the complex Danieli system (Fig. 18.1).

In literature, you can find plenty of definitions concerning the Project Manager role and the characteristics required to hold it.

In the job practice, the Project Manager role can be summed up as follows:

- *Contract manager*: the PM is the person who manages the contract and transforms it into clear directives to be given to the various company functions. Actions are taken up by the various company functions on the basis of their respective competence.
- *Resource manager*: the PM defines the organisation chart according to Project needs, by picking out the roles and the relevant competence. He is responsible for the correct coordination of the resources assigned and shared during the whole Project execution phase (the role representation must preserve the existing

This chapter has been written by **Flavio Cozzi**, actually Member of the Board of MME Gmbh, Director Projects, Sales and Marketing Manager—former Danieli's Vice President Projects, on behalf and with permission of Danieli Group.

S. Tonchia, *Industrial Project Management*, Management for Professionals,
https://doi.org/10.1007/978-3-662-56328-1_18

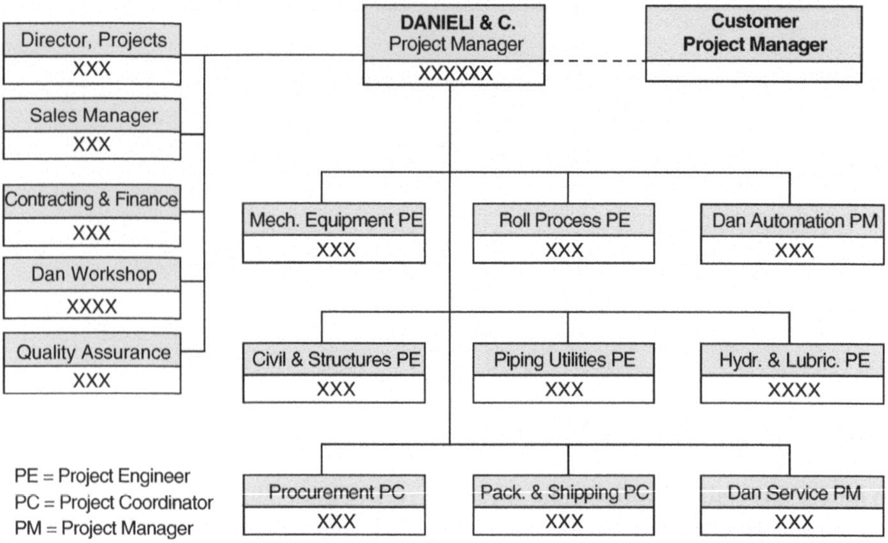

Fig. 18.1 Project organization chart—a common chart that gives the typical tree organisation of team key personnel for a turn-key project

structure, which generally tends to be a *weak matrix* structure, where therefore it is impossible to command the project resources and so their management, too).

- *Project planning manager:* with the support of the assigned functions, the PM defines the general job planning according to Project requirements, keeps monitoring the deviations and agrees on the necessary corrective actions with the various bodies making part of the operating cycle.
- *Person in charge of the profit and loss account for the job:* the PM is responsible for the profit and loss account for the job and for the relevant budget allocations. Together with the competent departments, he defines the purchasing strategies and participates in all the main decision-making processes for the achievement of the Project's economical objectives; he also carries out the project risk analysis.
- *Danieli representative vis-à-vis the Customer:* the PM is the Danieli representative vis-à-vis the Customer. He must therefore represent the Company's interests vis-à-vis the final Customer on the one hand, and take care of the relations with the Customer on the other, in order to obtain the Customer's satisfaction and esteem for Danieli.
- *Person responsible for the Project vis-à-vis the Company Management:* the PM reports to the Product Line managers and top managers for reviews, approvals and problems of transversal nature. The PM draws up concise reports on the job progress to be sent to the Company Management on a monthly basis.
- *Person in charge of the job-closing phase*: the PM collaborates with the sales department in the final negotiation phases by discussing the general conditions of the Contract with the Customer. The PM closes the job by drawing up the job closure report in order to point out its improvable aspects.

The PM plays a fundamental role within the company organisation, whose results mostly depend exactly on how the PM carries out its function.

From a know-how point of view, the PM's knowledge can be assimilated with that of an entrepreneur, whose activity is the *systematic management of a complex, single and short duration concern, aiming at the achievement of a predefined objective, through continuous planning and monitoring of differentiated and limited resources, with interdependent time–cost–quality constraints.*

The ability to organise everything in view of the result, to take decisions and to lead to the right direction, to negotiate with suppliers and/or customers and/or institutions, to understand the basic financial rules (L/C, bonds, guarantees, payments, etc.), as well as the knowledge of the project phases (i.e. until delivery to the Customer), are the essential requirements for a good PM.

The PM is the *Process Owner*, i.e. the person responsible for the performance, continuity and improvement of the process.

To understand on what is based the PM's work and which are the elements that he must strictly monitor, it will be necessary to explain the internal management tool that gathers together all the project management data.

Here follows a summary of the procedures to be executed for the project management.

As mentioned earlier, the JET (Job Enterprise Target) is the reference tool for all the business units at Danieli. It is also the updated document to obtain filterable *up-to-date* and *time-now* reports. All the data are withdrawn from one single IT system, One World, into which all the in-progress data are poured.

The PM is in charge of confirming the JET0 exploded that was issued during the first phase of initial planning. Today this phase involves different verification and checking phases prior to the final verification and confirmation of the project budget.

We already know that the origin is at offer stage. Then, at verification stage, the internal document is first compared with the contents of the contract itself, and secondly it is compared and agreed with the design and production technical departments, each for its own competence; finally, the updated document is checked as regards costing to confirm the project budget.

The macroeconomic figures shall be referred to the following:

- Internal Manufacturing Product Budget (IMPB)
- External Manufacturing Product Budget (EMPB)
- Technical Office Product Budget (UTPB)
- Service Budget (MACPB)
- Packing and Transportation Product Budget (PTPB)

Subsequent to the earlier, and to the internal agreement on the time schedule, the document is officially internally acknowledged and becomes the working basis in terms of time, costs and invoicing.

We would better though first deal with those process considerations that have led to the reorganisation of the PM micro-activities within the job process, considering the key activities, the possible monitoring of the performances and finally the identification of the low added value activities.

Let us define, for instance, the key activities as priorities for the PM. These can be aggregated into three macro activities (processes): Customer's satisfaction, follow-up and achievement of payment collection, job setup and follow-up.

On their turn, these three macroactivities are subdivided into several other ones; i.e. we can say that as regards customer's satisfaction, we will have certainly achieved the objectives when there will be timely answers to fax and e-mail messages; prompt decisions, leading the system to be receptive to the Customer's problems; actions directed at raising the level of familiarity and confidence, and therefore urging to act with professionalism and fairness; clearness in the presentation of the extra contract (time and costs) so as to make sure that its contents are understood and shared. Finally, the PM must represent the Customer within the Company and therefore must have very frequent contacts with him, by using any available communication means, such as e-mail, web cam, videoconference, meetings.

As far as payment collection is concerned, we know that the follow-up of the job milestones linked to payments is essential, as well as the ex ante analysis of the possible causes of delays in payment collection and the definition/implementation of the relevant countermeasures, besides the creation of all the elements that can make advance collections possible, if any.

As regards the job setup and follow-up, we have already largely pointed out the importance of the JET2 issuing, and then of the budget costs follow-up, as well as of the lead times follow-up (checking and updating of UT listings, order issuing, work done for UP, shipping, erection, startup, PAC, FAC, job closure meeting).

All these are related to the possible process performances (KPI) and their measurement, which could then be linked to deviations and payment collections, to recovery from the customer of the project extra costs due to changes made during project execution on Customer's demand, to JET issuing lead times, to recovery from suppliers of extra costs due to non-compliant supplies and consequently the relevant debiting for collection.

A series of activities that are carried out by the PM but turn out to be instead PE (Project Engineer) activities, or *low added value* activities, which occupy in an abnormal way the resource and its time, so affecting its actual effectiveness, have been analysed in detail in order to find implementable solutions. The activities considered as low added value ones *do not contribute significantly to the achievement of the PM performances* and more specifically, as mentioned earlier, they subtract time to the key activities, so hindering the attainment of the objectives.

For instance, the PM had to manage the listings of the contractual spare parts, and was in charge of the total activity management. So the whole was shifted to the existing Spare Parts Department (UCR), since UCR is the most appropriate body to draw up the list and forward it. Measures were taken also at JET level to correct the flow of activities.

In the initial planning analysis, the PM also managed the *flows* concerning production components. It was decided to pass this activity directly to the production department, which is specifically acquainted with these activities.

The PM was in charge of issuing all the Material Requests to the Purchasing Department, also in relation to Technical Specifications issued by the Technical Department. It was decided that this procedure could be launched automatically by the Record Office, after the PM had checked the UP dates and the contractual vendor list, also because the technical specification is drawn up by UT. Therefore the MR could be loaded automatically by UT, with at most a check made by the PM, since the technical specification may contain some terms that can be identified only by the PM, such as delivery of sub-suppliers' documentation and guarantees.

Others have been solved by streamlining the PM function, so that it could be dedicated to the actual management of the project in the strict sense of the word.

Different matters altogether are claims to suppliers, which fall under the PM's priorities since they have an influence on the job's economic result. A claim process has been defined that starts from the Nonconformity Report (NCR) as input, and must produce the debiting bill as output, through a step-by-step procedure involving the following: loading of the NCR onto the system, (documented) FAX of request for intervention or replacement under guarantee, evidence, if necessary, of the *prompt action* process in case of non-intervention on the supplier's part, notification to the supplier of the activation of the Danieli prompt action process to his charge, notification of the documented costs of the intervention, invoicing (agreed with the supplier). This carries the involvement of various resources, such as the job site issuing of the NCR, the DS secretary's office for the NCR loading, the PM for the notification to the supplier and the ATE/LE/PM departments for agreement on the invoice.

Quite interesting is the development and study at organisational design level of the Claim Manager figure whose position within the PM department could be as part of the staff of the various teams. The CM works according to job item and receives inputs from the PM (job references, item under discussion, intervention lead times). The CM notifies the NCR to the supplier (and sends copy to the ATE/LE department), so that the latter can take action and/or replace the piece under guarantee. Notifications are made by means of standard forms and these apply to all claims. The form is agreed with the ATE/LE and the legal departments. According to the supplier's answer, the CM discusses with the PM whether to activate the prompt action process and in parallel the claim one. The CM discusses with ATE/LE and the *prompt action executors* to define the (achievable) objectives of the recovery. If necessary, he meets the supplier and completes the debiting by charging the invoice on the relevant job item. The CM does not carry out prompt actions.

The first three points are easy to implement. As regards the fourth, the key factor that determines the introduction or not of the CM figure is the involvement or not of ATE/LE.

The CM profile requires experience in negotiating with suppliers, ability to use the IT system to find references for the claim and determination.

ATE/LE must be involved by introducing the total recovery from the supplier among the KPI (unless already included).

The tools used by the PM to tackle the project execution are therefore the JET, the Contract, the MPL and the general and detailed time schedule.

For the project execution, it will be necessary to take into account all the phases related to the inshore part, such as engineering, procurement, production, expediting, packaging and shipment. Moreover, the control procedures of all the planning phases shall be defined, the coordination of the resources of the team dedicated to the project shall be implemented and the interfacing and monitoring task shall be extended to the subsidiaries and their parallel internal structures.

The inshore phase shall end by checking the goods shipment and delivery on site where, according to contract, either the Customer or the Danieli job site structure shall be in charge of checking the compliance of the arriving goods with the documentation through the Open Package Inspection (OPI) procedure, on which basis the on-site receipt and storage certifying document shall be drawn up.

For the storage on site, the customer or the Danieli job site structure shall be given previously all the references concerning storage requirements and relevant procedures, in order to avoid damaging the goods before the installation phase (Figs. 18.2 and 18.3).

18.2 Project Monitoring

Each company, according to its own product, can develop suitable systems for collecting the necessary data so that they can be properly analysed. Therefore to analyse the project in all its main phases, some sub-phases have been created at a lower level, which fall directly within the competence of the single business units. At the same time these sub-phases are very important to determine the correct flow of project documents and to set the basis for the upper levels and be, if necessary, detail instruments. Of course we must avoid stiffening the bureaucratic procedure. So we shall only give an example of the data that must be monitored in progress by the PM during job development.

As far as the Cost Control plan is concerned, we shall have to focus attention on cost vs. financial accounting; job costs; input of costs and relevant descriptions; target cost, committed cost, comparative estimate at completion; extra cost; non-JET costs, meant as figures exceeding budget values; summarising job cost listing, as both analytic and summary statement.

As far as the Time Control plan is concerned, we shall make sure to observe strictly target dates—JET sub-phases; document planning for design; value and quantity progress; order issuing referred to the dates and requests of execution; expediter relations with suppliers, sub-phases for UPP (Production); release for shipment—relevant sub-phases; monitoring program; Project Review Meeting and final acceptance.

To meet the financial requirements, we shall have to be *on line* as regards the Cash Flow Control, i.e. cash in versus cash out, technical milestones for collection,

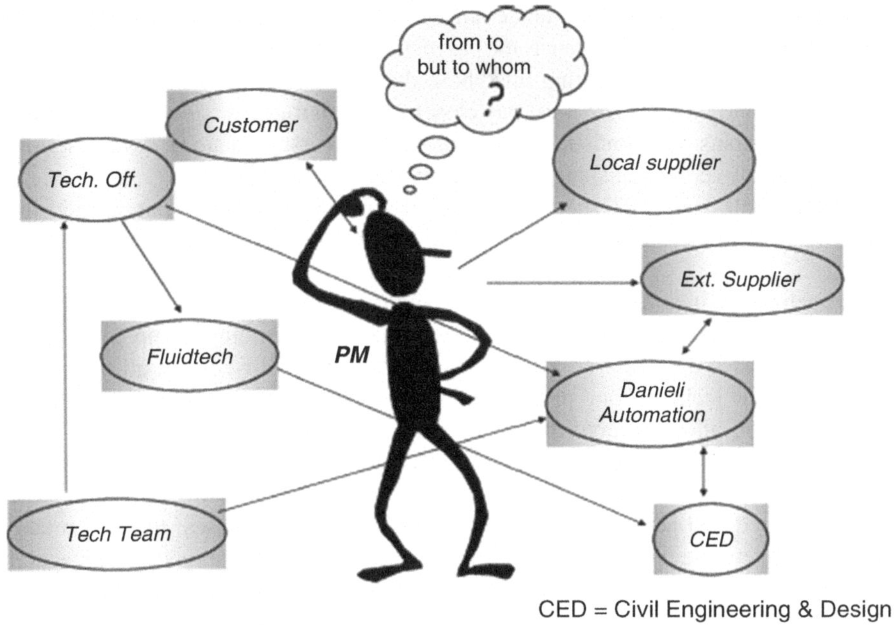

Fig. 18.2 Engineering document—from to but to whom

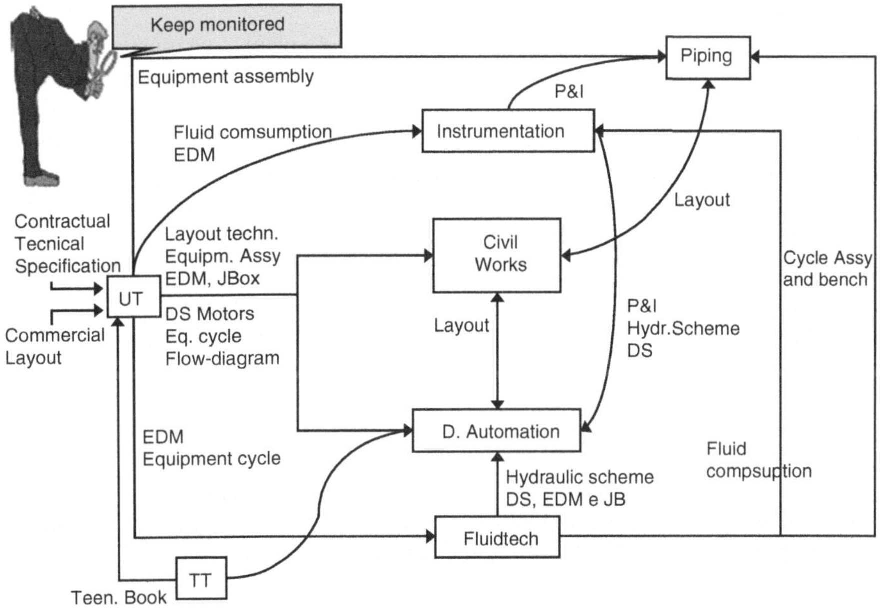

Fig. 18.3 Between technical offices—keep monitored

possible causes of postponement and countermeasures, follow-up with the Financial Management and invoice clearance for payment of suppliers.

We must not forget that during this phase the PM will be the sole interface with the customer, and therefore he must pay closer attention to forms of communication, cooperative versus antagonistic approach, budget versus customer satisfaction, progress control versus transparency and claim management.

The PM shall use the IT means to withdraw from the company database in the form of standardised reports whatever necessary to the project management activities.

For each single item described in the JET and in the MPL, the database list should consider the codified price group, which determines the invoicing group conditions; the filing status, if any; reference to the contractual WBS; of course, all the coded references of the technical and production departments that will handle the item and which therefore allow to confirm the coding of the determined flow; individual weights indicated in the JET; the target, expected and actual dates of the various design phases; the percentage progress status of the design of a single machine; the target, expected and actual dates of the various production phases; the percentage progress status of the production of a single machine; the target, expected and actual dates of the various works done and shipping phases; the shipment percentage; the JET cost; the invoicing price; the value of the invoice made out; the invoicing percentage.

Yet, this verification involves a detailed analysis of the single job, and it is certainly demanding if used as a reference for verification of the planning and of the project schedule. We shall therefore carry out, if necessary, a general monitoring via MPL.

For this reason, in case of complex projects or whole plants, we must create more standardised and more flexible management models by using the common planning tool named PRIMAVERA.

Let us therefore return to the master plant list that subdivides the plant into *areas* through a standardised coding. On each area we shall enter the equipment listed in the JET with all the relevant reference data, such as weights, costs and dates.

The progress monitoring concerns both the technical phases and the economic trend.

As regards the technical phases, the detailed monitoring concerns design, procurement, production, shipment, erection, startup and commissioning.

The Project Manager builds up the master plant list referred to the plant areas to link the machines to the plant.

The internal IT system will update the design document planning, and the monitoring will take place both at JET level for the single detail items, and in the total computation of the master plant list and of the relevant general job schedule.

The updating draws data from the company database according to the agreed needs, and then calculates and restates the various phases.

The replanning phase, if any, is highlighted by the single business units, which are in charge of updating the system and then informing the PM. In any case, replanning takes place at the end of the necessary verifications and consequent analyses and recovery attempts. However, we must pay attention to the risk involved in these type of activities assigned to the single business units, which might create a *discrepancy* between business units, and limit this activity to a *bureaucratic* system action without the necessary verifications and analysis of the impact that the replanning could have on the project or on one of its phases. This problem, to be considered in any case as relative, derives from the matrix structure and must be limited as much as possible through weekly meetings to be held within the company with the participation of the representatives of the different business units and always of the PM, who coordinates meetings and activities.

At planning stage we have created a document that is easily comprehensible within the company and for the customer as a report, and gives an effective description of the data we want to present.

The subject of reporting and project monitoring has been analysed within the company through a dedicated project work that pointed out the need of adapting reporting to the project manager activities.

Now let us explain briefly what is the internal IT tool utilised for the purpose.

Danieli started to implement the new One World IT tool developed by JD Edwards from January 2003. The tool target was providing an enterprise framework sufficiently flexible and configurable to face the continuously changing technology. In general, apart from the *adaptability* (to different languages, currencies, reporting provisions and technical standards) and *user-friendly* features that all software normally claim to have, this tool offers the possibility of multi-platform calculation; i.e. data exchange and calculation can be made across different platforms. Actually people in the company often import data from the system and work with the data in the way they feel comfortable with.

The interoperability function enables to take advantage of the existent network and reserves potential for integrating third-party products. After the implementation of One World, the previous system was utilised as a basis for the new structure. One World succeeded to inherit the functions of the previous system by keeping costs and weights on track.

The integrated supply chain allows One World to provide the ability to use the Internet or Intranet for internal and external communications. Information can be shared inside the company as well as with suppliers and customers.

Danieli had taken the first steps in this direction by providing standardised reports in a very small scope. With the launch of One World, this unique system became of common use throughout the company and its core part, i.e. the JET, became the basis of daily activities.

Of course, the monitoring system could end in a technicism applied to the project, and therefore limit itself to a cold data analysis.

On this basis, instead of the previous analysis, now the PM is responsible for the data analysis in order to get a time-now picture of the situation. Therefore, since he knows the project targets, he is the one who can activate warnings, if any, and propose and implement corrective solutions to possible malfunctioning over the course of the project itself, by always posing to himself three fundamental questions: where are we, where should we be, where are we going?

Then it will be possible to obtain an *S* curve of the project as shown in Fig. 18.4.

CLIENT AG SIDERURGICA BALBOA — OVERALL PROJECT — CUTOFF DATE 29JUL2005
JOB DP06 BALBOA — ENGINEERING-PROCUREMENT AND MANUFACTURING-PACKING/SHIPMENT/DELIVERY — DATE 23AUG2005

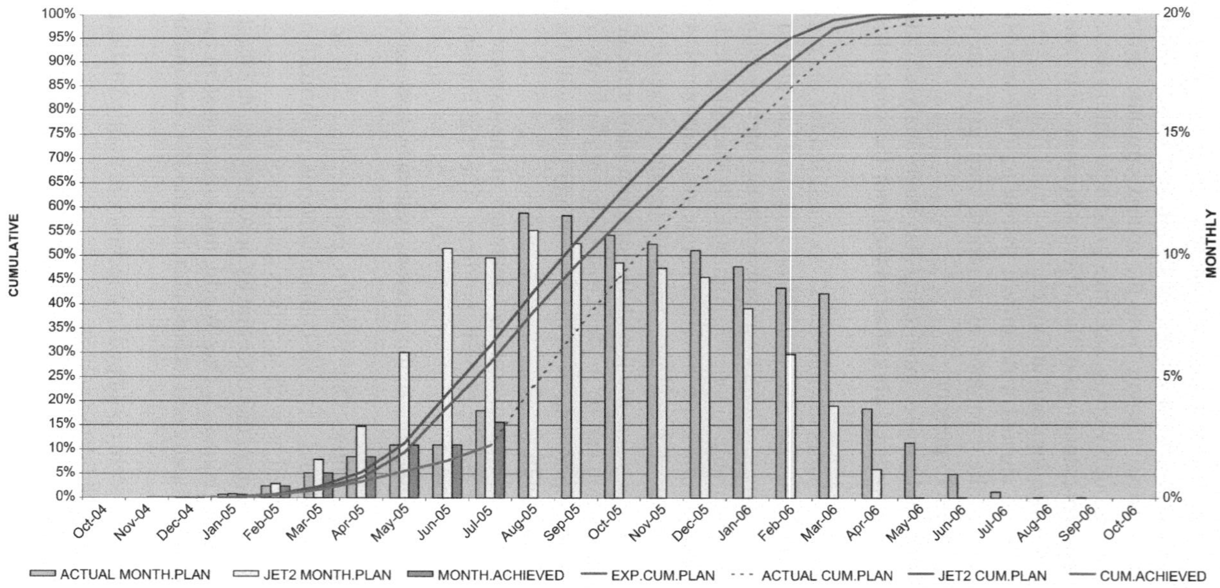

OVERALL			-2	-1	1	2	3	4	5	6	7	8
JOB DP006B	WF 100000	INC 100.0%	Oct-04	Nov-04	Dec-04	Jan-05	Feb-05	Mar-05	Apr-05	May-05	Jun-05	**Jul-05**
JET 2	MONTHLY PLAN			0.0%	0.0%	0.2%	0.6%	1.6%	2.9%	6.0%	10.3%	9.9%
	CUMULATIVE PLAN			0.0%	0.0%	0.2%	0.7%	2.3%	5.3%	11.3%	21.6%	**31.5%**
EXPECTED	MONTHLY PLAN			0.0%	0.0%	0.1%	0.4%	1.3%	2.3%	5.2%	9.4%	9.1%
	CUMULATIVE PLAN			0.0%	0.0%	0.1%	0.5%	1.9%	4.1%	9.4%	18.7%	**27.9%**
	MONTHLY ACHIEVED				0.0%	0.1%	0.5%	1.0%	1.7%	2.2%	2.2%	3.1%
	CUMULATIVE ACHIEVED				0.0%	0.1%	0.6%	1.6%	3.3%	5.5%	7.7%	**10.8%**
ACTUAL	MONTHLY PLAN				0.0%	0.1%	0.5%	1.0%	1.7%	2.2%	2.2%	3.6%
	CUMULATIVE PLAN				0.0%	0.1%	0.6%	1.6%	3.3%	5.5%	7.7%	**10.8%**

JOB DP006B		9	10	11	12	13	14	15	16	17	18	19	20	21	22	23
		Aug-05	Sep-05	Oct-05	Nov-05	Dec-05	Jan-06	Feb-06	Mar-06	Apr-06	May-06	Jun-06	Jul-06	Aug-06	Sep-06	Oct-06
JET 2	MONTHLY PLAN	11.0%	10.5%	9.7%	9.5%	9.1%	7.8%	5.9%	3.8%	1.2%	0.0%	0.0%				
	CUMULATIVE PLAN	42.5%	53.0%	62.7%	72.2%	81.3%	89.1%	95.0%	98.8%	100.0%	100.0%	100.0%				
EXPECTED	MONTHLY PLAN	10.5%	9.7%	8.9%	8.8%	8.8%	8.2%	7.5%	6.8%	2.0%	0.7%	0.3%	0.1%	0.0%		
	CUMULATIVE PLAN	38.4%	48.1%	57.0%	65.7%	74.5%	82.7%	90.1%	97.0%	99.0%	99.7%	99.9%	100.0%	100.0%		
	MONTHLY ACHIEVED															
	CUMULATIVE ACHIEVED															
ACTUAL	MONTHLY PLAN	11.7%	11.6%	10.8%	10.5%	10.2%	9.5%	8.7%	8.4%	3.7%	2.3%	1.0%	0.2%	0.0%	0.0%	0.0%
	CUMULATIVE PLAN	23.0%	34.7%	45.5%	56.0%	66.2%	75.7%	84.4%	92.8%	96.5%	98.7%	99.7%	100.0%	100.0%	100.0%	100.0%

Fig. 18.4 Progress calculation

This graph represents the progress according to the JET of the Actual Monthly Plan (AMP), Cumulative Plan on an expected basis (CPE), Monthly Plan (MP), Cumulative Plan (CP), Monthly achieved (MA) and Cumulative achieved (CA).

On the basis of the information, Actual is updated on the MP and the CP.

If we want to go into detail, then we shall refer to the MPL and the data contained in it and already represented in the initial planning picture, which shows in its graphic representation of the job progress all the typical elements of a time bar chart. It is therefore possible to check connections, define the critical path, and modify, if necessary, flows and time schedules.

This is a tool that is always available for use also vis-à-vis the final customer, so as to be able, if necessary, to show data to the customer in a professional way.

What must never be neglected, since it is essential for the success of the first part of the project, is the link between the MPL and the job site activities related to the erection of the equipment and systems, and to their startup.

Therefore, variations must be weighed up not only with regard to the workload, or in production or financial terms, etc., but also by analysing their consequences and impact on the future job site activities. It will be also necessary to check the possibility to modify the operating sequences and the various activities in time without missing the objective or failing to achieve the contractual target (Table 18.1).

As regards project monitoring and internal reporting, Table 18.2 later gives an idea of how to tackle the problem of company reporting with the consequent creation of standardised models that can be filtrated at various levels.

As a result, on the basis of these indications, a filterable report must be drawn up, based on the completeness of the requested and necessary data for the PM's reading.

The aforementioned should in any case be accompanied by all the necessary data to be filtered according to the PM's needs in the represented details (Fig. 18.5).

18.3 Onshore Phase

The site is organised in such a way as to carry out all the installation and startup phases of the supplied plant (Fig. 18.6).

To this end, Danieli has dedicated business units (MAC) that follow the process step by step and handle Danieli–customer relations on site.

We have already mentioned that the MAC PM takes part in the Job Order Review and initial planning. The MAC PM will be in charge of some operating stages that we can summarise in three macro activities, i.e. Executive planning, ongoing work management and closing of job site. To all intents and purposes, these activities will be managed under a job site *suborder*.

The Project Manager coordinates the activities of the MAC PM and maintains contacts with the Technical Departments to check the installation, engineering and plant systems in general, contacts with the production department for machine quality and with the purchasing department for orders with sub-suppliers. Furthermore, the PM will act as watchdog for extra costs deriving from modifications to be

Table 18.1 Derogation form—check and authorize

| DANIELI | CHANGE / DER
OGATION REQUEST FORM | |
|---|---|---|
| | Project name : | Project ref : |
| | Date of opening : | Sheet # : |

Modification demanded by	☒ Customer	☐ Sub-contractor: :....................	☐ Danieli internal : ....
	☐ Customer's engineer	☐ Co-contractor :......................	☐ Other: ....

MODIFICATION REQUEST SOURCE with references of the documents : mail, meeting minutes...
REMINDER OF SERVICES PROVIDED FOR IN CONTRACT with references of contractual documents :
DESCRIPTION OF THE REQUESTED MODIFICATION:

TECHNICAL/COST CONSEQUENCES (list each time the references of affected documents and, for installation, the concerned zone)				
Studies	Supplies / Spare parts	Installation	Tests / Performances	Delay

WRITER of the sheet		PROJECT MANAGER SIGNATURE	
Name :	Entity :	Name :	Entity :
Function :	Signature :	Function :	Signature :

Project manager's decision	External circulation to : Date :...............	For obser-vation	For accep-tance	Internal circulation to : Date :...............	Actions to be conducted :
		☐	☐		
		☐	☐		
		☐	☐		

Acceptance decision / External addressee's observations	Company : Name : Function : Date : Signature :

Table 18.2 Project report requirement

Stakeholders	What they care	What they need
Top management and executives	Company development, profitability, reputation	Project status report, issues
Director	Project performance	Project status, cash flow, earned value, issues
Executive Project Manager	Time, cost, quality, cash flow	Task status, cost, problems, issues
Project Manager (support staff)	Time, cost, quality, cash flow	Task level details, overall project status
Team members from various departments	Schedule, workload	Overall project status and department details
Client	Schedule, quality	Project status, schedule, quality, selected details
Suppliers	Exchange of information, schedule	Delivery target dates, technical details
Information Management	Database update	Dedicated information on project

done on site, making sure any anomalies are properly documented to prevent them from occurring again in the future and, if necessary, to contest any non-compliant supplies, to safeguard the activities the company carries out for the customer in full compliance with the contract and respect for the relationship of trust with the customer.

There could be differently structured organisation charts depending on the type of supply and the type of contract signed with the customer. The most complete and complex organisation chart is in the case of turnkey projects. Therefore, these types of documents will always be used to indicate the activities, the requirements and the management parameters for on-site work.

This is always based on budgeting that is also indicated in the JET. The PM must continuously update any ongoing variations that could impact on planning in terms of deadlines and budget.

The budget is determined by analysing a series of standard parameters such as the machines in the scope of supply; the machines with external technology, which will consequently have erection and supervision *packages*; the contract time schedule with timing of engineering, manufacturing, purchasing, erection, commissioning and performance tests; the regulations in force in the customer's country; any standards that may differ from Danieli standards and that have to be applied; special logistics dictated by the environment in which we will be operating (desert, equatorial or Siberian environments must be carefully studied to assess work execution); specific necessary means, quantities and weights of machines, electrical and automation equipment and piping, and, if possible, according to the various types of fluids; quantities expressed in weights, lengths (if necessary) of cables and cable trays with all the specified piping conduits.

Main data	Contract N°		Client		Country		PM		Month	
	Supply						Currency			
	CIF		PAC		FAC		EPM			
	Payment form									
	Gurantee				Penalty					

Cash In			CIF	Eng LC	LC	Sequence Mtg1	Sequence Mtg2	Cold test	PAC	FAC	Gran Total
	Exp Cashin	Date									
		Amount									
	Act. Cashin	Date									
		Amount									
	Delay (days)										
	UTD Variance		%								

Project performance	Tach Group	PV	CIF	AC	CV		BV		CN	BN	
		budget	Coming intoforce	Actual Cash	Ant. CIF-AC	%	Ant. CV-BV	%	BV/AC		%
	Month										
	total										

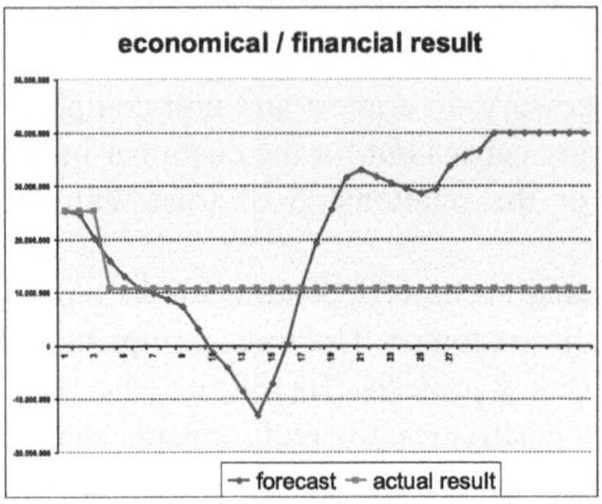

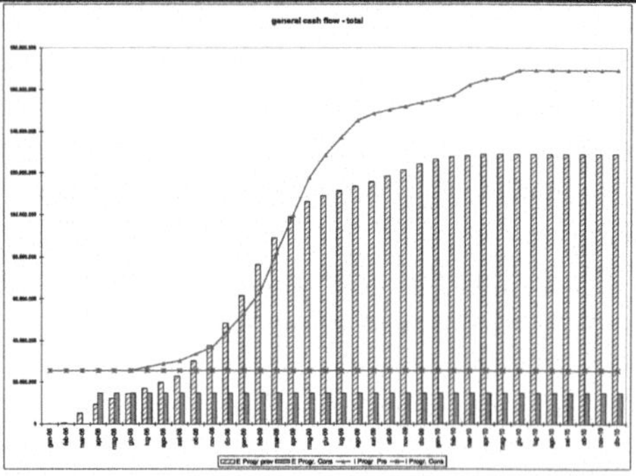

JOB	COSTS(€/1000)				
	TOT. BUDGET	USED	TO GO	TO COMPLETE	DELTA %
DP02DH01	2601,80	2355,00	248,60	2603,60	0,07
DP033J01	4785,90	401,90	4389,40	4791,30	0,11
DP004U01	6410,10	5503,10	246,10	5749,20	-10,31
DP02AN01	5778,10	4462,30	755,70	5218,00	-9,69
DP00Z701	2537,10	2072,20	270,50	2342,70	-7,66
DP00Z801	234,00	292,70		292,70	25,09
DP02X801	2981,60	192,00	2788,30	2980,30	-0,04
DP00D701	3731,30	3173,60	277,10	3450,70	-7,52
DP00D702	3214,00	2534,40	120,90	2655,30	-17,38

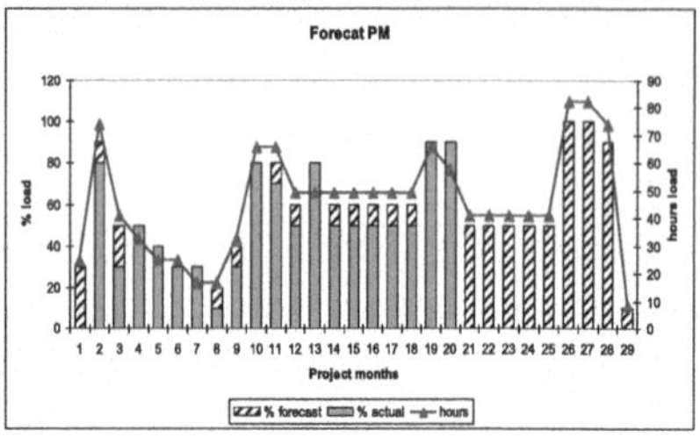

Project Status	□ On track	□ Some issue	□ Out of control
Opportunities			
Issues			
Action to be taken			
Action follow up			

Fig. 18.5 PM report

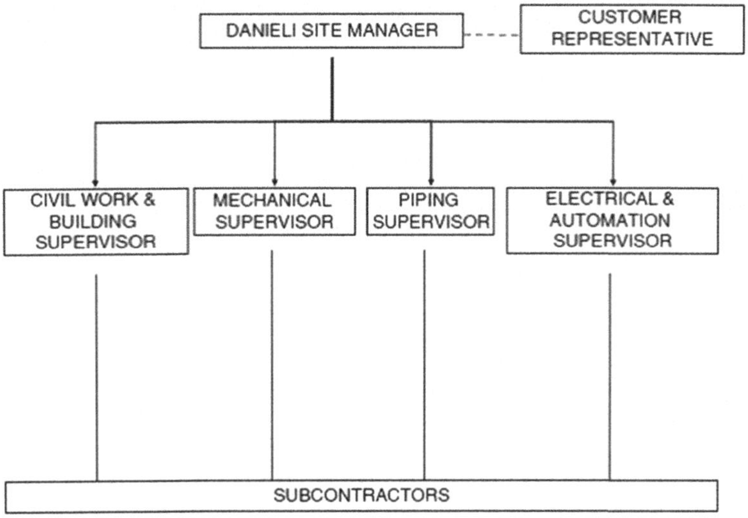

Fig. 18.6 Onshore OBS

To enter in the executive planning stage, before starting the on-site activities, the PM has to make sure that the original master schedule for the site drawn up in the initial planning stage is updated and then has to create a site organisation chart.

Therefore, executive planning comprises specific activities such as the issue of a project schedule to be defined with the customer, or with the erection companies involved in the project; the Site Manager must collect all the necessary information to properly carry out the project on site, transfer the information concerning changes to the project during design or manufacturing, information specific to the delivery of materials to the job site and means of transportation; it consists of preparing a Summary Budget Cost Diagram (SBCD) that needs to be updated constantly to keep the expected installation and supervision costs under control.

If plant erection is not assigned to Danieli's Construction International company (DCI), an invitation to tender must be published to select the erection company. This is normally done long in advance the on-site activities are to begin for obvious reasons of preparation and organisation.

If we analyse how to properly carry out the initial planning, we also have to consider simple elements, which are often neglected and considered at the last minute when a budget warning has already been given. Therefore we will try to provide a guide for these specific activities that refer to management of on-site activities and then to job order management. It is interesting to note that detail activities can impact on the contract with the risk of extra costs. For example, in some contracts a delayed arrival on site of the technical supervisor is penalised economically, and this creates a type of management that focuses on availability of personnel, which always has to be agreed with the customer.

The job site OBS has to define the use of personnel based on the type of machine in order to use the more specialised resources specific to the product. We can thus

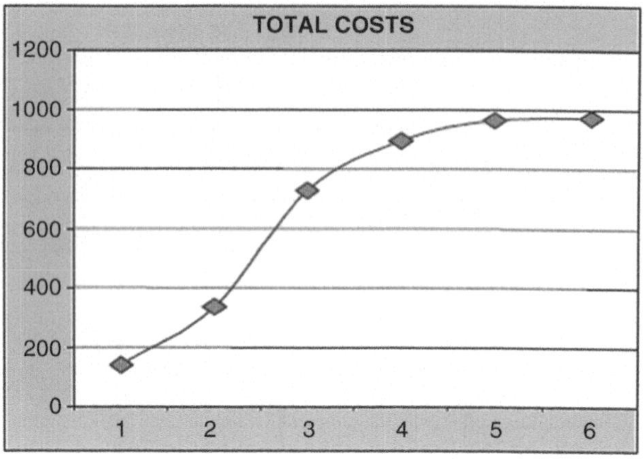

Fig. 18.7 Total costs on site

draw up a site organisation chart that we will submit to the customer as the structure that represents Danieli's activities within the timeframe defined for installation and startup.

We will collect useful, specific data such as costs for room and board; we will define all personnel mission planning data and consequently the type of airline tickets, any entry and residence visas, work permits, hotels and accommodation, including the organisation of local transportation, making sure they are correct and complete. Based on the data collected we can then come up with specific agreements to reduce costs.

The PM ensures that all the necessary documentation is gathered according to a standard checklist. The documentation will be studied and submitted to MAC for the preparatory stage of the site manager. The PM will also make the necessary arrangements to equip the site with all the possible connections to the central system so that the site is part of the company network. He will also make sure that project data such as drawings, technical specifications, non-conformity reports, etc. can immediately be transferred via computer. Finally, the PM makes sure that the Work Schedule is drawn up and that it has the typical S-shaped curve indicating the forecast cost trends that are discussed and agreed with the erection and startup business unit (Fig. 18.7).

The Work Schedule, which is based on the contract schedule, shows the macroactivities in detail and therefore contains all the elementary activities entered in the units that are part of a sub-plant, which in turn are elements of a plant unit and already in the MPL.

It is essential that in the planning stage, the availability of the equipment and its supply be carefully examined in order to optimise the availability of resources and means. For this reason, planning according to MPL is an indispensable tool because it highlights the variables that impact on manufacturing, making it possible to identify detailed corrective actions. So the PM manages the critical points that could arise, with the help of various operating units on the job order staff and must have the ability, once the critical path is determined, to guide the project with the necessary authority in order to avoid the heavy impact of possible variations, based on forecasts and not on the final balance.

The job site is required to manage the work schedule and, in complex cases, is assisted by planners who collaborate with the site manager to update the work progress report and keep the PM informed. Managing the work schedule is not a simple operation, as it requires the use of means and resources that have to stay within a cost and time budget. It is also a tool used with the customer to check the progress of the project and to determine the cash flow if the contract specifies payments related to Work Progress. The schedule must show the milestones such as availability of utilities, plant completion, beginning and end of cold tests, beginning and end of hot and performance tests.

The SBCD drawn up by the MAC PM with the JET 2 reference data contains the two main erection, supervision and startup costs with specific details and budget concerning technical services and the services of specialists, since today it has significant costs within the framework of job site costs and therefore must be carefully considered.

This planning schedule is updated monthly, directly by the site and is one of the monthly reports that the site manager and the MAC PM send to the job order PM. The data are then compared with those provided by the central system to see if there are any inconsistencies, especially in terms of planning *to go* and planning of extra costs (Table 18.3).

It is important to set up the system to receive the non-conformity reports (NCR) from the site with the procedure monitored by the PM. Thanks to the job site network connection—the NCRs can be processed in the company's integrated system, which means we always have real-time detailed critical points and the corresponding extra costs. In the planning stage, communication with the customer takes on importance as the customer must collaborate and take cognisance of the on-site project.

The company can propose project-planning schedules according to its own direct experience. An important database is the one used for returns and resources for the execution of the project.

Figure 18.8 can give you an idea of the extent of erection activities in a complex plant.

Proper planning is therefore a must but the difficulty generally encountered is complying with the planning schedule, especially in consideration of the agreed contract target, in terms of putting the plant in production and therefore return on investment. The schedule is not *fossilised*; through our resources and experience, we have to be able to modify actions *on-line*, thereby limiting the economic and time impacts, and look for the opportunities created by the changes, aiming for continuous improvement and giving feedback to the technical design departments and the purchasing department.

Management of work in progress is also specific because it requires the preparation of Danieli startup personnel and the transfer of specific information on the types of machines and processes, so that the activities can be carried out properly until the contractual performances are reached.

Table 18.3 Extra-cost customer rank—make experiences

Client	Looking for claims and back charges	Requires strictly the respect of the schedule	Requires the respect of STD much different from ours	Yield of the country very low	Presence of Union	Safety rules very restrictive	Low capability in the management of Customer activities with impact on DANIELI activities (as sample Civil Works)	Requires the strictly respect of the contract	Looking for the opportunity
C1		3	1	1				2	1
C2									1
C3									
C4	2	2	1		3	1	1	2	
C5	1	1	1	1	1	1	1	1	
C6	1	2		1				2	1
C7				1			1	1	1
C8				1			1		
C9									
C10									
C11									
C12		1					1	1	1
C13	1	2	1	1			1	1	1
C14									
C15									
C16									
C17						1	1		
C18				1			1		
C19	1	1	1			1	1	1	1
TOT	6	12	5	7	4	4	9	11	7

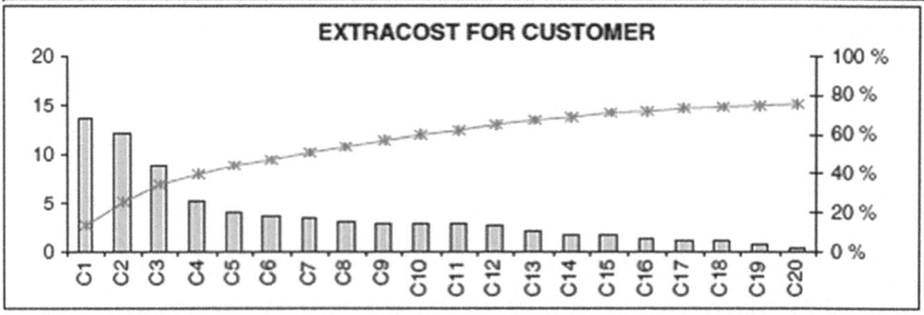

The job order PM must manage and coordinate the resources involved in the project such as MAC, technical departments, shipping staff and financial departments.

Making sure the various stages are progressing properly is not limited to a functional checklist but also includes specific technical meetings to understand the problems that could arise. It is up to the PM's organisational capacity, together with the company's standard and structural elements, to make sure that a list of

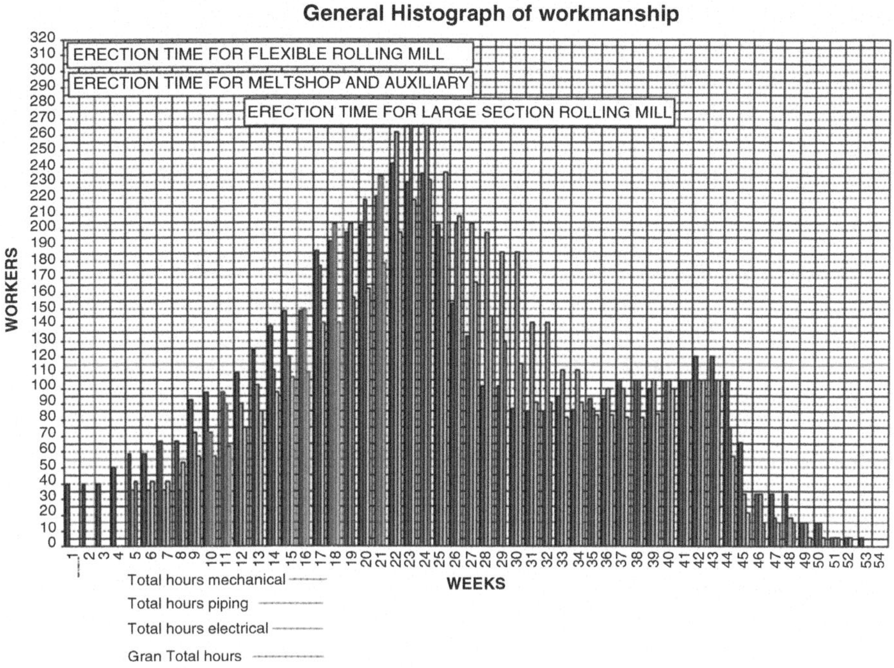

Fig. 18.8 Total man-hours

outstanding points is maintained and updated as a document of exchange with questions and answers being dealt with, requests for clarifications, the people in charge of solving problems and time schedule.

This document becomes a list of critical points and an aid in creating a job order history. It is the tool that determines the three WWWs (Who, What, When), which can be extended to include concepts and information—WWH (Where, Why, How): And finally, it makes the PM and the Site Manager responsible for solving problems.

In a job order, the opening of a job site is one of the milestones included in the project-planning schedule, because as of that moment it officially gives the green light to the actual work of the job order and the project. It is an important time for the Buyer and the Seller and must therefore be made official and monitored so that everything is done correctly and that startup complies with what has been specified, prepared and planned.

With the opening of the job site, the Site Manager becomes the operative interface between the company and the customer and therefore must be in contact with the PM, who in turn is also the customer's representative in his dealings with the company. The Site Manager and the PM must work together to prevent these problems and to quickly solve them making sure they do not go over the budget.

The Site Manager will monitor all the production activities on the job site.

With the earlier input the PM receives the output from the site, i.e. updating the installation schedule or the issue of new specific and detail schedules; the certification of the activities such as the beginning and end of erection, cold tests, hot tests, etc. and consistent accurate updating of the Summary Budget Actual Cost (SBAC) to be compared with the SBCD.

This is referred to the MAC PM since it is a management task specific to the job site, which requires a detailed knowledge of the activities involved, knowledge of direct operative resource management. In any case, this important suborder is coordinated by the job order PM, since the economic result is entered in the overall economic result of the job order.

Specifically, job site order management includes the settling in of the site manager, instruction of erection and startup personnel, monitoring of production activities and the running of the site itself.

Important aspects have to be considered in this settling in, with respect to the country and its regulations.

Depending on the contract, there can be various starting conditions depending on whether it is a greenfield plant, an extension of an existing plant or a revamping.

We have to be aware of local laws and regulations, especially those referring to safety and safe working conditions. Our experience in this respect is wide ranging and highlights the fact that there are still no international standards in place. It is true that more and more countries are standardising, using international technical standards as a basis, but a lot is still tied to local or state policies. With these variables, we will undoubtedly use our previous experience and the lessons we learned to reduce margins of error to a minimum in carrying out these activities. Considerable support is provided by our representative offices located all over the world, which know the local fabric and network, definitely useful on the job site.

When the job site opens, Danieli will take over the technological areas and all the temporary areas required to carry out the work through inspection and delivery of the area of erection.

The kick-off meeting with the customer and the subcontractors determines the provisions and management rules of the job site in all its aspects. When the OBS is presented and the detailed erection and startup work schedule is agreed, the site manager will organise the management of materials and draw up the site reports to confirm the work progress report. It is important to plan periodical meetings with the customer to decide on any actions that need to be taken and to keep both sides informed on work progress (the involvement of the customer is fundamental in turnkey plants as well so that the customer can become familiar with the technology as soon as possible and start managing his investment right away).

In addition to site management tasks, there are also specific activities such as monitoring and coordination of the activities of the company's technical supervisors and any sub-suppliers. The schedule plans the work of the technicians thus creating the best rotation of resources. We will have to determine and agree the criteria to be used to estimate work progress, with both the erection company and the customer. We know that there are various ways of applying this estimation, the most common being the one related to mounted weight. We will therefore draw up a document to officialise the start of activities, which is the civil works start certificate for civil works and the start erection certificate for buildings and machines in general.

Turnkey plants in particular require complex local organisation and, therefore, entail thorough preparation of site documentation, identification and qualification of possible local and backup suppliers, organisation of a technical site file, logistics and checking of staff residences.

As regards the instruction of site personnel, the various specialists need to produce mechanical, hydraulic, electronic and automation, medium- and high-voltage electronics documentation. The PM shall ensure that the documents have been prepared properly. A typical example is the automation FAT test for specialised personnel in charge of starting up the plant. This consists of checking the application software prepared according to the mechanical and operating specifications. Returns will be requested to make sure that the assigned task is being done. Please note that MAC is set up in such a way that all specific activities are carried out, as it is responsible for managing its own budget for the training of personnel at Danieli headquarters.

We can define the main monitoring phases of site production activities such as erection, cold tests, hot tests and performance tests.

The PM may have to be directly involved in specific site activities, and the matrix-oriented organisational structure means that the PM has to make sure that the various steps have been carried out according to the time schedule, procedures and costs.

The PM's *forecasts* are possible if the reports and work progress report are clear and can be compared with other experiences and therefore suitably assessed.

An erection start certificate must be issued at the beginning of erection guided by the site manager, followed by a technical follow-up of work performance to be indicated in the monthly reports. It must be backed by specific site progress reports enabling the outstanding list to be updated. The last stage of erection will come to an end when the erection end certificate is issued. The management of the erection company and the personnel involved is then referred to the figures of the Site Manager and the MAC PM.

Erection management is a *parallel world*, which today uses the latest technologies, including IT. Today this field requires high managerial and operational qualifications, and it is for this reason that the company has created its own specialised company (Danieli Construction International). Not only is this company able to perform the work—thanks to its specific skills—but its feedback leads to design and equipment improvements, and to targeted planning schedules in line with the company's needs.

At the end of the erection phase comes the cold test phase, which, on the basis of a plant and equipment checklist, is done by specialised technicians. These tests confirm that the erection work has been done precisely and determine an important work progress report for the project because they basically *start moving* the equipment, thereby giving life to the project. All these activities are planned during plant startup, and resources are monitored and organised. Determining a timeframe to carry out all the cold tests is the task of MAC, which has experience related to the type of plant and the detailed knowledge required to complete the task. Once the work is completed a certificate will be issued and a report prepared in order to deal with the next step based on all the necessary information and feedback, including any flaws that have to be eliminated.

Hot tests: This is the important commissioning phase where all the equipment starts to *work* in load conditions and the material is processed in order to make the product specified in the contract signed with the customer. In this phase, the customer's staff becomes part of the plant to be started up. This is also the beginning of transfer of the know-how that the company places at the disposal of its customers.

The contract already defines the milestones ascertaining that the test results have been reached, which are generally necessary to prove the plant's capacity to produce what is specified in the contract.

The company guides the customer in familiarising himself with the supply and the operating process in preparation for future production.

Naturally, in this phase as well, a certificate is required that is generally attached to the technical part of the contract and that in some cases has economic–financial milestones. However, all the documents qualifying the supply are drawn up while the activities are being carried out, including the specific technical reports and the technological reports now required since they concern the customer's final commercial product. Process experts and technologists are part of the startup team; they make sure that the technical design data have been applied properly and they test the plant, gathering experience and data for the purpose of continuous improvement. Today, knowledge of products such as stainless and special steels is important as they may be transformed into equipment, and the possibility of satisfying the specific requirements of customers that need to focus on market demands.

We can even say that up to this phase the contents of the report are standardised and based on a model that uses *time now* calculation. The principal items monitored by the PM are the Gantt diagram with the PPC obtained for each operating unit; the progress curves and baseline comparison system specified in the schedule and the progress actually reached up to the processing date; the list of critical points encountered with an indication of their consequences, such as for example the postponements and economic costs together with any suggested actions; a list of non-conformities encountered during the month with a complete list in the network, highlighting the corresponding site costs; the amount of present and future technical supervisors required based on time-now planning; comments on work progress and relations with subcontractors. Finally, the report can also include photographs of the work areas to have a visual idea of work progress. Attached to the descriptive document is the SBAC analytical document for erection, supervision to erection and startup and any additional costs.

The job order PM decides whether or not to call meetings to analyse problems and solve them.

The final or performance phase is the one where the plant's production capabilities are proven and where, because of its complexity, it requires a specific type of management where the PM handles relations with the customer, already an integral part of the plant that needs to be taken over with confidence and ability. The contractual aspect of this activity is very important because as mentioned earlier, the result of the performances is strictly related to the economic part for the release of the final payment, because it is connected to a possible application of economic penalties for failure to reach technical or technological parameters. Notwithstanding the contractual fulfilment the PM starts to consider the negotiations between the technical and economic part.

Reaching Final Acceptance of the plant is an important goal that proves that the contractual requirements have been fulfilled and that the customer agrees that the plant is in compliance.

18.4 Contract Closing

The final onshore process involves closing the job site and ending all the corresponding activities; it must also include monitoring of technical job site activities, as well as the related logistic–administrative activities; it will end with the job site mobilisation phase and the drawing up of an End of Job Site Technical Report.

Among other things, specific activities are carried out that are not the PM's responsibility but that need to be identified in the job site procedure used to do them. The job site documentation has to be gathered for filing, and the various technical requirements need to be known as well as the analysis of any materials that advanced during project execution and that now can be delivered to the customer, a confirmation that the equipment has been picked up or delivered, closing of anything open with local institutions, mobilisation of temporary facilities used during on-site work performance and drawing up a punch list with any claims to be settled. The punch lists are agreed with the customer and the site manager and, in addition to technical–economical information, must also include the deadlines.

The positive definition of the site and therefore the making of the plant may require a subsequent assistance stage. The company has a service centre at the disposal of customers even after the cancellation of the contract. Service means continuity in a relationship of trust between customer and supplier and also improves the exchange of information between the parties.

There are considerable growth margins in this specific field as well with the possibility of developing networks that use innovative technology and transfer of know-how.

As regards the closing of site activities, a commissioning report will be drawn up, which, as a basis for a *closeout* meeting, contains the main critical elements of the project and the positive elements to be considered to improve the result. This will then be attached to the PM's final job order report in his end-of-job-order memo. It is clear that the standard site report will contain a summary of the monthly reports.

The PM's tasks in the project-closing phase are closely tied to the contractual elements and to the relationship of trust with the customer. The recurring elements in this stage that require negotiations leading to a final agreement are an analysis of any extra costs charged to the customer and to Danieli and technical and economic claims between the parties.

Once these points are settled, the final acceptance certificate will be signed.

The project is not only closed with respect to the customer, but is also closed with respect to the company; for the analysis of project closing, the stages of its execution are: the concepts of the project, analysing feasibility, the offer and the launch of the project, the planning concepts in terms of both preliminary and detailed planning; the production aspect through its complex procedure; analysis and detection through actual production and costs; checking efficiency through deviation analysis, corrective actions and change management.

Going through a critical analysis and historisation of results is our goal for the final job site report.

We already know that we cannot define a *closed* project until all the contractual and administrative obligations have been fulfilled, including a critical analysis of results highlighting the causes for success or failure, management strengths and weaknesses and historisation of main data in order to turn it all into a reference basis for future projects.

It could be somewhat subjective to determine the highlights of a project and therefore, based on experience and feedback, a model is used that can be changed depending on market, technical or company requirements.

We provide a general application basis in a summary containing a series of items and by-products that already appeared during production but that need to be gathered in a document.

It defines the project in terms of scope of supply and dimensions; it indicates the complexity of the environment, possible variations in progress, the real expectations of the customer in terms of time, costs and quality; it shows estimates and time-based calculations based on the initial planning schedule and actual completion dates, estimates and final budget balances; it analyses costs and revenue comparing the expected job order Gross Operating Margin with the one that was actually obtained; it considers external suppliers, evaluating their behaviour and either confirming or rejecting the saving clause.

It also analyses internal human resources in relation to skills used, type of training received, performance parameters according to types and characteristic activities; it analyses yield parameters and ends with an analysis of the Risk Plan through its components, its variables and the correctness of the metrics used to calibrate any models.

The main critical points that were dealt with are useful as lessons learned.

How positive was the result of the project? We have to have the capacity to read the result as a company, which, using a management system, has had the advantage of an initial planning schedule, consistently effective management, a system to support decision-making, the creation of a database and a company image.

We are in a position to give a specific opinion on functionality through a typical checklist showing the main phases.

Initial planning schedule: a structured project with an unequivocal allocation of units to save time in searching for information.

Effective management: to reduce information flow time by means of homogenous information and a common vocabulary, keeping the general or particular situation under control by a better use of resources and by anticipating negative events.

Support of decision-making: more effective management to bring about a reduction in penalties, anticipation of payment collection, objective claims justification, creating a less-expensive management system.

Creation of a database: for greater planning objectivity, effective historical support in drawing up the offer, determines a consolidation of company and sector know-how.

Company image: quick development of a large amount of documentation; attention to image by means of painstaking graphics and representation.

So, even if the result can be measured and assessed in numerical terms we have to consider not only the result of each project but also its complexity and evolution.

18.5 Risk Monitoring

The specific analysis of technical risks must refer to the supply; project data must be checked, such as output, types of steels, types of finished products, standards and non-standard supplies to identify details, vendor list, the features of the foundations and buildings, type of installation, supervision and startup.

Risk or project analysis is continuously evolving and requires continuous improvement in project risk management. So to keep the process and the analysis structure active, one strategy is to hold regular project review sessions on the analysis and management of project risks. The lessons learned aim to improve the company production cycles and project management methods and cross-referencing of management experiences for the project.

Typical project review sessions are as follows: Project staff meetings are held regularly every 1 or 2 months during the life of the project; it is attended by General Management (DG), the project manager (PM) with some people of the project group (GdP) and the managers of all company structures. In this session, the PM explains the executive status of the project and the risks, and requests any exceptional interventions.

This is the *moment of truth* avoiding among the liars around the table, where any unclear situations (lies) emerge, on the status of the production cycles and management strategies in order to define realignment actions.

The information exchanged and the decisions made are recorded in minutes of meeting, which are distributed to all the concerned sectors and which will eventually make up the daily project log, which can be consulted at the end for a final review.

Project risk checkpoints are specific project review sessions requested by the project manager when he/she feels that the status of the project requires a cold-eye analysis by a group of experts outside the Project Group made up of other PMs and company managers.

The post-project review meeting is held at the end of the project where the PM reassumes the risk areas found during project development, expresses his/her opinion on the final results and uses them as lessons learned, in terms of both positive examples to be copied and negative examples to be avoided. The post-project review is organised according to an agenda that allows you to feed a company database with the risks managed during projects.

In this database we also have to consider the levels at which these subjects have been handled, with high and medium risks generally discussed in detail, while for the minor risks, a periodically updated *watch list* would be sufficient.

Often, in the execution of particularly complex projects, there are risks triggered by production processes that are no longer suitable and by improper management of customer relations. When these issues become generalised or occur repeatedly, a specific session is held before the project is completed. Only management times are analysed in these sessions for the purpose of identifying general solutions that could entail modifications to the production processes and company organisations. The decisions that are made become operative for all new projects.

These extraordinary meetings identify general solutions for risks that tend to become repetitive, improve company processes through lessons learned and support the exchange of experience among GdPs.

The database referred to in the analysis becomes a register of the risks that arise during identification, which are mainly as follows:

- Identified risks, description of risks, concerned project areas, relative causes and influence on project goals
- Risk holders and assignment of responsibilities
- Output from quantitative and qualitative analyses with priorities and probabilities of occurrence
- Agreed response strategies
- Specific actions to develop the selected response strategy
- Establishing symptoms and warning signals for possible risk occurrence
- Budget and planned activities required to fulfil action strategies
- Contingency reservations in terms of time and costs with corresponding plans
- Reserve plans to be used as risk reaction for unsuitable response
- Residual risks together with those accepted together
- Collateral risks occurring after the application of a solution
- Contingency reservations calculated in terms of project quality

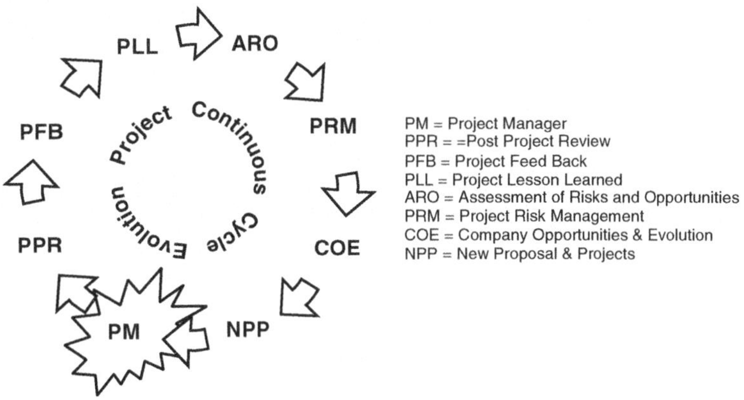

Fig. 18.9 The virtuous circle

In conclusion, the goal of the described process is to feed a virtuous corporate circle that can be summarised as follows.

Improve the company's ability to identify and manage project risk and increase confidence in pursuing new opportunities. This virtuous circle is simplified in Fig. 18.9 in which, starting any point of the circle, we can transfer each other's experience, having a continuous improvement in the risk evaluation and minimising as much as possible the impacts.

Annex: How to Use Microsoft Project Effectively[1]

By Microsoft Corp.

Index

A.1 An Introduction

A.1.1 What Is Microsoft Project?

With the ever-growing complexity of projects and their schedules, it is not exaggeration to say that no project today is managed without using a Project Management Software. Microsoft Project is one of the industry-leading tools for creating, managing and tracking project schedules and generating related reporting that could be used to make insightful decisions about projects.

[1]This Annex has been written by **Prasanna Adavi** on behalf of Microsoft Corp.

S. Tonchia, *Industrial Project Management*, Management for Professionals,
https://doi.org/10.1007/978-3-662-56328-1

MS Project serves as Project Management Information System, by allowing you to model your schedule including dependencies between various activities, any constraints, leads and lags and helps forecast the outcome of current progress of activities in your schedule.

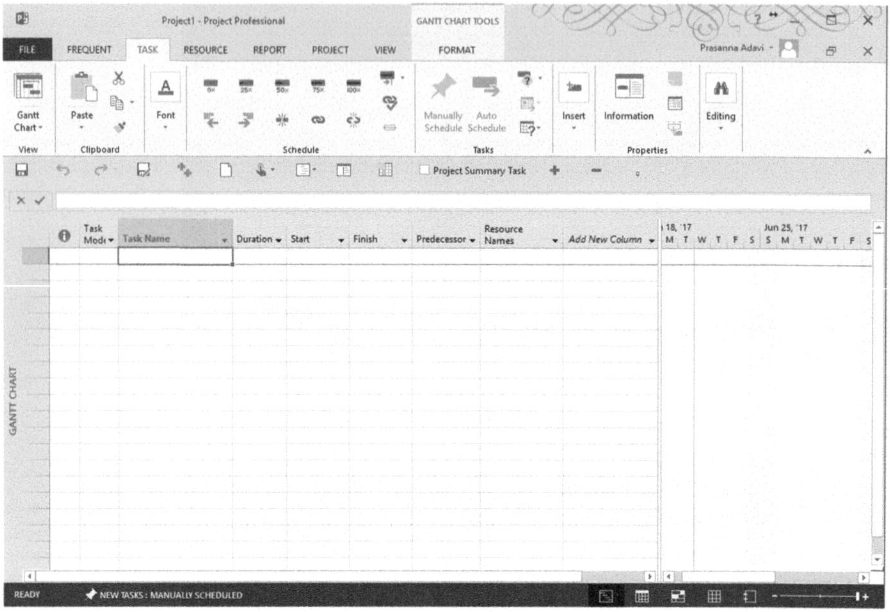

Fig. A.1 MS Project input area

A.1.2 Versions of Microsoft Project

Microsoft Project comes in many editions and versions. In addition, with the advent of cloud based products, MS Project us also available in a cloud based version.

In a nutshell, the following are the versions available currently for MS Project.

ON-premises solutions	Cloud-based solutions
Project Standard	Project Online Professional
Project Professional	Project Online Premium

The following link provides more detailed information about each of these products: https://products.office.com/en-us/project/compare-microsoft-project-management-software?tab=tabs-1

A.1.3 Features and Navigation

Microsoft Project uses the Fluent UI interface, which is seen across all Microsoft Office product family. The ribbon menu provides an easy to find mechanism for various menu options.

Each of the tabs under the ribbon group the relevant commands for easier access and management.

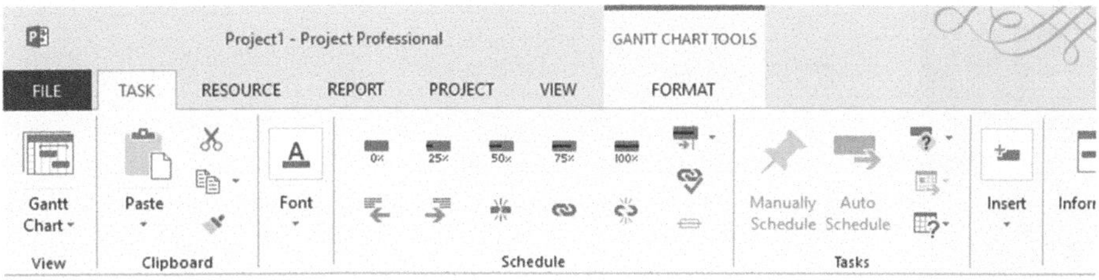

Fig. A.2 MS Project Fluent UI ribbon

A.1.4 Enterprise Project Management

Each organization is different and while smaller organizations manage projects on an individual basis, the need often is to manage not just projects, but also programs and portfolios. Organizations need information about these initiatives rolled up in various dimensions, and also look at them in a holistic perspective to be able to make business decisions. This is nothing but Enterprise Project Management.

Microsoft Project by itself is a stand-alone desktop product, and great for the management of individual project schedules. However, paired with either Project Server or Project Online, it becomes a powerful EPM software, providing the ability to manage thousands of project schedules at a given time and be able to report and track them.

A.2 Project Set-Up and Basics

A.2.1 Project Options

Microsoft Project allows the users to specify several options like date formats, default views, how the schedule calculations are performed etc. This provides the users a powerful control over how their schedules are built and calculated.

You can navigate to the project options from the backstage by navigating to ***File*** >> ***Options***.

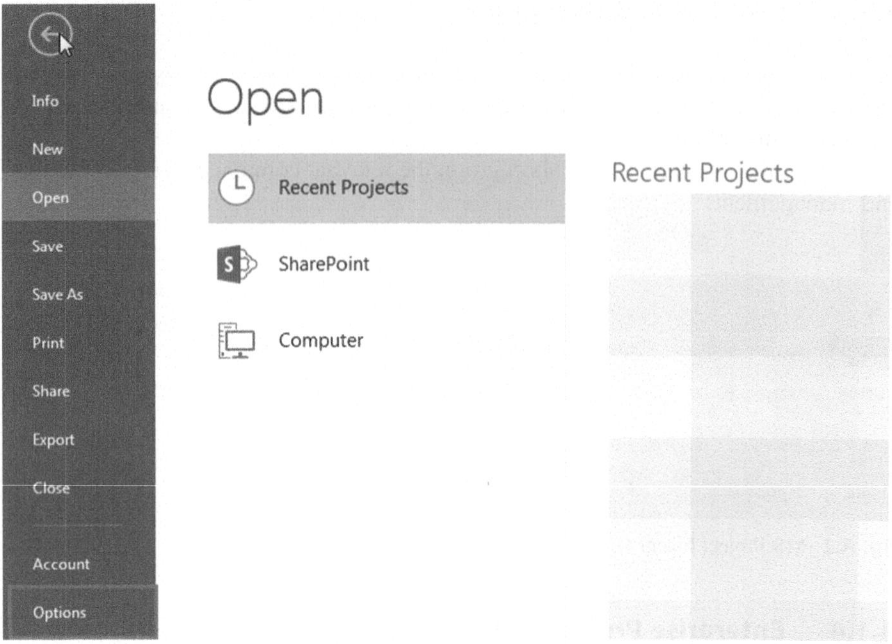

Fig. A.3 Accessing MS Project options

The options are divided into the following categories:

Fig. A.4 MS Project options

Since discussing each option in detail is outside the scope of this material, the following table provides the recommended option settings. Please feel free to change them as necessary to suit your project.

General Project Options

- Default View—Gantt Chart with Timeline
- Date Format—Set it to the locale of your project

Display Options

- Calendar Type—Gregorian Calendar
- Currency—Set it to match the currency of your project

Schedule Options

Please note that changes made to the settings in this section, do not apply to existing projects.

Calendar Options

- Calendar Options for this project: Change to ***All New Projects***
- Modify the remaining settings in this section to match the project you are working with. Please keep in mind that these settings are not used in calculating the schedule, rather in the timescale of how Project displays weeks, months and years.

Scheduling Options

- Schedule Options for this Project—Change to ***All New Projects***
- New Tasks created—Change to ***Auto Scheduled***
- Default Task Type—This depends on your scheduling technique, we recommend ***Fixed Duration*** task type as this the most common.
- New Tasks are Effort Driven—Leave this box unchecked.
- Split In-Progress Tasks—Check this box to allow MS Project to split tasks across weeks when work happens out of sequence.
- Tasks will always honor their constraint dates—Leave this box checked, so that MS Project always schedules in relation to the constraints.

Calculation

- Calculate project after each edit should be set to ***On***.

As mentioned before, there are several other options for MS Project that could influence how a project schedule is developed and calculated. The user is encouraged to work with these options and find what suites best for their scenario.

A.2.2 Project Information

The Project Information dialog box is access from the tab ***Project*** >> ***Project Information*** and provides quick information on the project.

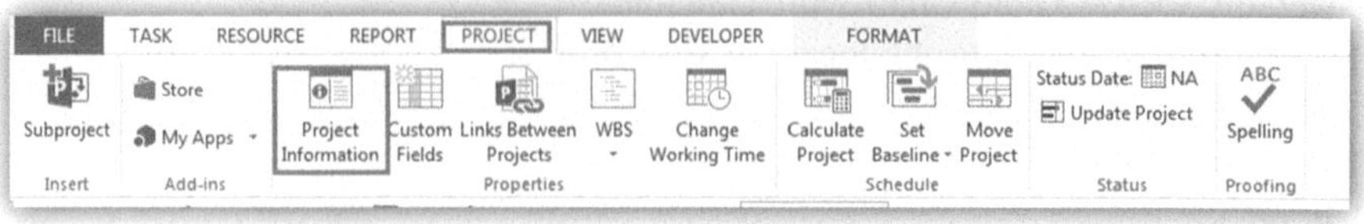

Fig. A.5 Accessing project information

This dialog box allows you to specify the Project Start Date, the Project Status Date, Project Calendar and the Project Priority number (used for Resource Leveling).

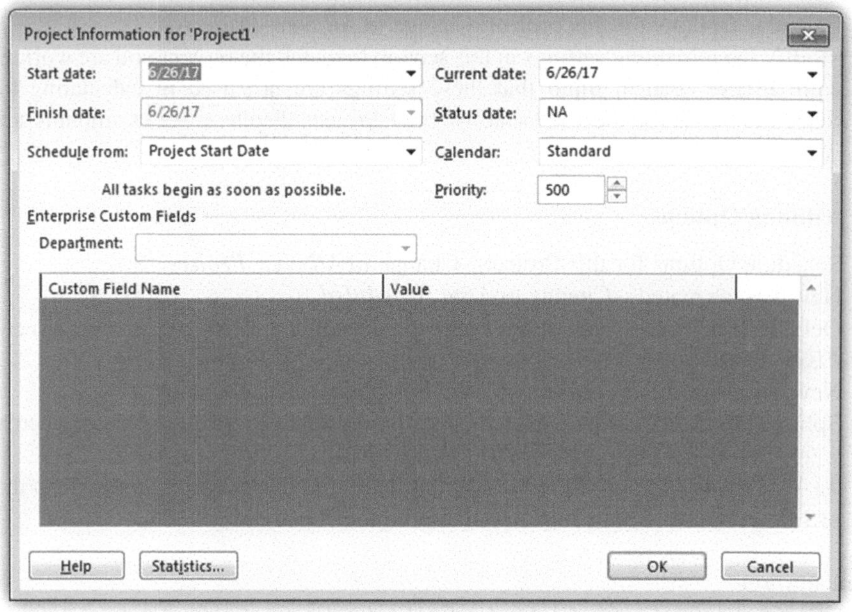

Fig. A.6 Project information dialog box

A.3 Scenario

Avameri Inc. is an international Product Development company that designs, manufactures and distributes hardware tools. The organization has standardized on several Project Management processes, but with the growing size, it is increasingly getting difficult to keep track of the various initiatives happening across the company from a schedule, budgeting and resourcing perspective.

The leadership team of Avameri has decided to acquire a tool for creating and managing project schedules and to be able to easily report on them, namely MS Project.

The rest of the chapters and exercise will focus on how to build project schedules for Avameri Inc.

A.4 Building Initial Schedule and WBS

The first step in building a project schedule is to build the Work Break-Down Structure (WBS), A WBS defines the complete scope of work that the project is expected to deliver. This work, is then ***decomposed*** into several levels called work packages, until it does not make any more sense to split the lowest level into further detail.

A typical project at Avameri, Inc. could be listed as below.

Identify the launch team
Determine sales objectives
Define launch goals (launch timing and publicity objectives)
Identify channel partners
Identify retail partners
Identify online opportunities
Identify budget requirements
Obtain launch budget approval
Kickoff product launch
Define creative brief
Prepare messaging framework
Review current literature and plan new requirements
Define press release requirements
Define product specification literature
Define sales presentations
Define internal communication needs
Align product release timing with marketing plan
Prepare for volume manufacturing according to sales goals
Plan sales force staffing and training to support sales goals
Plan team staffing to support sales goals
Plan field service staffing to support sales goals
Supply updated requirements and budgets based on departmental plans

(continued)

Obtain approval for updated plans and budgets
Evaluate market and refine messaging
Design and order final packaging
Create press releases
Create product specification literature
Create sales presentations
Communicate product launch details to internal organization
Create sales, field, and product support training
Update product forecasts based on market feedback and analysis
Update launch plan based on forecast
Complete and test product
Create prototype products
Prepare pipeline
Establish sales channels
Establish product support mechanisms
Establish field service organizations
Implement change control
Finalize support policy
Certify product
Manufacture the volume of product planned to meet sales objectives.
Hire and train sales staff
Hire and train product support personnel
Hire and train field service personnel
Conduct final quality review
Needed materials are on order or in stock
Launch product

Now let us see how this could be developed into a project schedule. As discussed before, our first step is to decompose the list of activities into deliverables based work packages. This can be in done in several different ways, based on the nature of the project. One such way is shown below (Phase 1 shown).

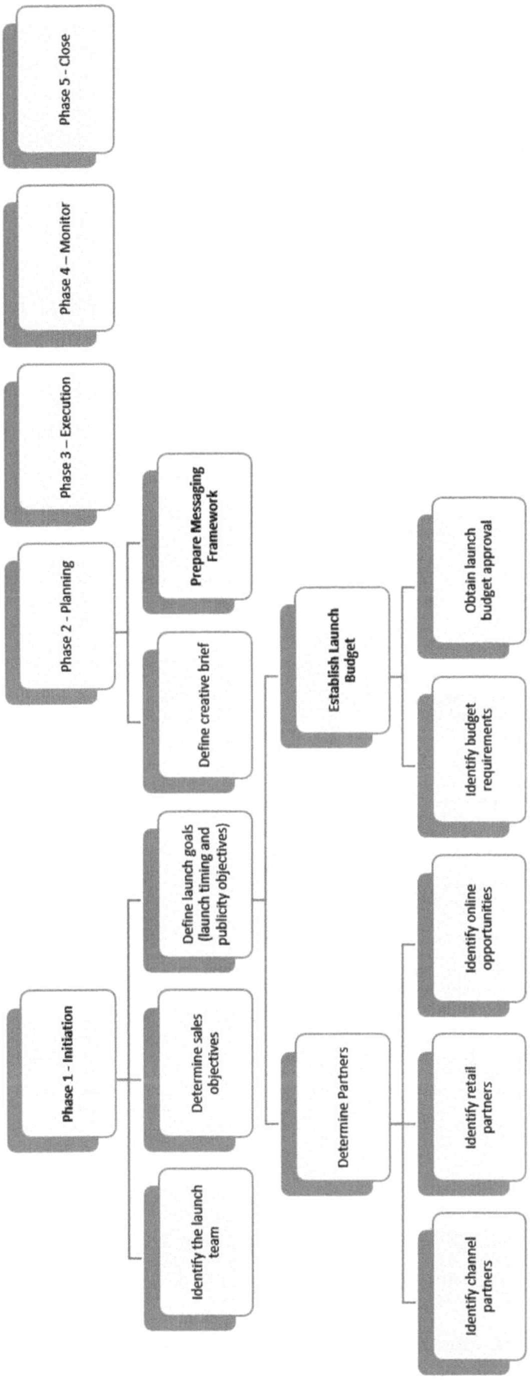

Fig. A.7 Work break down structure

A.4.1 Setting Up Tasks and Building WBS in MS Project

Once you have the outline of your activities broken down by work packages to the required level, you can start building the WBS in MS Project.

A.4.2 Instructions

To begin building the WBS and start adding tasks.

- Open MS Project, with a blank project.
- Navigate to ***Project*** >> ***Project Information*** and set the ***Project Start Date***
- Now navigate to ***File*** >> ***Save as*** and save the file with a name of your choice.
- Navigate back to the ***Gantt Chart*** view and start adding tasks, simply by clicking in the Gantt Chart area.
- Once all tasks have been added, your schedule would look something similar to this (not complete picture).

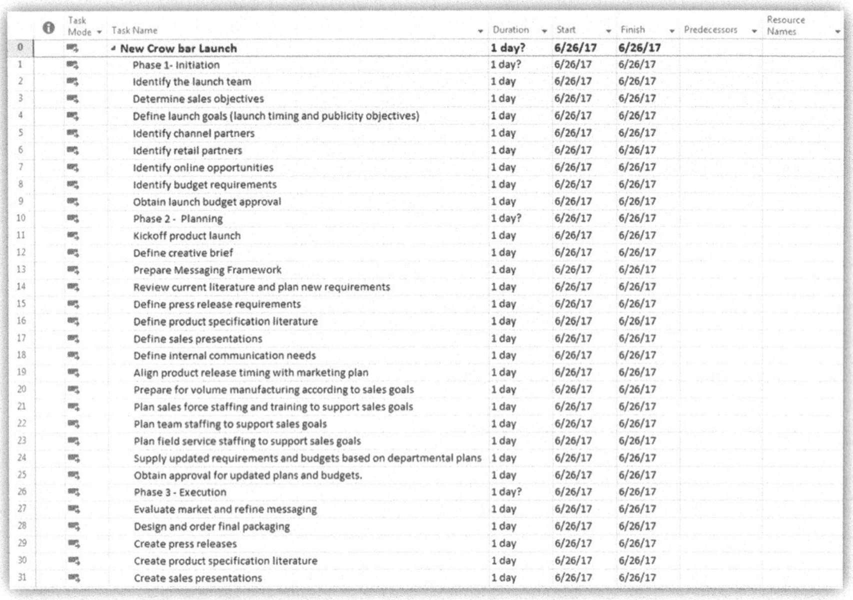

	Task Mode	Task Name	Duration	Start	Finish	Predecessors	Resource Names
0		New Crow bar Launch	1 day?	6/26/17	6/26/17		
1		Phase 1- Initiation	1 day?	6/26/17	6/26/17		
2		Identify the launch team	1 day	6/26/17	6/26/17		
3		Determine sales objectives	1 day	6/26/17	6/26/17		
4		Define launch goals (launch timing and publicity objectives)	1 day	6/26/17	6/26/17		
5		Identify channel partners	1 day	6/26/17	6/26/17		
6		Identify retail partners	1 day	6/26/17	6/26/17		
7		Identify online opportunities	1 day	6/26/17	6/26/17		
8		Identify budget requirements	1 day	6/26/17	6/26/17		
9		Obtain launch budget approval	1 day	6/26/17	6/26/17		
10		Phase 2 - Planning	1 day?	6/26/17	6/26/17		
11		Kickoff product launch	1 day	6/26/17	6/26/17		
12		Define creative brief	1 day	6/26/17	6/26/17		
13		Prepare Messaging Framework	1 day	6/26/17	6/26/17		
14		Review current literature and plan new requirements	1 day	6/26/17	6/26/17		
15		Define press release requirements	1 day	6/26/17	6/26/17		
16		Define product specification literature	1 day	6/26/17	6/26/17		
17		Define sales presentations	1 day	6/26/17	6/26/17		
18		Define internal communication needs	1 day	6/26/17	6/26/17		
19		Align product release timing with marketing plan	1 day	6/26/17	6/26/17		
20		Prepare for volume manufacturing according to sales goals	1 day	6/26/17	6/26/17		
21		Plan sales force staffing and training to support sales goals	1 day	6/26/17	6/26/17		
22		Plan team staffing to support sales goals	1 day	6/26/17	6/26/17		
23		Plan field service staffing to support sales goals	1 day	6/26/17	6/26/17		
24		Supply updated requirements and budgets based on departmental plans	1 day	6/26/17	6/26/17		
25		Obtain approval for updated plans and budgets.	1 day	6/26/17	6/26/17		
26		Phase 3 - Execution	1 day?	6/26/17	6/26/17		
27		Evaluate market and refine messaging	1 day	6/26/17	6/26/17		
28		Design and order final packaging	1 day	6/26/17	6/26/17		
29		Create press releases	1 day	6/26/17	6/26/17		
30		Create product specification literature	1 day	6/26/17	6/26/17		
31		Create sales presentations	1 day	6/26/17	6/26/17		

Fig. A.8 Task list in MS Project

Now to start bringing the outline structure to this task list,

- Navigate to the ***Task*** tab and use the ***Indent*** and ***Outdent*** buttons

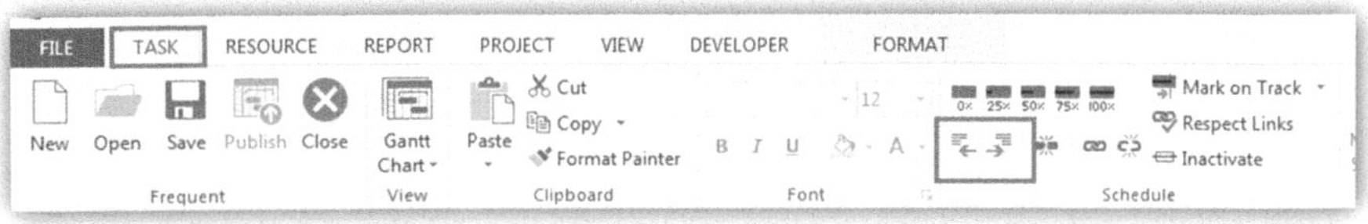

Fig. A.9 Indenting and outdenting

- To ***Indent*** a task, select the task and click on the ***Indent*** button on the ribbon. Similarly for ***Outdenting***, you can click on the ***Outdent*** button, after selecting the task.
- You can select multiple tasks, by holding the ***Shift*** key while selecting the tasks.

The tasks that represent a group of sub-tasks are called ***Summary Tasks*** in MS Project nomenclature.

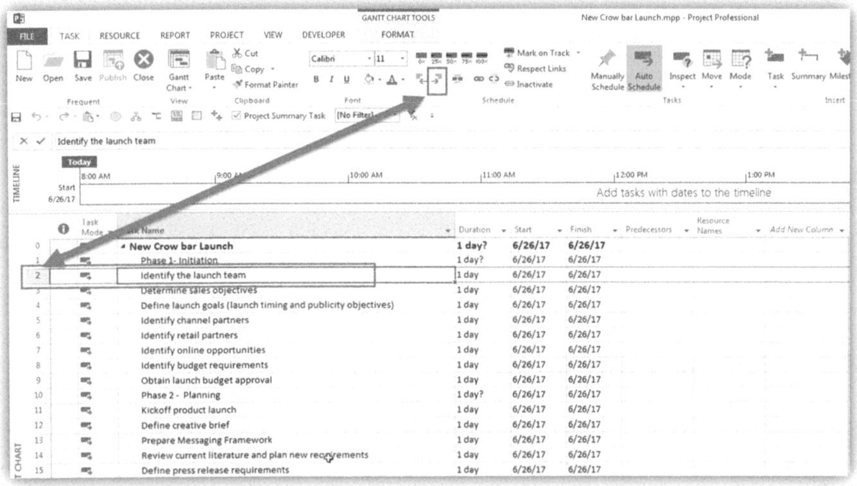

Fig. A.10 Building WBS in MS Project

Once you go through the task list, building your WBS outline as above, your schedule will start to look as shown.

	Task Mode	Task Name	Duration	Start	Finish
0		**New Crow bar Launch**	**1 day**	**6/26/17**	**6/26/17**
1		**Phase 1- Initiation**	**1 day**	**6/26/17**	**6/26/17**
2		Identify the launch team	1 day	6/26/17	6/26/17
3		Determine sales objectives	1 day	6/26/17	6/26/17
4		Define launch goals (launch timing and publicity objectives)	1 day	6/26/17	6/26/17
5		Identify channel partners	1 day	6/26/17	6/26/17
6		Identify retail partners	1 day	6/26/17	6/26/17
7		Identify online opportunities	1 day	6/26/17	6/26/17
8		Identify budget requirements	1 day	6/26/17	6/26/17
9		Obtain launch budget approval	1 day	6/26/17	6/26/17
10		**Phase 2 - Planning**	**1 day**	**6/26/17**	**6/26/17**
11		Kickoff product launch	1 day	6/26/17	6/26/17
12		Define creative brief	1 day	6/26/17	6/26/17
13		Prepare Messaging Framework	1 day	6/26/17	6/26/17
14		Review current literature and plan new requirements	1 day	6/26/17	6/26/17
15		Define press release requirements	1 day	6/26/17	6/26/17
16		Define product specification literature	1 day	6/26/17	6/26/17
17		Define sales presentations	1 day	6/26/17	6/26/17
18		Define internal communication needs	1 day	6/26/17	6/26/17
19		Align product release timing with marketing plan	1 day	6/26/17	6/26/17
20		Prepare for volume manufacturing according to sales goals	1 day	6/26/17	6/26/17
21		Plan sales force staffing and training to support sales goals	1 day	6/26/17	6/26/17
22		Plan team staffing to support sales goals	1 day	6/26/17	6/26/17
23		Plan field service staffing to support sales goals	1 day	6/26/17	6/26/17
24		Supply updated requirements and budgets based on departmental plar	1 day	6/26/17	6/26/17
25		Obtain approval for updated plans and budgets.	1 day	6/26/17	6/26/17
26		**Phase 3 - Execution**	**1 day**	**6/26/17**	**6/26/17**
44		**Phase 4 - Release**	**1 day**	**6/26/17**	**6/26/17**

Fig. A.11 MS Project with outline structure

Depending upon your project need, this could be further elaborated, and decomposed.

A.4.3 Assigning WBS Codes

MS Project automatically assigns outline codes for each level in the WBS. However, you could create a custom outline code for your schedule, if your project demand it. WBS codes are useful when you are addressing or reporting on specific items or work packages.

For further information on assigning custom WBS codes, please refer to this link: https://support.office.com/en-us/article/Create-WBS-codes-bb6a61aa-debd-4e38-8c04-8e2c1ae3cbda

A.5 Entering Estimates

A.5.1 Work, Duration and Units

Once you have a defined WBS structure, it is now time to enter the estimates for the tasks. In MS Project, you can enter the estimates in terms of duration, work or units of a resource's time.

Duration is the number of units of calendar time, that a task will require to complete.
Work is the actual effort it takes to complete a task, usually measured in hours.
Units represent the % of a resource(s) that are needed to work on a task.

However, before entering estimates, one needs to understand how MS Project calculations work. MS Project expects you to enter two of the three variables (Duration, Work, Units), and will calculate the third variable for you. This allows MS Project to dynamically re-estimate your project schedule when tasks slip or progress differently than planned.

The equation that MS Project uses to calculate is as ***Duration * Units = Work***

For example, if there is a task that requires **5 days** to complete, with a resource working on it only **half time**, the amount of work on that task is calculated as **5 days * 0.5 = 2.5 days (or 20 hours if it is a 40-hour work week)**.

A.5.2 Task Types

Another concept that influences how MS Project schedules and recalculates a schedule model, is ***Task Types***.

MS Project provides three different task types:

- **Fixed Duration** Tasks where you can estimate the duration more accurately, and it generally remains constant for a given task. For example, it is easy to estimate the number of days a specific training will take. In this case you would set this task type to be of Fixed Duration, so that MS Project schedules the task to have a fixed duration and changes work or units to accommodate the model.
- **Fixed Work** Tasks where you can estimate the work (or number of business hours) more accurately, and it generally remains constant for a given task. For example, it is easy to estimate the number of hours it takes to paint a wall. In this case you would set this task type to be of Fixed Work, so that MS Project scheduled the task to have a fixed work and changes duration or units to accommodate the model.
- **Fixed Units** Tasks where you can estimate the amount of resources available for a given task, and it generally remains constant for a given task. Microsoft Project in this case tries to manipulate the duration of the task or the work hours, to accommodate the schedule model. An example of this scenario would be where

you have a resource that can only work half time. In this case you are restricted by 50% availability, so inevitably the tasks would have increased duration.

An additional parameter that is provided for these task types is whether the tasks are **effort driven**. This setting dictates in scenarios where more than one resource is assigned, if the work is divided between the resources or the additional work is added to the new resource. Fixed Work tasks are Effort Driven by default.

For our exercise, we will treat all tasks as of type **Fixed Duration**. To confirm this setting,

A.5.3 Instructions

- Insert a column called ***Type*** by clicking on ***Add New Column***

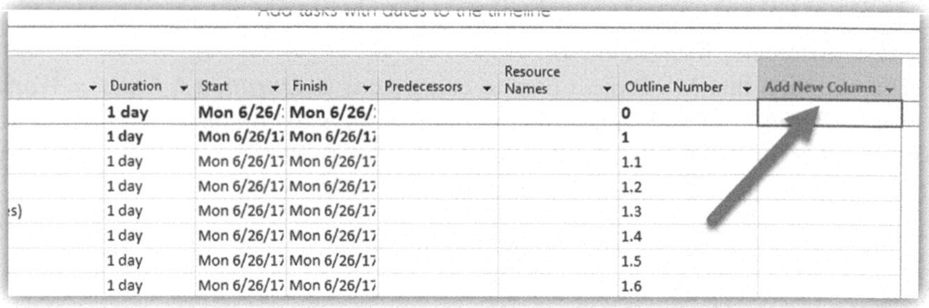

Fig. A.12 Adding task type

- If the Task type is not **Fixed Work**, you can change it by clicking in the column and selecting **Fixed Work** from the drop-down, and applying the same to all tasks.

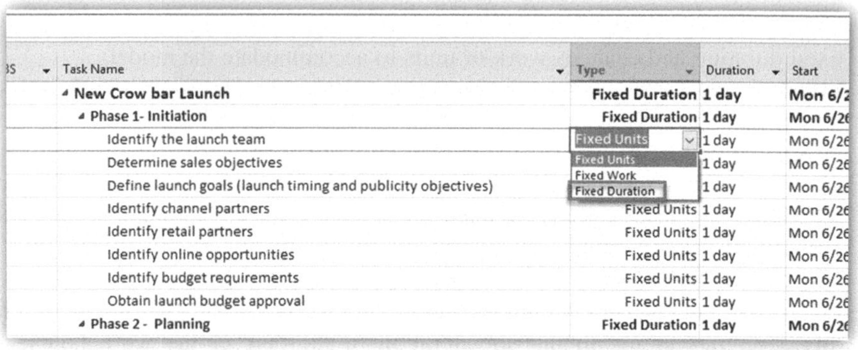

Fig. A.13 Changing task type

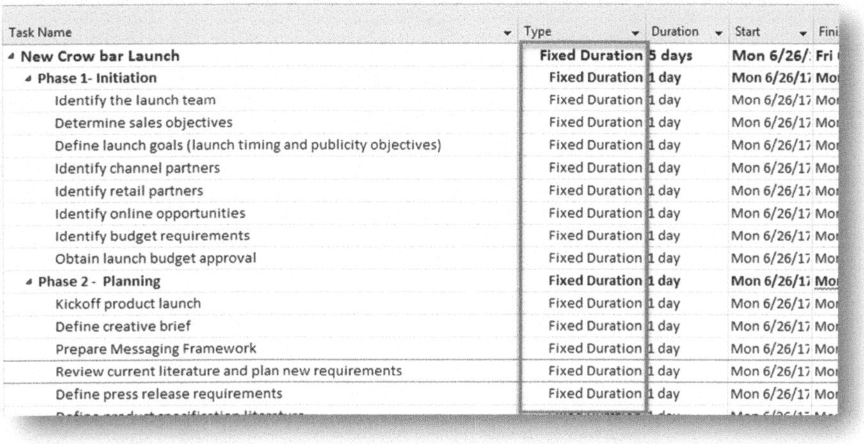

Task Name	Type	Duration	Start	Fini
New Crow bar Launch	Fixed Duration	5 days	Mon 6/26/	Fri
Phase 1- Initiation	Fixed Duration	1 day	Mon 6/26/1	Mo
Identify the launch team	Fixed Duration	1 day	Mon 6/26/17	Mo
Determine sales objectives	Fixed Duration	1 day	Mon 6/26/17	Mo
Define launch goals (launch timing and publicity objectives)	Fixed Duration	1 day	Mon 6/26/17	Mo
Identify channel partners	Fixed Duration	1 day	Mon 6/26/17	Mo
Identify retail partners	Fixed Duration	1 day	Mon 6/26/17	Mo
Identify online opportunities	Fixed Duration	1 day	Mon 6/26/17	Mo
Identify budget requirements	Fixed Duration	1 day	Mon 6/26/17	Mo
Obtain launch budget approval	Fixed Duration	1 day	Mon 6/26/17	Mo
Phase 2 - Planning	Fixed Duration	1 day	Mon 6/26/1	Mo
Kickoff product launch	Fixed Duration	1 day	Mon 6/26/17	Mo
Define creative brief	Fixed Duration	1 day	Mon 6/26/17	Mo
Prepare Messaging Framework	Fixed Duration	1 day	Mon 6/26/17	Mo
Review current literature and plan new requirements	Fixed Duration	1 day	Mon 6/26/17	Mo
Define press release requirements	Fixed Duration	1 day	Mon 6/26/17	Mo

Fig. A.14 Changed task types for all tasks

- Click on ***File*** >> ***Save*** to save the project.
- To enter the estimates of Duration for these tasks, you can enter them directly in the Duration column, . Another way to enter them is to use the ***Task Information*** dialog box.
- To access the ***Task Information*** dialog box, click on the Task Name and then select ***Information***. You can also access the task information dialog box by double clicking on the task.

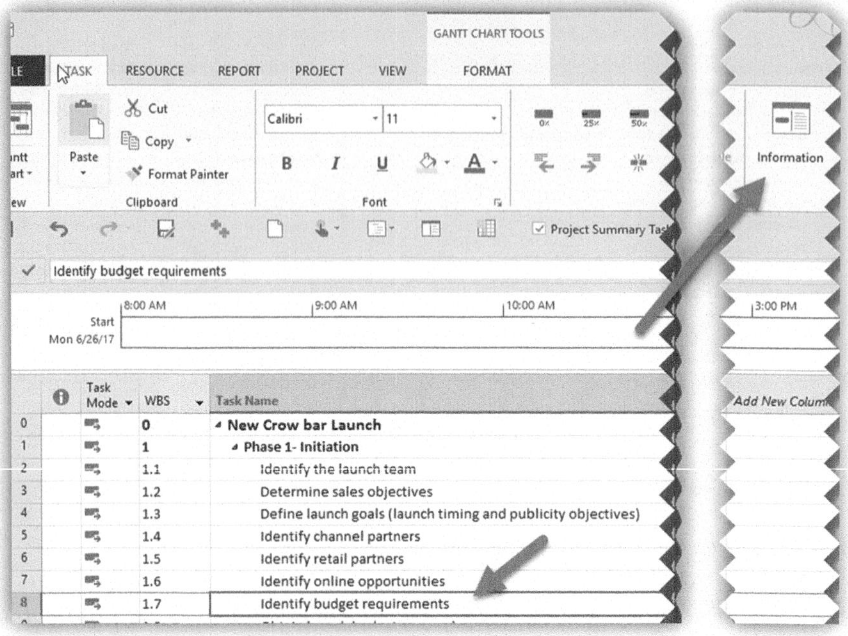

Fig. A.15 Accessing the task information dialog box

- This brings up the ***Task Information*** dialog box.

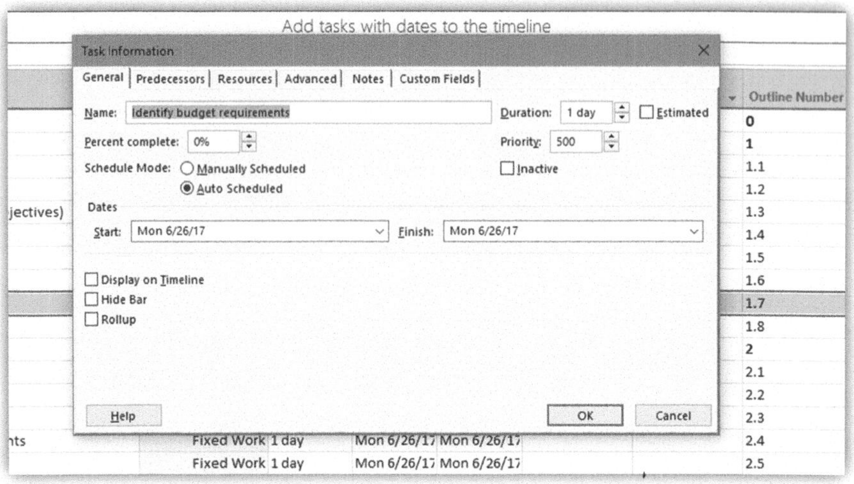

Fig. A.16 Task information dialog box

- In the ***Duration*** box, enter the estimated duration of the task, and click ***OK***. Repeat for all other tasks.

The schedule model with estimates will look something like the following picture.

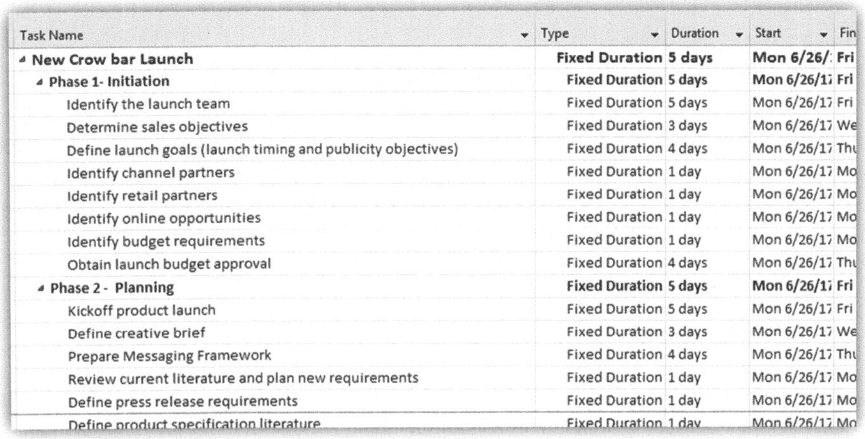

Task Name	Type	Duration	Start	Fin
New Crow bar Launch	**Fixed Duration**	**5 days**	**Mon 6/26/**	**Fri**
Phase 1- Initiation	**Fixed Duration**	**5 days**	**Mon 6/26/17**	**Fri**
Identify the launch team	Fixed Duration	5 days	Mon 6/26/17	Fri
Determine sales objectives	Fixed Duration	3 days	Mon 6/26/17	We
Define launch goals (launch timing and publicity objectives)	Fixed Duration	4 days	Mon 6/26/17	Thu
Identify channel partners	Fixed Duration	1 day	Mon 6/26/17	Mo
Identify retail partners	Fixed Duration	1 day	Mon 6/26/17	Mo
Identify online opportunities	Fixed Duration	1 day	Mon 6/26/17	Mo
Identify budget requirements	Fixed Duration	1 day	Mon 6/26/17	Mo
Obtain launch budget approval	Fixed Duration	4 days	Mon 6/26/17	Thu
Phase 2 - Planning	**Fixed Duration**	**5 days**	**Mon 6/26/17**	**Fri**
Kickoff product launch	Fixed Duration	5 days	Mon 6/26/17	Fri
Define creative brief	Fixed Duration	3 days	Mon 6/26/17	We
Prepare Messaging Framework	Fixed Duration	4 days	Mon 6/26/17	Thu
Review current literature and plan new requirements	Fixed Duration	1 day	Mon 6/26/17	Mo
Define press release requirements	Fixed Duration	1 day	Mon 6/26/17	Mo
Define product specification literature	Fixed Duration	1 day	Mon 6/26/17	Mo

Fig. A.17 Schedule model with duration estimates

A.6 Setting Up the Dependencies

A project schedule model is not a static task list. The key element that makes the schedule model dynamic are the interdependencies between the tasks. MS Project allows the users to set up the following types of dependencies between the tasks.

A.6.1 Types of Dependencies

Finish to Start (FS) This type of dependency dictates that the successor task cannot start until the predecessor task is complete. If you set up this dependency between two tasks, MS Project will schedule the successor task to start after the finish date of the predecessor task.

Start to Start (SS) This type of dependency dictates that the successor task starts at the same time as the predecessor task. If you set up this dependency between two tasks, MS Project will schedule the successor task to start on same date as the start of the predecessor task.

Finish to Finish (FF) This type of dependency dictates that the successor task finishes at the same time as the predecessor task. If you set up this dependency between two tasks, MS Project will schedule the successor task to finish on same date as the start of the predecessor task.

Start to Finish (SF) This type of dependency dictates that the successor task finishes when the predecessor task starts. If you set up this dependency between two tasks, MS Project will schedule the successor task to finish on same date as the start of the predecessor task.

A.6.2 Lead or Lag

Sometimes, it is necessary to separate a predecessor or successor task by a predefined amount of time. In MS Project, this is accomplished by Leads and Lags. A lead or Lag is added as a positive or negative number of business days that is specified as part of the dependencies between the tasks.

A.6.3 Instructions

As an example, in our project schedule, the following tasks have dependencies as defined below.

1. **Task 3—Determine sales objectives task** cannot start until the **Task 2—Identify Launch Team task** is complete.
2. **Task 5—Identify Channel Partners** task can start simultaneously with **Task 6—Identify Retail partners task**.
3. **Task 9—Budget Approval** generally takes 5 days after **Task 8—budget requirements are identified and submitted**.
4. **Task 13—Prepare Messaging Framework** can start 2 days earlier than **Task 12—Defining Creative Brief**

To model the above scenarios, follow the steps below:

- Open your project schedule, if it is not already open.
- To model the first scenario above, which is a ***Finish-To-Start*** dependency, in the Predecessors column for Task 3, enter ***2FS*** (the default dependency is Finish to Start, so you could also just enter the number 2). Notice how the start date of Task 3 is scheduled to start after Task 2.

		Task Mode	Task Name	Duration	Start	Finish	Predecessors
0			New Crow bar Launch	**8 days**	**Mon 6/26/17**	**Wed 7/5/17**	
1			Phase 1- Initiation	**8 days**	**Mon 6/26/17**	**Wed 7/5/17**	
2			Identify the launch team	5 days	Mon 6/26/17	Fri 6/30/17	
3			Determine sales objectives	3 days	Mon 7/3/17	Wed 7/5/17	2
4			Define launch goals (launch timing and publicity objectives)	4 days	Mon 6/26/17	Thu 6/29/17	
5			Identify channel partners	1 day	Mon 6/26/17	Mon 6/26/17	
6			Identify retail partners	1 day	Mon 6/26/17	Mon 6/26/17	
7			Identify online opportunities	1 day	Mon 6/26/17	Mon 6/26/17	
8			Identify budget requirements	1 day	Mon 6/26/17	Mon 6/26/17	
9			Obtain launch budget approval	4 days	Mon 6/26/17	Thu 6/29/17	
10			Phase 2 - Planning	**5 days**	**Mon 6/26/17**	**Fri 6/30/17**	
11			Kickoff product launch	5 days	Mon 6/26/17	Fri 6/30/17	
12			Define creative brief	3 days	Mon 6/26/17	Wed 6/28/17	
13			Prepare Messaging Framework	4 days	Mon 6/26/17	Thu 6/29/17	

Fig. A.18 Finish to start link in MS Project

- To model the second scenario above, which is a ***Start-To-Start*** dependency, in the Predecessors column for Task 6, enter ***5SS***. Notice how the start date of Task 6 is scheduled the same as the start of Task 5.

		Task Mode	Task Name	Duration	Start	Finish	Predecessors
0			New Crow bar Launch	**8 days**	**Mon 6/26/17**	**Wed 7/5/17**	
1			Phase 1- Initiation	**8 days**	**Mon 6/26/17**	**Wed 7/5/17**	
2			Identify the launch team	5 days	Mon 6/26/17	Fri 6/30/17	
3			Determine sales objectives	3 days	Mon 7/3/17	Wed 7/5/17	2
4			Define launch goals (launch timing and publicity objectives)	4 days	Mon 6/26/17	Thu 6/29/17	
5			Identify channel partners	4 days	Mon 6/26/17	Thu 6/29/17	
6			Identify retail partners	1 day	Mon 6/26/17	Mon 6/26/17	5SS
7			Identify online opportunities	1 day	Mon 6/26/17	Mon 6/26/17	
8			Identify budget requirements	1 day	Mon 6/26/17	Mon 6/26/17	
9			Obtain launch budget approval	4 days	Mon 6/26/17	Thu 6/29/17	
10			Phase 2 - Planning	**5 days**	**Mon 6/26/17**	**Fri 6/30/17**	
11			Kickoff product launch	5 days	Mon 6/26/17	Fri 6/30/17	
12			Define creative brief	3 days	Mon 6/26/17	Wed 6/28/17	
13			Prepare Messaging Framework	4 days	Mon 6/26/17	Thu 6/29/17	
14			Review current literature and plan new	1 day	Mon 6/26/17	Mon 6/26/17	

Fig. A.19 Start to start link in MS Project

- To model the scenario # 3 above, which is a ***Finish-To-Start*** dependency with a Lag of 5 days, in the Predecessors column for Task 9, enter ***8FS+5d***. Notice how the start date of Task 9 is scheduled to start after 5 days after finish of Task 8.

		Task Mode	Task Name	Durati	Start	Finish	Predecessors
0			**New Crow bar Launch**	**10 days**	**Mon 6/26/17**	**Fri 7/7/17**	
1			**Phase 1- Initiation**	**10 days**	**Mon 6/26/17**	**Fri 7/7/17**	
2			Identify the launch team	5 days	Mon 6/26/17	Fri 6/30/17	
3			Determine sales objectives	3 days	Mon 7/3/17	Wed 7/5/17	2
4			Define launch goals (launch timing and publicity objectives)	4 days	Mon 6/26/17	Thu 6/29/17	
5			Identify channel partners	4 days	Mon 6/26/17	Thu 6/29/17	
6			Identify retail partners	1 day	Mon 6/26/17	Mon 6/26/17	5SS
7			Identify online opportunities	1 day	Mon 6/26/17	Mon 6/26/17	
8			Identify budget requirements	1 day	Mon 6/26/17	Mon 6/26/17	
9			Obtain launch budget approval	4 days	Tue 7/4/17	Fri 7/7/17	8FS+5 days
10			**Phase 2 - Planning**	**5 days**	**Mon 6/26/17**	**Fri 6/30/17**	
11			Kickoff product launch	5 days	Mon 6/26/17	Fri 6/30/17	
12			Define creative brief	3 days	Mon 6/26/17	Wed 6/28/17	
13			Prepare Messaging Framework	4 days	Mon 6/26/17	Thu 6/29/17	
14			Review current literature and plan new requirements	1 day	Mon 6/26/17	Mon 6/26/17	

Fig. A.20 Adding lag in MS Project

- To model the scenario # 4 above, which is a ***Start-To-Start*** dependency with a Lead of 2 days, in the Predecessors column for Task 13, enter ***12SS-2d.*** Notice how the start date of Task 13 is scheduled to start 2 days before the start of Task 12.

		Task Mode	Task Name	Durati	Start	Finish	Predecessors
9			Obtain launch budget approval	4 days	Tue 7/4/17	Fri 7/7/17	8FS+5 days
10			**Phase 2 - Planning**	**32 days**	**Mon 7/10/17**	**Tue 8/22/17**	
11			Kickoff product launch	5 days	Mon 7/10/17	Fri 7/14/17	9
12			Define creative brief	3 days	Mon 7/17/17	Wed 7/19/17	11
13			Prepare Messaging Framework	4 days	Thu 7/13/17	Tue 7/18/17	12SS-2 days
14			Review current literature and plan new requirements	1 day	Wed 7/19/17	Wed 7/19/17	13
15			Define press release requirements	2 days	Thu 7/20/17	Fri 7/21/17	14
16			Define product specification literature	1 day	Mon 7/24/17	Mon 7/24/17	15
17			Define sales presentations	1 day	Tue 7/25/17	Tue 7/25/17	16
18			Define internal communication needs	4 days	Wed 7/26/17	Mon 7/31/17	17
19			Align product release timing with marketing plan	5 days	Tue 8/1/17	Mon 8/7/17	18
20			Prepare for volume manufacturing according to sales goals	3 days	Tue 8/8/17	Thu 8/10/17	19
21			Plan sales force staffing and training to support sales goals	4 days	Fri 8/11/17	Wed 8/16/17	20

Fig. A.21 Adding lead in MS Project

- To have a dynamic schedule, it is important for every task in the project schedule to have a predecessor or successor linkage. This allows MS Project to translate the effect of a change in one of the tasks to all the tasks in the project schedule.

- There are many ways in MS Project to enter the dependencies, and the above method covers only one of those methods.

A.7 Setting the Constraints

Sometimes it is necessary in the project schedule to constrain some of the tasks to either start no earlier than or finish no later than or simply say that a task must start or finish on a specific date.

MS Project six types of constraints to suit various scenarios. Four of them are what we call **Soft Constraints** and two are **hard constraints**.

- **Soft Constraints**
  - **Start No Earlier Than**
  - **Start No Later Than**
  - **Finish No earlier than**
  - **Finish no later than**
- **Hard Constraints**
  - **Must Start on**
  - **Must Finish on**
- **Other**
  - **As soon as possible**
  - **As late as possible**

Note

- Setting constraints in a project schedule reduces the dynamic nature of schedule, and hence must be done only when it is necessary.
- Typing a Start date in MS Project directly, automatically sets the ***Start No Earlier Than*** constraint. Similarly typing a finish date in directly, automatically sets the ***Finish No Later Than*** constraint. This is the reason why it is generally advised not to type the dates in MS Project, but rather work with dependencies to arrive at a good dynamic schedule model.

A.7.1 Instructions

To set a constraint on a task in MS project

- Double click on the Task Name to bring up the ***Task Information*** dialog box.
- Click on ***Advanced*** tab

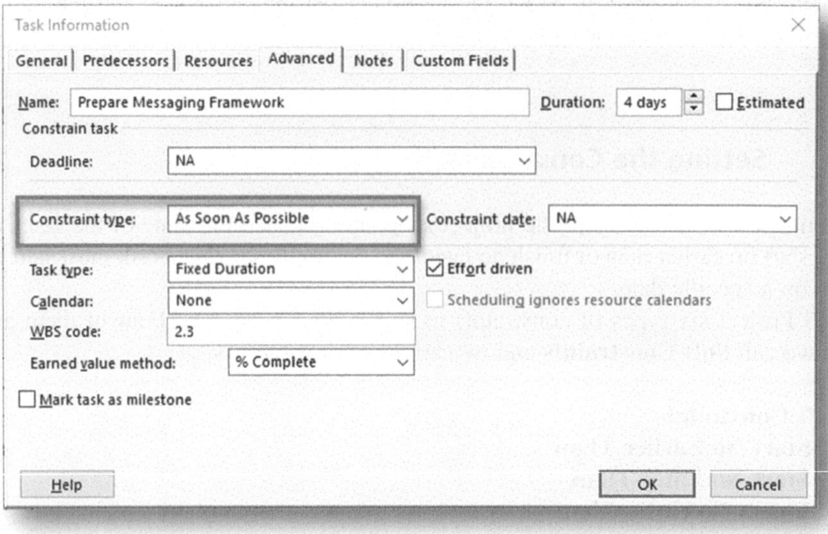

Fig. A.22 Task information dialog box

- Under the constraint type, select the constraint you want apply. For our exercise, select the ***Start No Earlier than*** constraint, and set a date for the constraint.

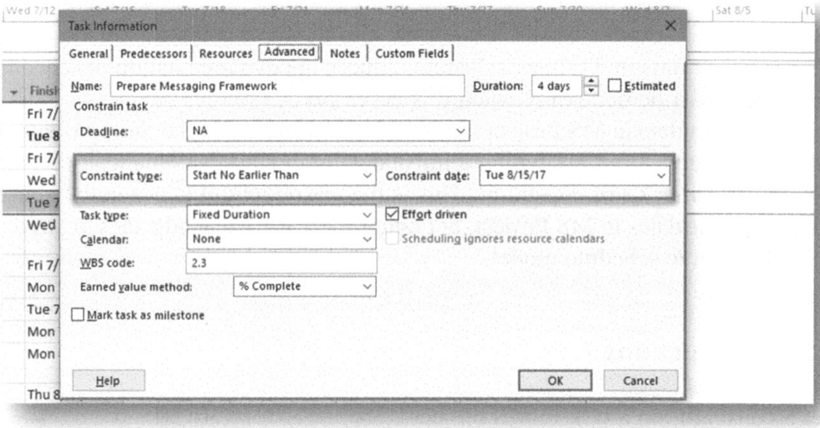

Fig. A.23 Adding a constraint

- Click ***OK***.
- Notice that date for the task we applied the constraint to has now moved to start on the constraint date, instead of honoring the dependency we set up earlier.

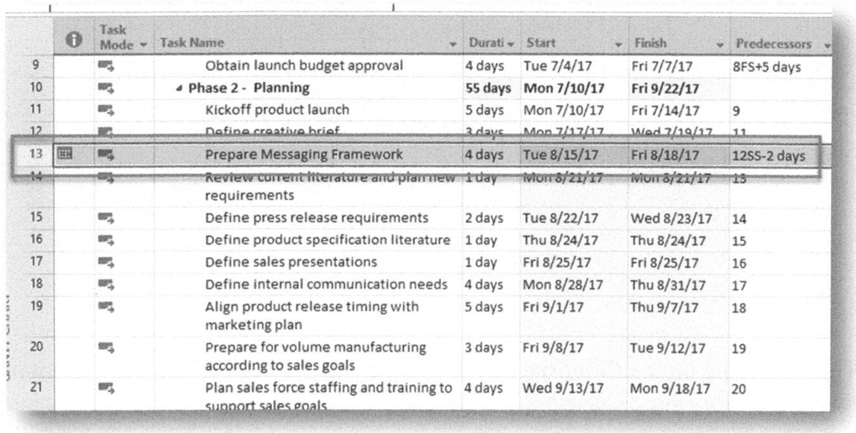

Fig. A.24 Constraint on a task

A.8 Assigning Resources

With the tasks, estimates, dependencies and constraints in place, the next step is to assign resources to the tasks in the schedule to make it a resource-loaded schedule.

MS Project provides three different types of resources to allow for various modeling scenarios

Work Resources which represent the human resources that will perform work on the project tasks

Cost Resources that can be used to capture costs on Project Tasks, that are not the result of work performed on the task

Material Resources which allow to model scenarios where the resource is not a person, but rather a material or equipment.

A.8.1 Instructions

To assign resources to tasks in MS Project, the first step is to build the project team.

- Navigate to the ***Resource Sheet*** view by clicking on ***View*** >> ***Resource Sheet*** view on the ribbon.

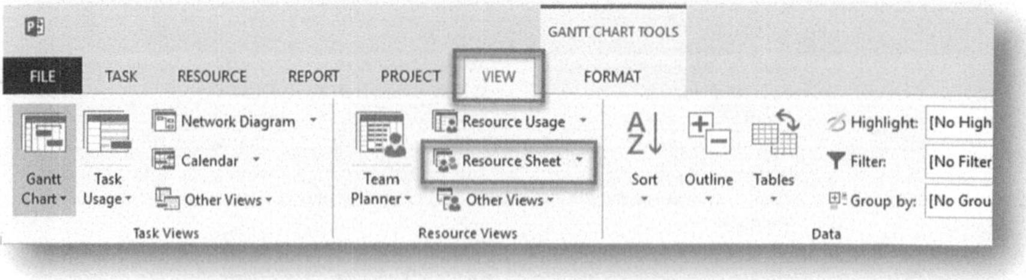

Fig. A.25 Resource sheet view

- This will bring up the ***Resource Sheet*** view. In this view, you can enter Resource Names and Resource Costs.

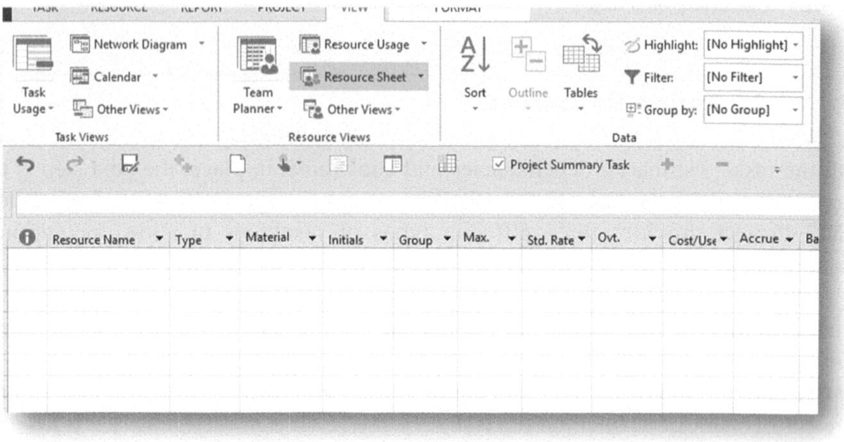

Fig. A.26 Resource sheet view

- The following screenshot shows various types of resources entered into the resource sheet view. The type of resource could be controlled by the ***Type*** column.

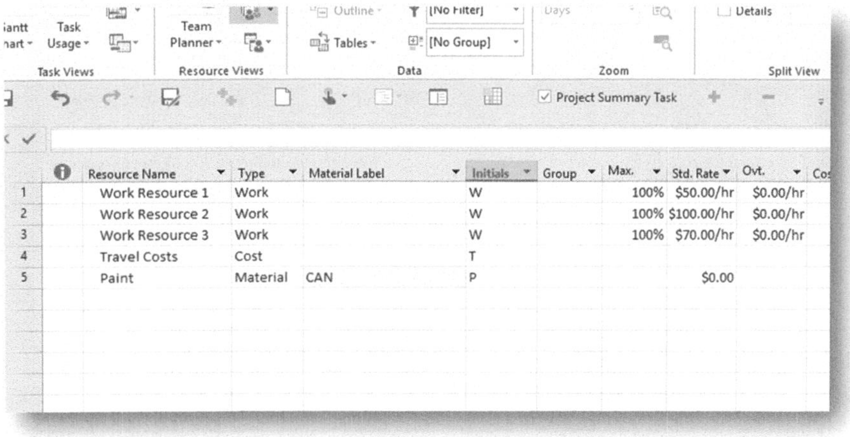

Fig. A.27 Adding resources

As you can see there is a standard rate applied to the ***Work*** resources, where there is no standard rate for the ***cost resources*** and the ***Material Resources***. For these two types of resources the costs are entered while assigning to a task.

Now to assign resources to your tasks, follow the steps below:

- Navigate back to the Gantt Chart View by clicking on ***View*** >> ***Gantt Chart***

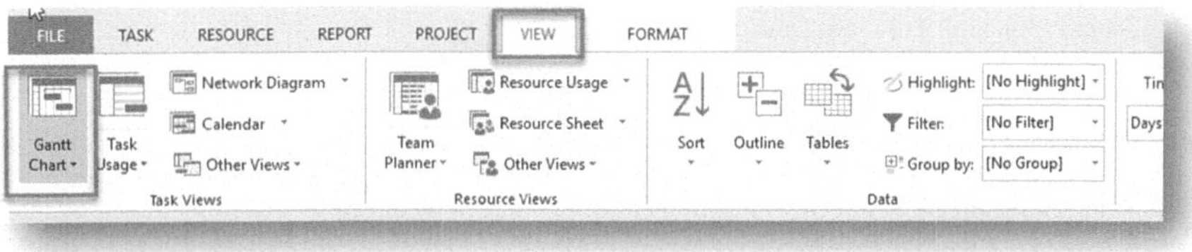

Fig. A.28 Accessing Gantt chart view

- In the Gantt Chart view, select a task you want to assign the resource to, and click on ***Resource*** >> ***Assign Resources***

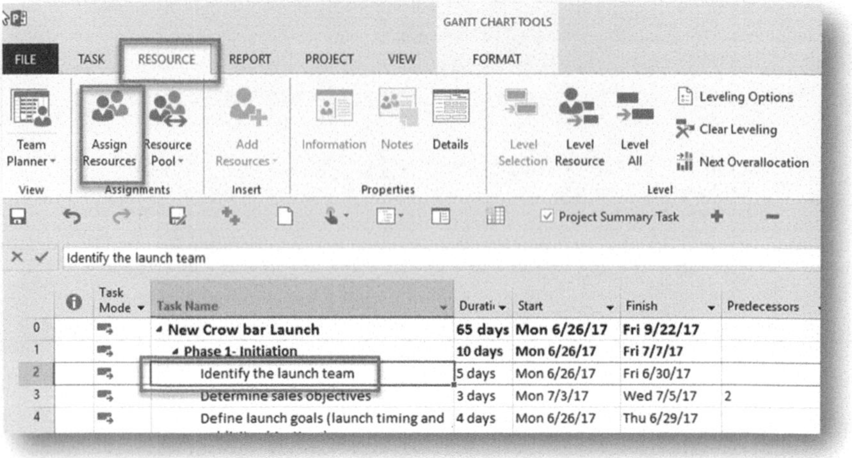

Fig. A.29 Gantt chart view

- This will bring up the ***Assign Resources*** dialog boxes

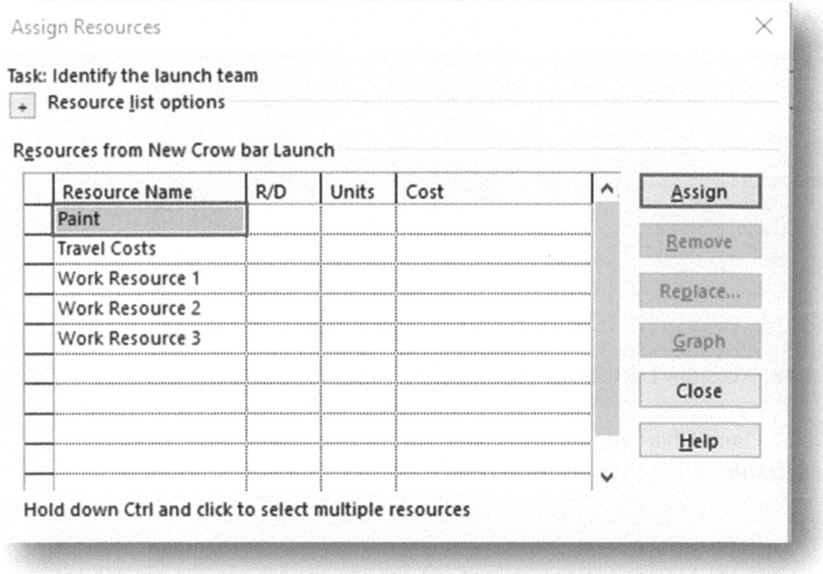

Fig. A.30 Assign resources dialog box

- From this screen, you can select the resource you want to assign and click ***Assign.***

- For Work Resources, you can specify the % of time the work resource is available for the task, by typing the number in the ***Units*** column.

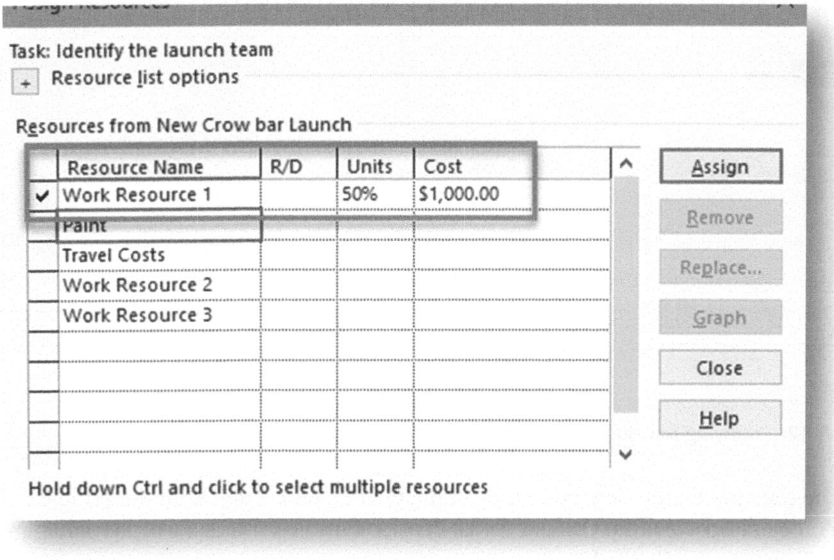

Fig. A.31 Assign resources dialog box

Once you assign resources to your tasks, the work hours are automatically calculated using the formula ***Duration*** * ***Units*** = ***Work*** and the corresponding cost is calculated by applying the hourly cost to the work hours.

As mentioned before, there are many ways in MS Project to assign resources, and the above covers one of the many ways.

A.9 Optimizing and Finalizing the Schedule Model

A.9.1 Identifying the Critical Path in MS Project

After you finish estimating and developing your resource-loaded schedule model, you may want to see if the schedule model is fully optimized or has unnecessary slack somewhere, that could be eliminated for a quicker finish date or if there is a possibility to crash the schedule to reduce the overall duration.

MS Project provides ways to view the critical path, critical tasks, Total Slack and Free Slack.

A.9.2 Instructions

- To view the Critical Path, Navigate to ***View*** >> ***Gantt Chart*** >> ***Tracking Gantt***

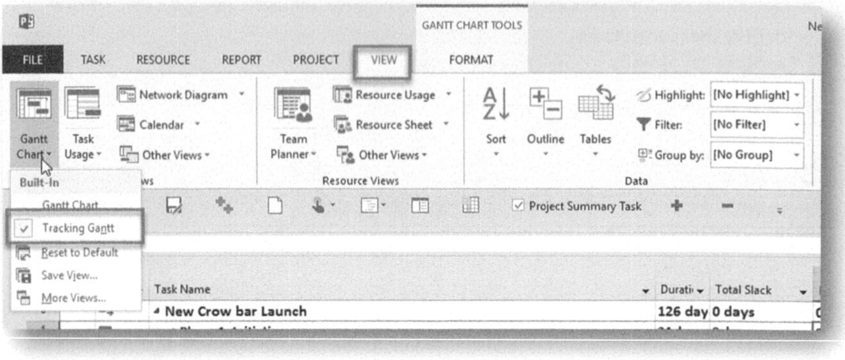

Fig. A.32 Applying tracking Gantt view

- The tracking Gantt view will show in the Gantt Chart Area, with the critical path.
- You can also view the Critical Tasks (Tasks with Free Slack = 0), by navigating to ***Format*** >> ***Critical***
- Project also calculates the **Total Slack** and the **Free Slack** for each task in the project schedule. To display the total and free slack, Click on ***Add new Column***, and add the columns.

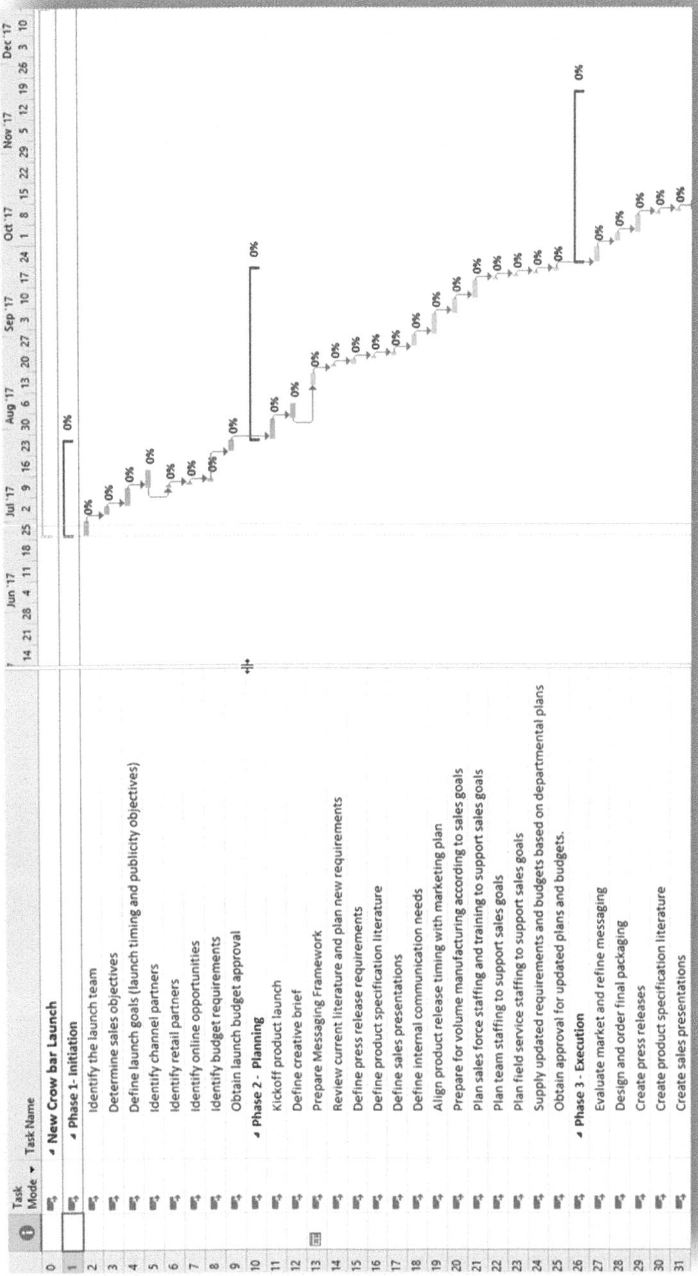

Fig. A.33 Viewing critical path

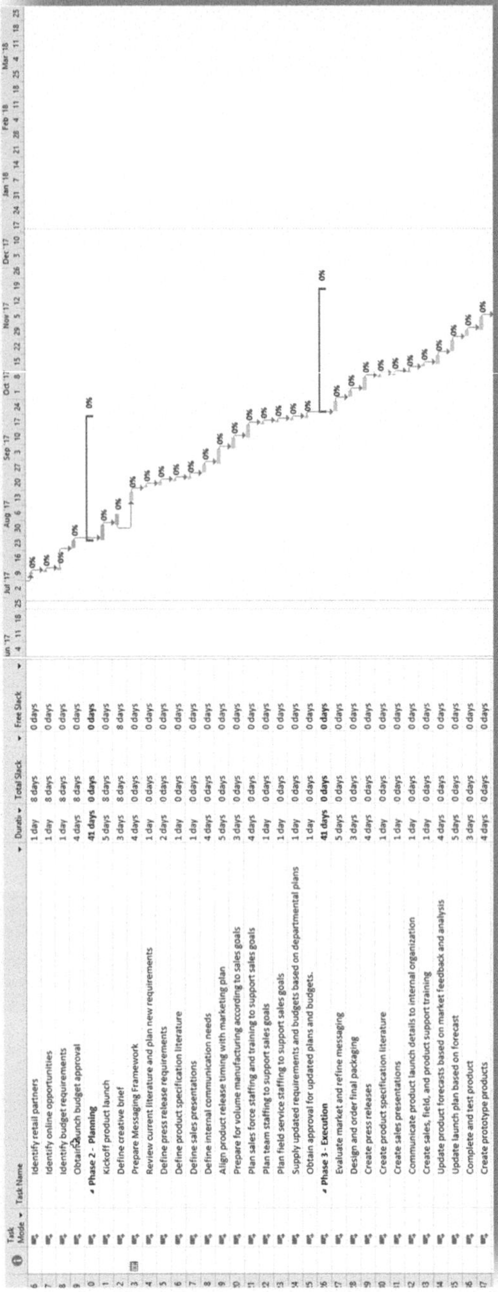

Fig. A.34 Viewing critical tasks, total slack and free slack

To optimize the schedule for time, it is important to see if the durations on the Critical Tasks could be modified. This will have an overall impact on the schedule duration, costs and timeline.

A.9.3 Setting Baseline

Once the project schedule is modelled and optimized, before any actual data is input into the schedule, it is important to take a snapshot of the schedule. This snapshot in time is called a Baseline for the Project Schedule.

This is important because, MS Project is a dynamic tool. It consumes the data from the actual work, actual costs etc., and tries to recalculate the schedule model based on the dependencies, constraints etc., In that scenario, it is important for us to have the original schedule to compare with.

MS Project offers 11 different Baseline slots to store snapshots of the project schedule data.

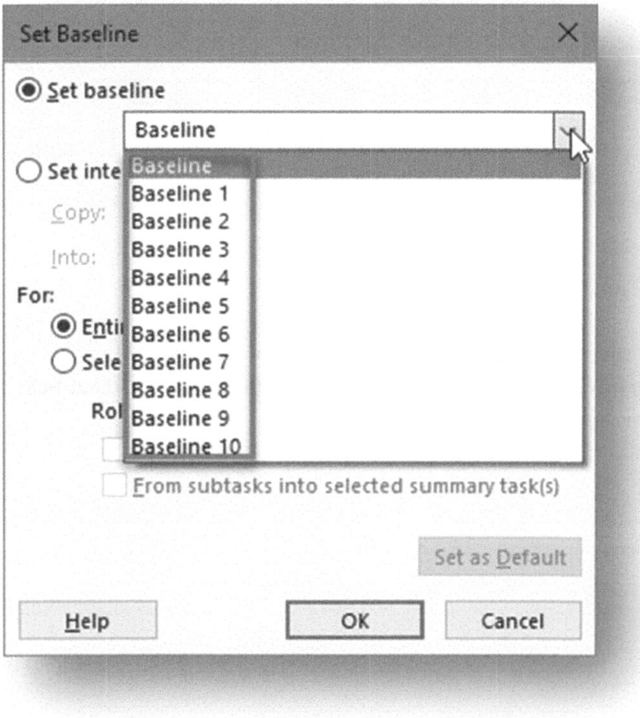

Fig. A.35 Baselines in MS Project

When you set a baseline in MS Project, the following data is saved:

- Start date
- Finish Date
- Work
- Duration
- Cost

MS Project also provides something called ***Interim Plans***, which you can use to save the start and finish dates at interim stages.

A.9.4 Instructions

- To set a baseline in MS Project, navigate to ***Project*** >> ***Set Baseline*** >> ***Set Baseline***

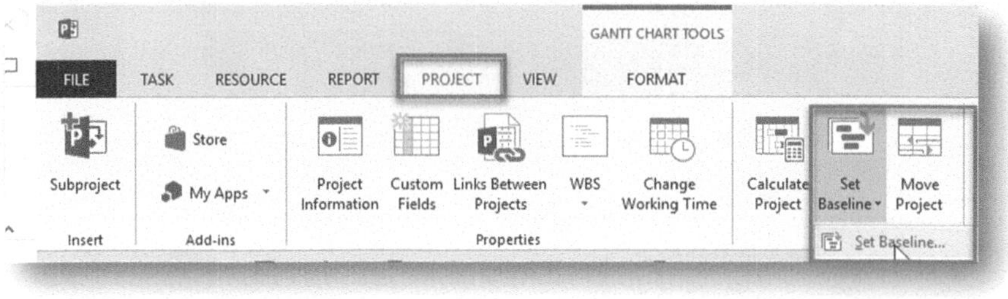

Fig. A.36 Accessing baselines in MS Project

- This will bring up the ***Set Baseline*** dialog box.
- Select ***Baseline*** and then select ***Entire Project*** (you can also set baseline for few specific tasks) and click ***OK***

Fig. A.37 Setting baselines in MS Project

- The ***Tracking Gantt*** chart will show that the Baseline has been set, by a gray bar.

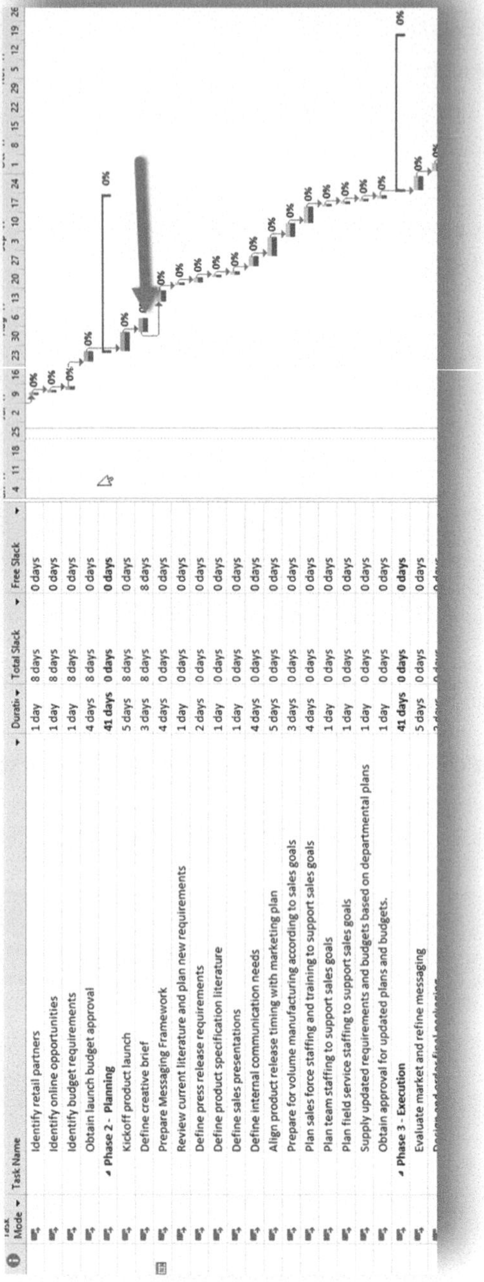

Fig. A.38 Viewing baselines in MS Project

A.10 Tracking and Managing Schedule

A project schedule is not supposed to a static model that is just a plan. It is important to update the project schedule with actual work happening, so that MS Project can calculate the rest of the schedule model to provide you with predictions on the rest of the tasks.

The process of updating the tasks in MS Project consists of many steps. The details of each step are discussed below.

A.10.1 Set Project Status Date

The Project Status Date is the date as of which the progress is being updated.

- To Set the project status date, navigate to ***Project*** >> ***Status Date*** and click on the Calendar Icon to set the Status Date. Then Click ***OK***

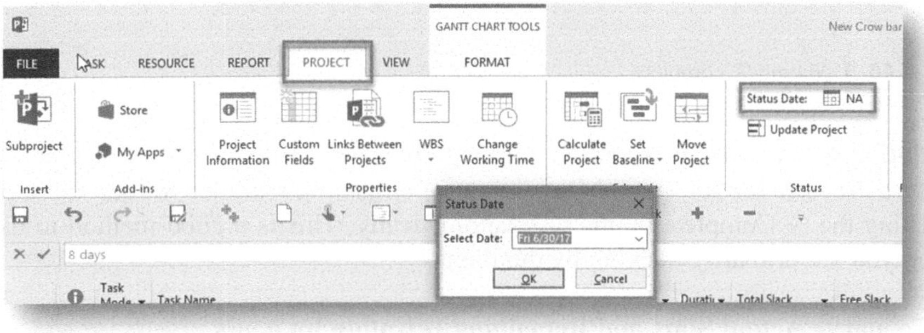

Fig. A.39 Viewing status date in MS Project

A.10.2 Update Task Progress

There are several methods to update the task progress in project schedule. We will cover some of them here.

Updating Percent Complete This method allows for users to update just the % Complete on the task. However, this approach assumes that the task happened according to the Plan, and the Planned Start and Finish Dates are treated as the Actual Start and Finish Dates, which may not be accurate in reality.

To update the Percent complete for a task,

- Apply the ***Tracking*** table, by clicking on ***View*** >> ***Tables*** >> ***Tracking***

- Type the % Complete Number you wish to update with in the ***% Complete*** Column

Task Name	Act. Start	Act. Finish	% Comp	%
New Crow bar Launch	**Mon 6/26/17**	**NA**	**3%**	**0%**
Phase 1- Initiation	**Mon 6/26/17**	**NA**	**15%**	**0%**
Identify the launch team	NA	NA	0%	0%
Determine sales objectives	NA	NA	0%	0%
Define launch goals (launch timing and publicity objectives)	NA	NA	0%	0%
Identify channel partners	NA	NA	0%	0%
Identify retail partners	NA	NA	0%	0%
Identify online opportunities	Thu 7/13/17	Thu 7/13/17	100%	0%
Identify budget requirements	Fri 7/14/17	NA	50%	0%
Obtain launch budget approval	Mon 7/24/17	NA	50%	0%
Phase 2 - Planning	**NA**	**NA**	**0%**	**0%**
Kickoff product launch	NA	NA	0%	0%
Define creative brief	NA	NA	0%	0%
Prepare Messaging Framework	NA	NA	0%	0%
Review current literature and plan new requirements	NA	NA	0%	0%
Define press release requirements	NA	NA	0%	0%
Define product specification literature	NA	NA	0%	0%
Define sales presentations	NA	NA	0%	0%

Fig. A.40 Updating % complete

Updating the Actual Start and Remaining Duration This method allows you to capture the Actual Start, Actual Duration and Remaining Duration values, there by updating the % Complete on the task automatically. This is a good method to use when you are primarily tracking by duration.

To update **Actual Start and Remaining Duration** for a task,

- Apply the ***Tracking*** table, by clicking on ***View*** >> ***Tables*** >> ***Tracking***
- Type the **Actual Start, Actual Duration and Remaining Duration** in the corresponding columns.

Task Name	Act. Start	Act. Finish	% Comp	Phys. %	Act. Dur.	Rem. Dur.	Act. Cost	Act. Work
Define creative brief	NA	NA	0%	0%	0 days	3 days	$0.00	0 hrs
Prepare Messaging Framework	Mon 6/26/17	NA	48%	0%	15 days	16 days	$0.00	0 hrs
Review current literature and plan new requirements	NA	NA	0%	0%	0 days	1 day	$0.00	0 hrs
Define press release requirements	NA	NA	0%	0%	0 days	2 days	$0.00	0 hrs
Define product specification literature	NA	NA	0%	0%	0 days	1 day	$0.00	0 hrs
Define sales presentations	NA	NA	0%	0%	0 days	1 day	$0.00	0 hrs
Define internal communication needs	NA	NA	0%	0%	0 days	4 days	$0.00	0 hrs

Fig. A.41 Updating durations

Updating Actual Start, Actual Work and Remaining Work This method allows you to capture the Actual Start, Actual Work and Remaining work values, there by updating the % Work Complete on the task automatically. This is a good method to use when you are primarily tracking by Work.

To update **Actual Start, Actual Work and Remaining Work** for a task,

- Apply the ***Tracking*** table, by clicking on ***View*** >> ***Tables*** >> ***Tracking***
- Type the **Actual Start, Actual Work and Remaining Work** in the corresponding columns.

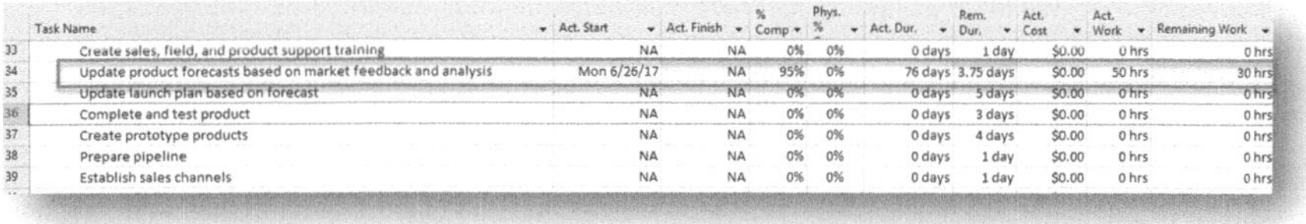

	Task Name	Act. Start	Act. Finish	% Comp	Phys. %	Act. Dur.	Rem. Dur.	Act. Cost	Act. Work	Remaining Work
33	Create sales, field, and product support training	NA	NA	0%	0%	0 days	1 day	$0.00	0 hrs	0 hrs
34	Update product forecasts based on market feedback and analysis	Mon 6/26/17	NA	95%	0%	76 days	3.75 days	$0.00	50 hrs	30 hrs
35	Update launch plan based on forecast	NA	NA	0%	0%	0 days	5 days	$0.00	0 hrs	0 hrs
36	Complete and test product	NA	NA	0%	0%	0 days	3 days	$0.00	0 hrs	0 hrs
37	Create prototype products	NA	NA	0%	0%	0 days	4 days	$0.00	0 hrs	0 hrs
38	Prepare pipeline	NA	NA	0%	0%	0 days	1 day	$0.00	0 hrs	0 hrs
39	Establish sales channels	NA	NA	0%	0%	0 days	1 day	$0.00	0 hrs	0 hrs

Fig. A.42 Updating actual start, actual work and remaining work

A.10.3 Reschedule Incomplete Work

After you have captured the actuals information, it is possible that some of the work has not been complete and is now showing in the past. You will need to move the work to the future so that it can be performed to a plan.

To reschedule incomplete work to new date,

- Navigate to ***Project*** >> ***Update Project***

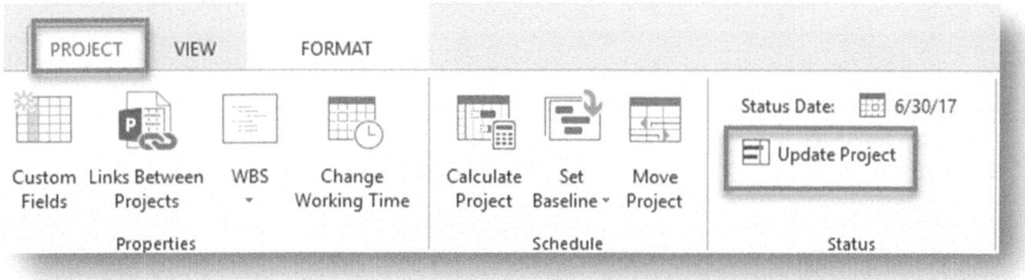

Fig. A.43 Reschedule incomplete work

- On the Update Project dialog box, select ***Reschedule Uncompleted work to start after*** and set a date. You can select some specific tasks or the entire project for this change. Click ***OK***

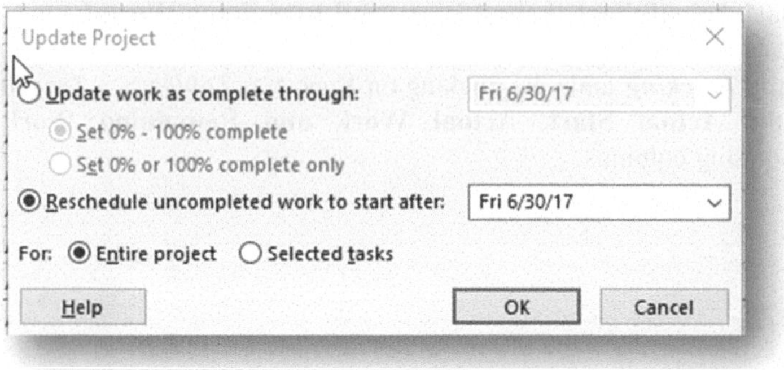

Fig. A.44 Reschedule incomplete work

This will reschedule all the incomplete work in the past to the new date you choose.

A.10.4 Reviewing Variances

At this point, you can review your schedule and compare it with the baseline to look at the variances, and take necessary corrective actions.

To view the variances in the project schedule, to the baseline:

- Navigate to ***View*** >> ***Tables*** >> ***Variance***. This will bring up the variance table which shows the variance between the current start and finish dates to the baseline dates.

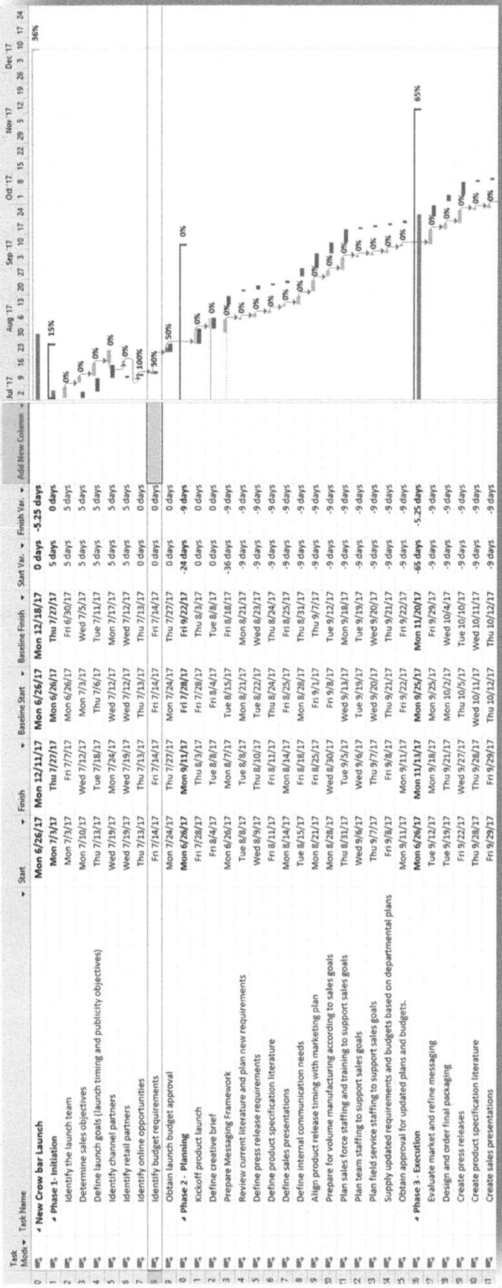

Fig. A.45 Reviewing variances

A.11 Reporting

MS Project provides very rich graphical reporting to be able to quickly generate reports on your project schedule data.

The following reports come inbuilt:

Dashboard reports	Burndown
	Cost overview
	Project overview
	Upcoming tasks
	Work overview
Resource reports	Overallocated resources
	Resource overview
Cost reports	Cash flow
	Cost overruns
	Earned value report
	Resource cost overview
Tasks reports	Task cost overview
	Critical tasks
	Late tasks
	Milestone report
	Slipping tasks

You can build your own reports, if needed from the provided templates.

To access the reports

- Navigate to ***Reports*** and select a reports section, to view the report

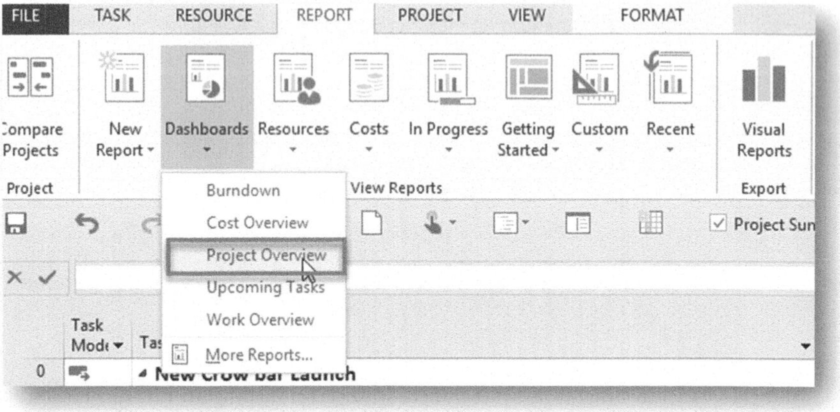

Fig. A.46 Viewing reports

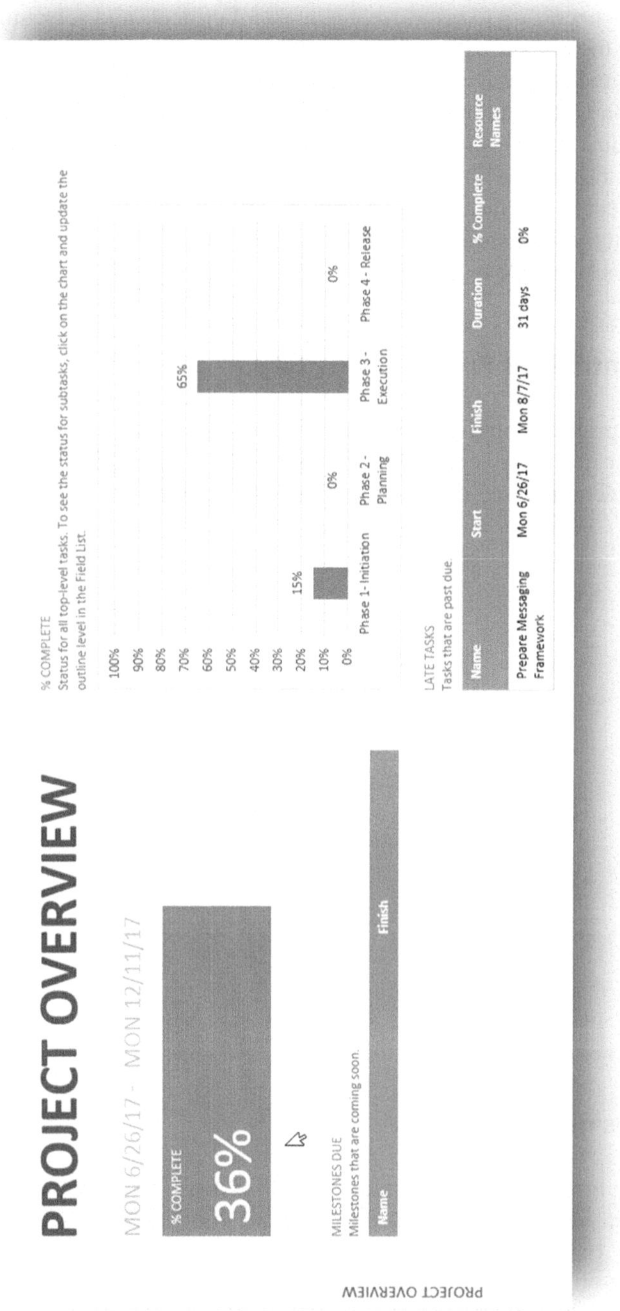

Fig. A.47 Project overview report

A.12 Project Portfolio Management with Project Online

When there are multiple projects competing for a limited budget and constrained time, it becomes necessary to prioritize the intake of projects and select the very best of them, which most align with your strategic objectives.

This management of projects jointly, to allow for the maximization of efficiency and optimization of constraints is known as Project Portfolio Management.

While a manual method could be built and used to prioritize and select projects, Project Online, the cloud offering for Microsoft PPM, and provides a more robust set of features to decide and manage your project portfolio.

The following sections offer a high-level overview of using Project Online for Portfolio Analysis and Management.

A.12.1 Centralized Project Repository

Before a set of projects could be analyzed together, they need to be located centrally in a single repository. Project Online provides this functionality, where projects could be created, stored and managed from the Project Center.

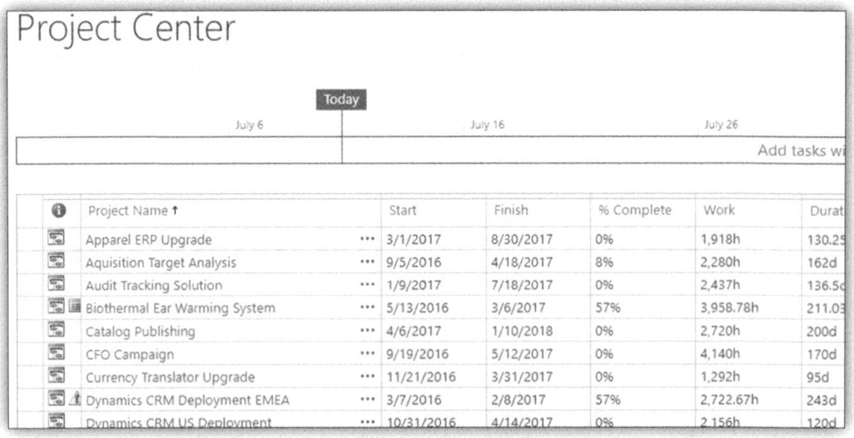

Fig. A.48 Project center

A.12.2 Business Drivers

When there are multiple projects competing for priority, one of the key factors that needs to be considered is how well they align to your business strategy i.e. their strategic value. Using Business drivers, and related prioritization technique allows business users to be objective about prioritize the business drivers themselves, and then also objectively eliminate the bias on selecting projects based on these business drivers.

Fig. A.49 Business drivers

It is recommended to define between, 5 and 12 drivers for each portfolio. Any less than five drivers are typically inadequate to capture the complexities of the organization. Any more than 12 often become unwieldy in the prioritization session.

To create a new business driver,

- Navigate to ***Driver Library*** and click on ***New***.
- Complete the details to create the business driver, and click ***Save***.

Fig. A.50 Creating a new business driver

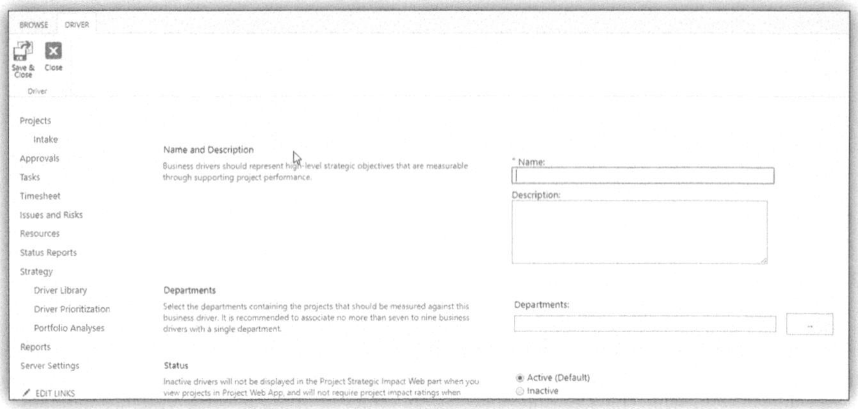

Fig. A.51 Creating new business driver

The business driver set up form includes specifying the ***impact statements*** that the project requestors will have to use to rate their project against that particular driver. These impact statements should be as quantifiable as possible to make sure that there is no ambiguity in the rating of the project.

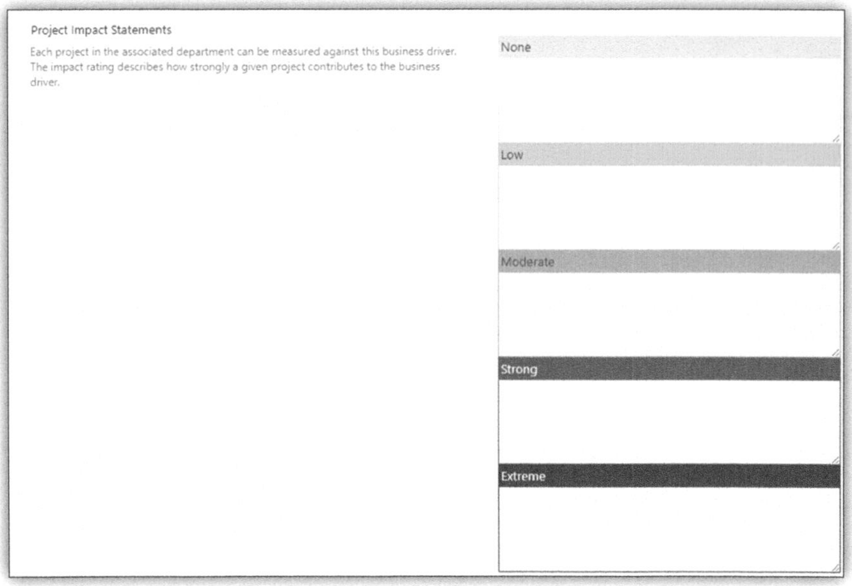

Fig. A.52 Business drivers impact statement

A.12.3 Business Driver Prioritization

To help remove the subjectivity of people rating their projects against business drivers, it is important to weigh the business drivers against each other and determine which driver is better than the other. Project Online uses a technique called ***Pair-Wise*** analysis to achieve this.

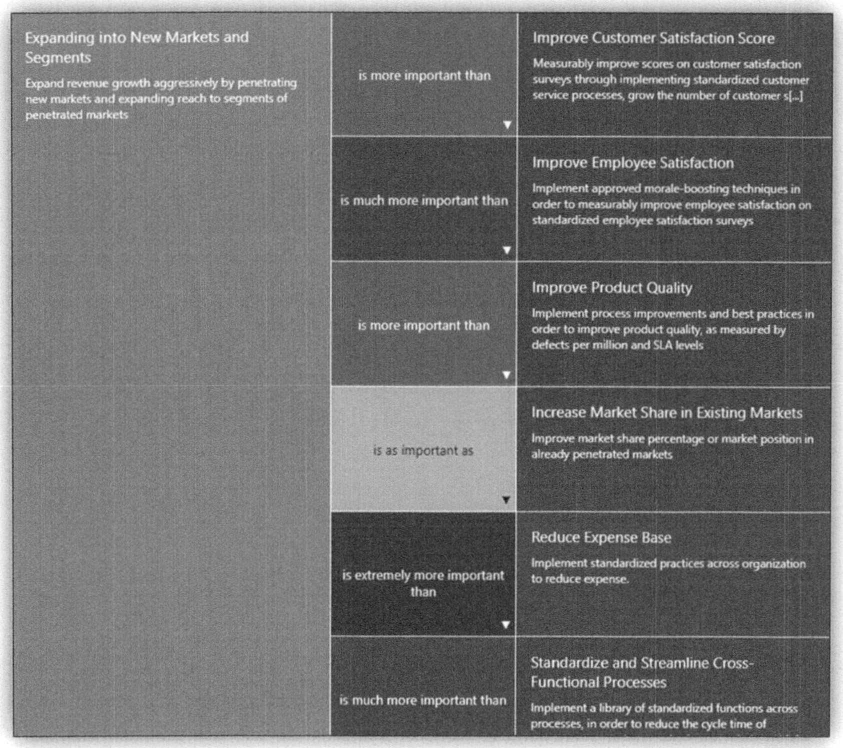

Fig. A.53 Business drivers pair wise analysis

Once you rate the business drivers against one another as more important or less important, the system is able to provide a prioritized list of Business Drivers in that prioritization scenario.

Driver Name	Priority ↓
Increase Market Share in Existing Markets	33.67%
Expanding into New Markets and Segments	30.92%
Standardize and Streamline Cross-Functional Processes	10.55%
Improve Product Quality	10.49%
Improve Customer Satisfaction Score	7.18%
Reduce Expense Base	3.74%
Improve Employee Satisfaction	3.45%

Fig. A.54 Business drivers prioritized list

Since the priorities within an organization might vary based on the perspective of the team conducting the pairwise analysis, Project Online allows in this module to set up several prioritizations to suit various needs.

Name ▲	Department ▲	Type	Complete
CIO Priortization		Calculated	Yes
COO Prioritization		Calculated	Yes
Executive Consensus		Calculated	Yes
HR Prioritization		Calculated	Yes

Fig. A.55 Business drivers prioritizations

In addition to this, the prioritization could be set to be calculated automatically or you can assign manual weights to each of the business drivers if you need.

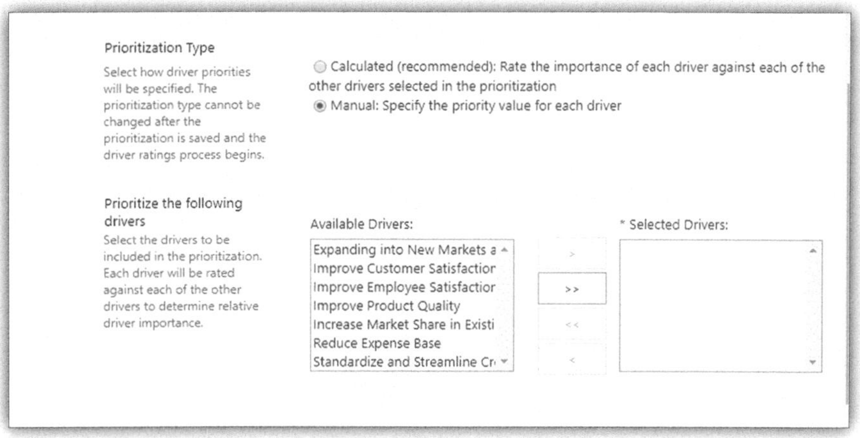

Fig. A.56 Business drivers automatic vs. manual

Fig. A.57 Business drivers manual prioritization

A.12.4 Project Rating Against Business Drivers

The business drivers we set up in earlier steps, will appear along with the impact statements, at the project level, under the title, ***Strategic Impact***. This is where the project requestors get to rate their projects alignment with the business strategy of the organization.

Once all the projects that are part of the current analysis have been rated using the Strategic Impact form, the next step is to set up and run the Portfolio Analysis.

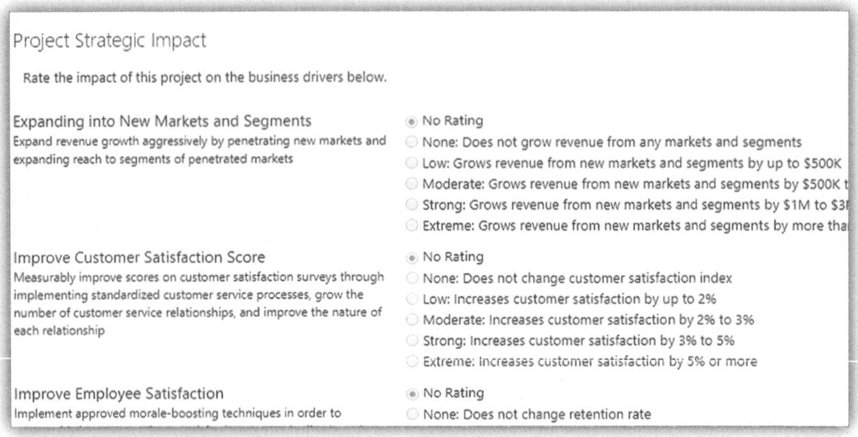

Fig. A.58 Strategic impact page

A.12.5 Setting Up the Portfolio Analysis

The portfolio analysis module in Project Online allows users to use all the information like the business drivers, business driver prioritization, and strategic alignment of projects to the business drivers, and analyze them together. It also allows users to build and run several what-if scenarios to optimize for the best possible combination of projects subject to constraints like Cost and Resources.

To get started with building an analysis,

- Navigate to ***Portfolio Analyses*** and click ***New***

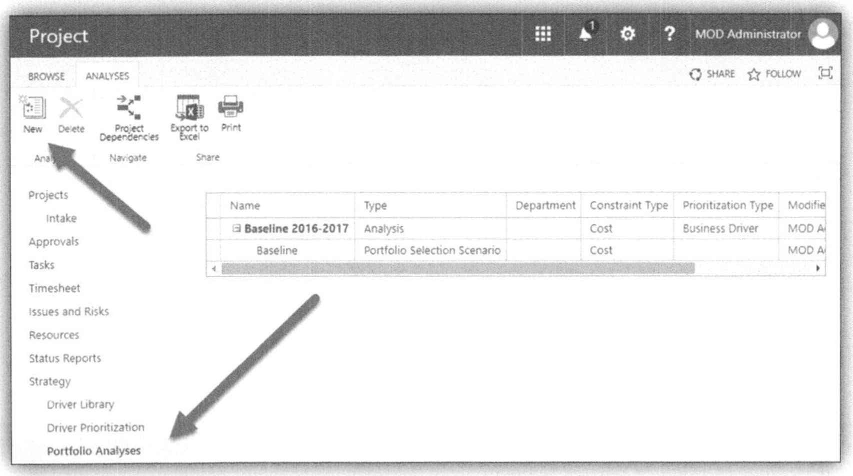

Fig. A.59 Creating new portfolio analysis

Set up your analysis by filling in the required fields for the portfolio analysis. The picture below shows a sample scenario setup.

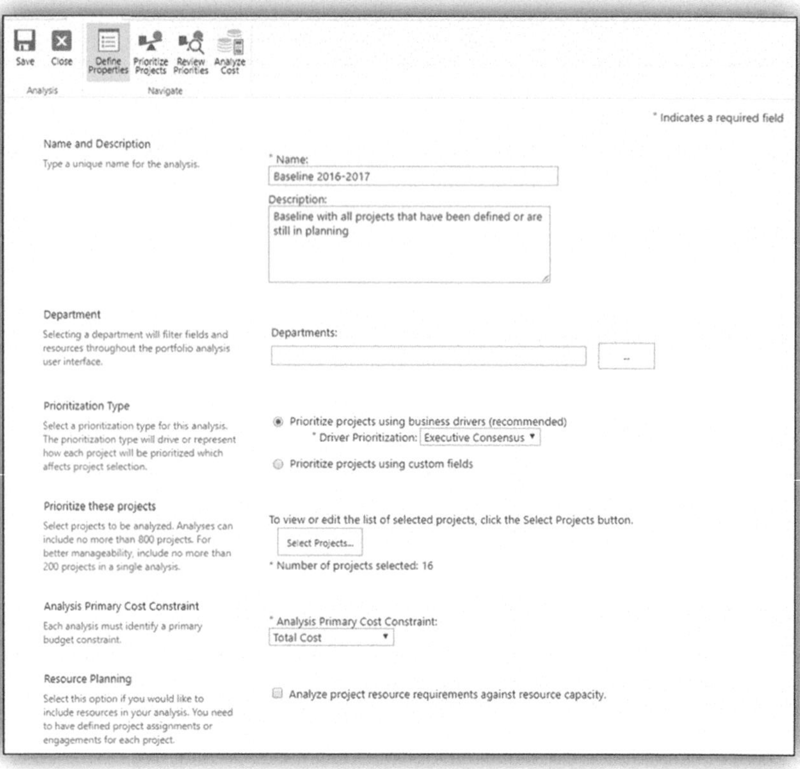

Fig. A.60 Creating new portfolio analysis page

- Once you are done setting up the portfolio analysis, the next step is to ***Prioritize Projects***. To move to this step, click on ***prioritize projects*** on the ribbon.

Fig. A.61 Prioritize projects

- This brings up a screen, where you can view the strategic impact ratings for each of the project selected. This screen also allows you to adjust the ratings, allowing for more what-if scenario modeling.

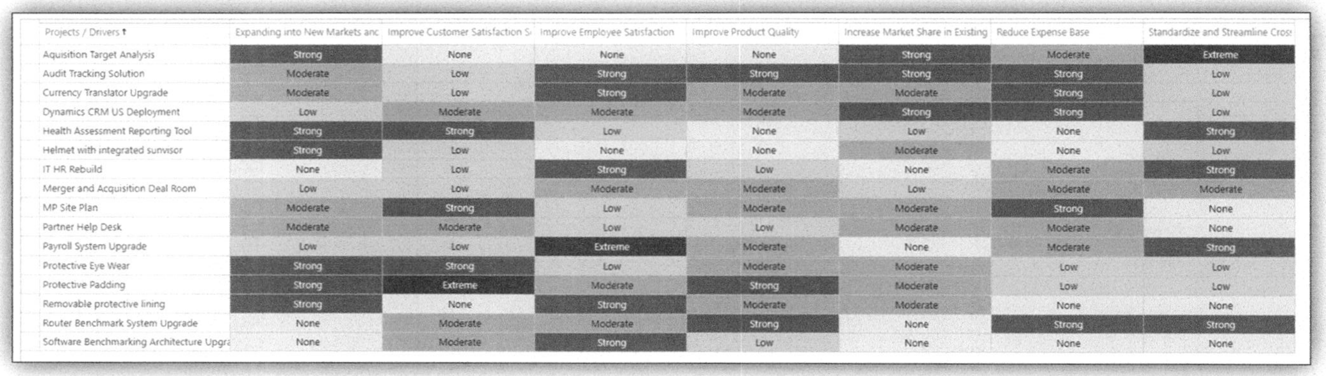

Projects / Drivers ↑	Expanding into New Markets anc	Improve Customer Satisfaction S	Improve Employee Satisfaction	Improve Product Quality	Increase Market Share in Existing	Reduce Expense Base	Standardize and Streamline Cross
Aquisition Target Analysis	Strong	None	None	None	Strong	Moderate	Extreme
Audit Tracking Solution	Moderate	Low	Strong	Strong	Strong	Strong	Low
Currency Translator Upgrade	Moderate	Low	Strong	Moderate	Moderate	Strong	Low
Dynamics CRM US Deployment	Low	Moderate	Moderate	Moderate	Strong	Strong	Low
Health Assessment Reporting Tool	Strong	Strong	Low	None	Low	None	Strong
Helmet with integrated sunvisor	Strong	Low	None	None	Moderate	None	Low
IT HR Rebuild	None	Low	Strong	Low	None	Moderate	Strong
Merger and Acquisition Deal Room	Low	Low	Moderate	Moderate	Low	Moderate	Moderate
MP Site Plan	Moderate	Strong	Low	Moderate	Moderate	Strong	None
Partner Help Desk	Moderate	Moderate	Low	Low	Moderate	Moderate	None
Payroll System Upgrade	Low	Low	Extreme	Moderate	None	Moderate	Strong
Protective Eye Wear	Strong	Strong	Low	Moderate	Moderate	Low	Low
Protective Padding	Strong	Extreme	Moderate	Strong	Moderate	Low	Low
Removable protective lining	Strong	None	Strong	Moderate	Moderate	None	None
Router Benchmark System Upgrade	None	Moderate	Moderate	Strong	None	Strong	Strong
Software Benchmarking Architecture Upgra	None	Moderate	Strong	Low	None	None	None

Fig. A.62 Prioritize projects—view strategic impact ratings

- The next step is to ***Review priorities*** of the projects. In this step, the tool uses **Analytical Hierarchical programming technique** to measure the alignment of the projects to the business drivers and assign them a relative priority value.

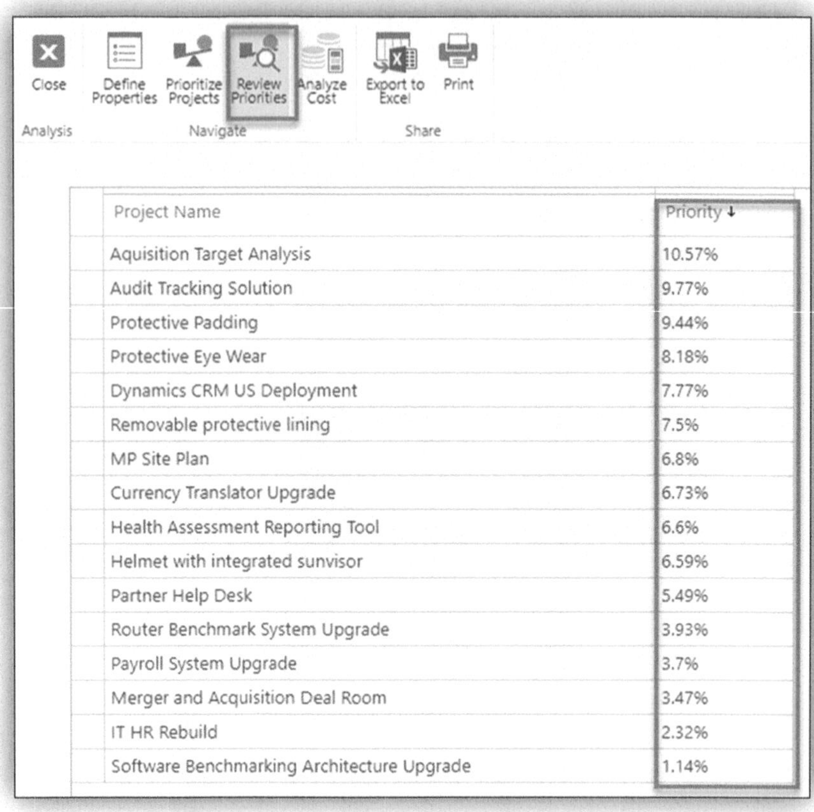

Fig. A.63 Review project priorities

- Once you have the relative priorities for the projects, it becomes easier to perform the cost constraint analysis. To do this, click on ***Analyze Cost*** on the ribbon.

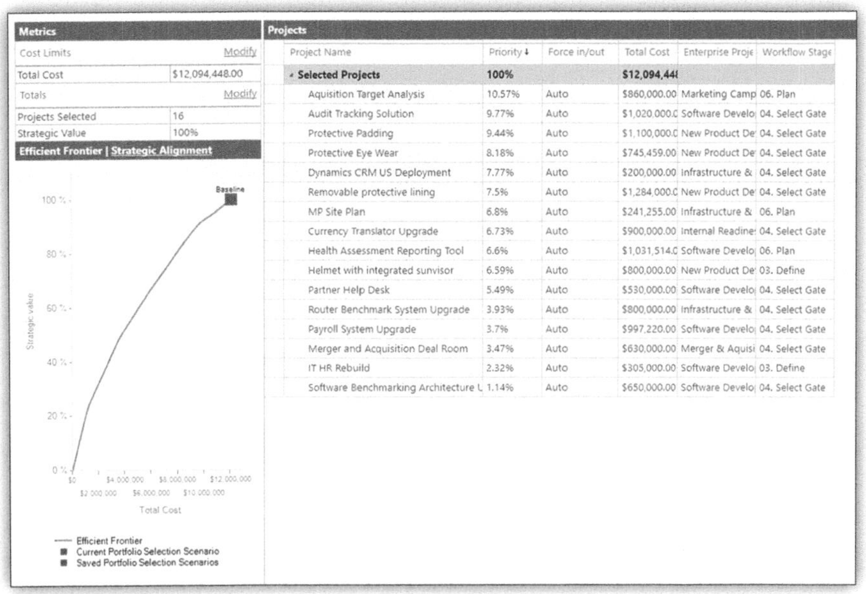

Fig. A.64 Cost constraint analysis

- Initially all the projects are selected because, there is no cost constraint applied yet. You can also see the efficient frontier is at 100%.
- To apply a cost constraint, in the ***Cost Limits*** section, enter a value for the constraint on ***Total Cost*** and click ***Recalculate***.

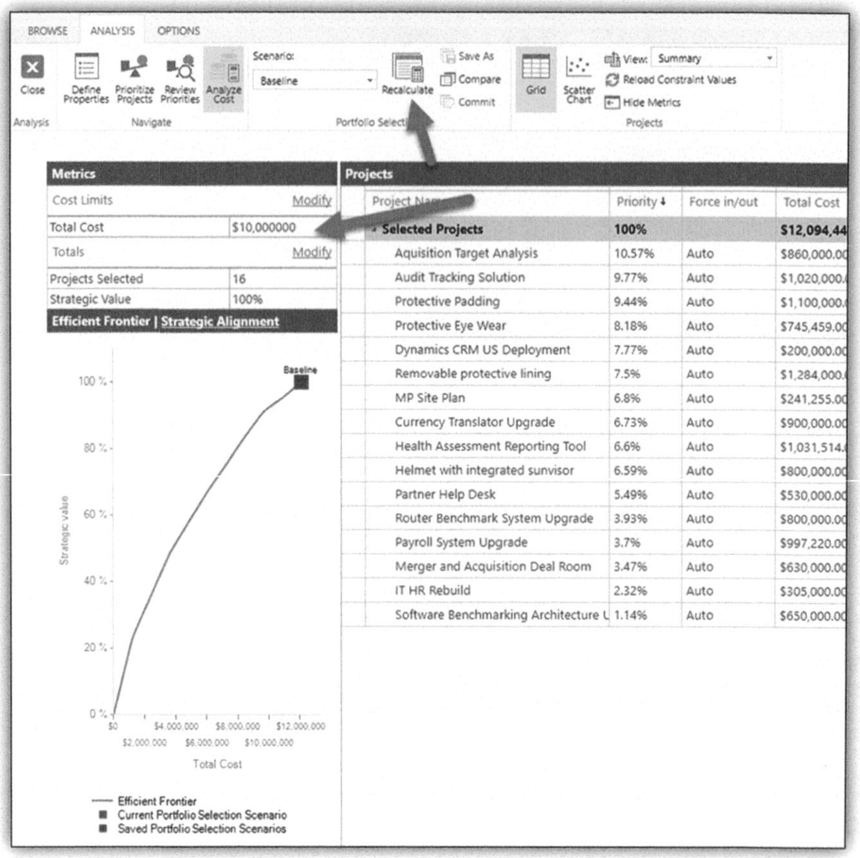

Fig. A.65 Cost constraint analysis

- Since the constraint limit we set is less than the total budget necessary for all of the projects, the tool will move out some projects that do not contribute to the strategic priority.

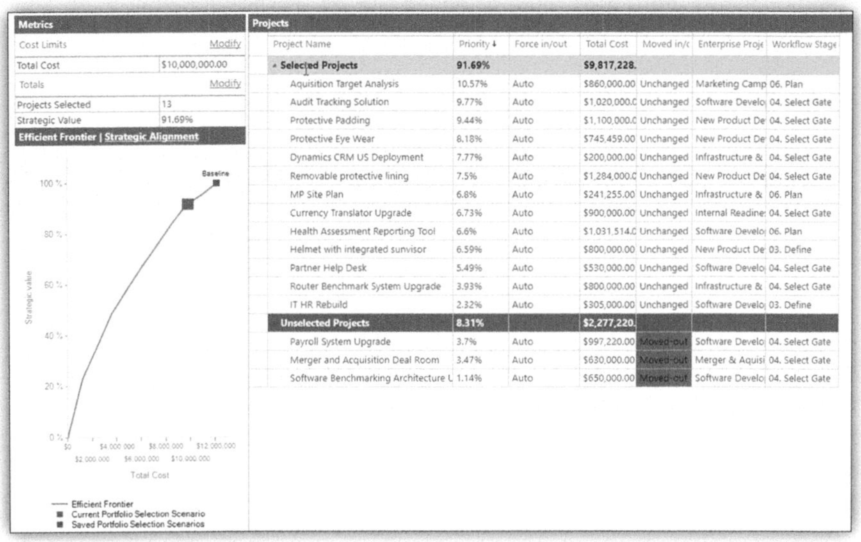

Fig. A.66 Cost constraint analysis

- If any of the projects that have been moved out are needed to be in the portfolio due to any reason (like a regulatory compliance reason), those projects could be ***Forced-in***. Obviously, if a project is forced in, the analysis will move one or more projects out of the selected portfolio to honor the cost constraint set.

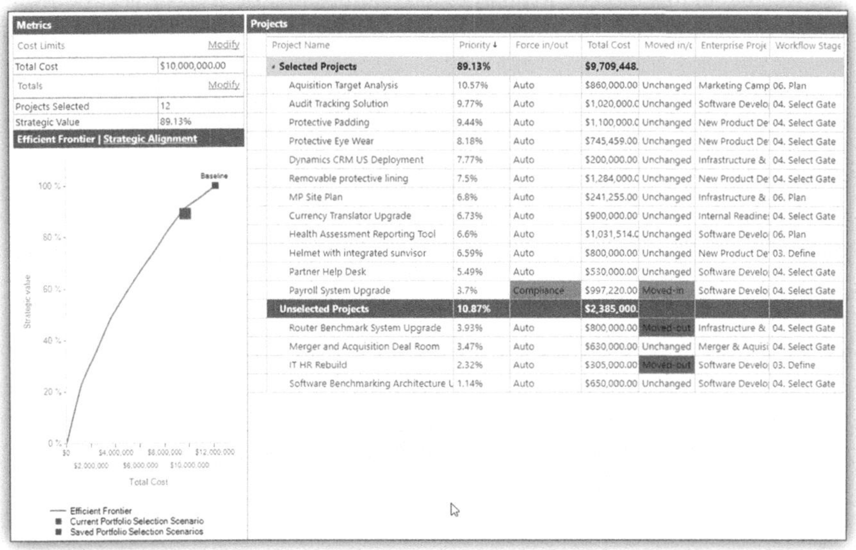

Fig. A.67 Cost constraint analysis

- Additional cost limits also could be added and set, as shown below.

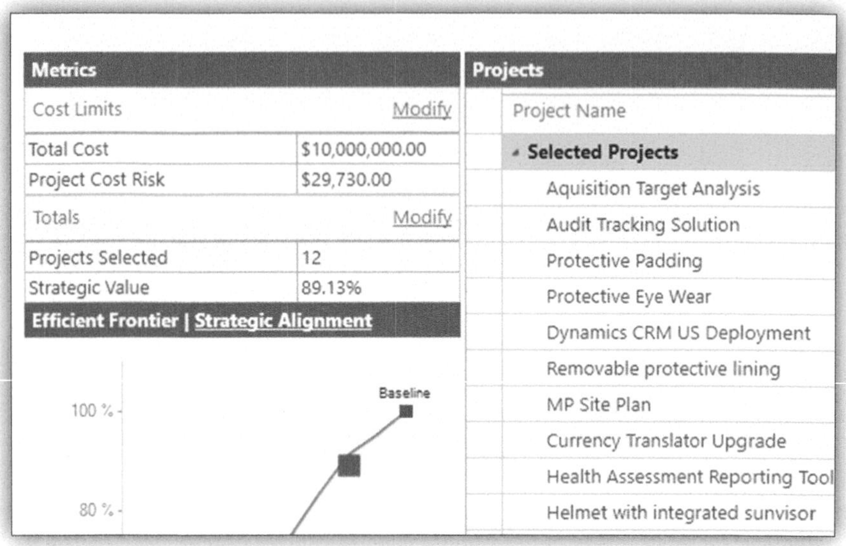

Fig. A.68 Cost constraint analysis

The goal of this analysis is to maximize the efficient frontier for a given set of projects, while keeping the strategic alignment high. You can use various reports and views available in this module to analyze the optimization of your project portfolio.

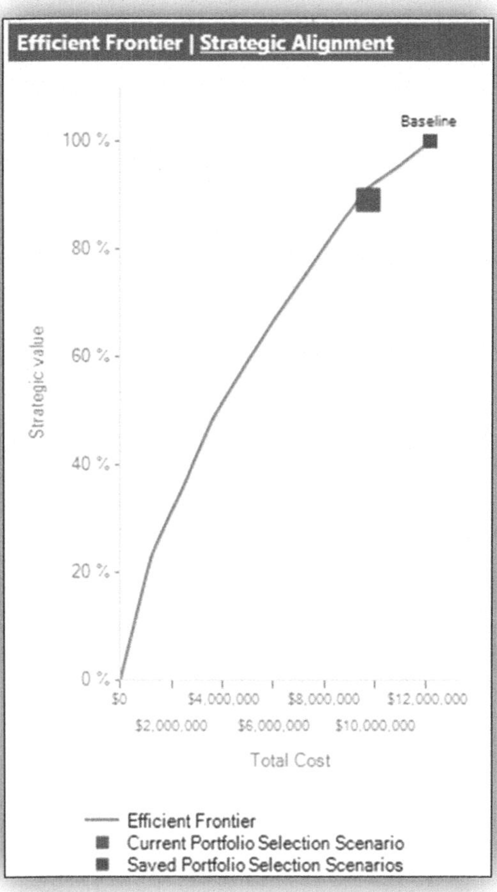

Fig. A.69 Efficient frontier

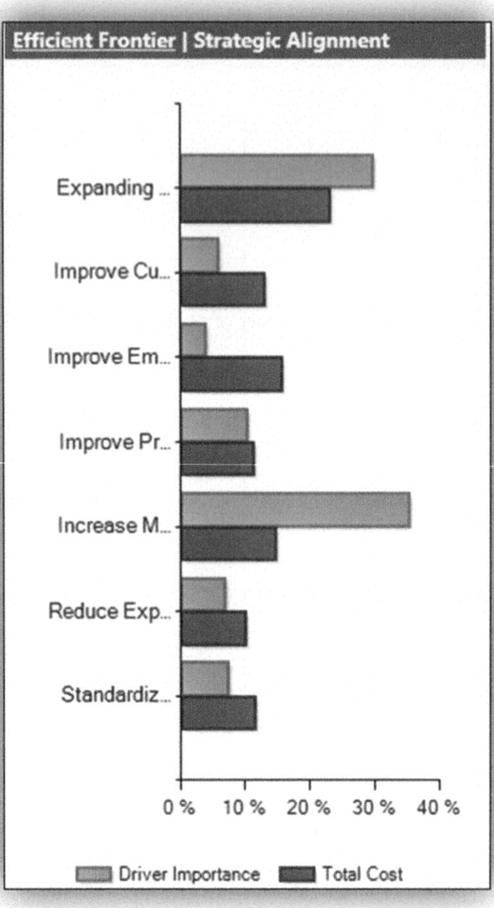

Fig. A.70 Strategic alignment chart

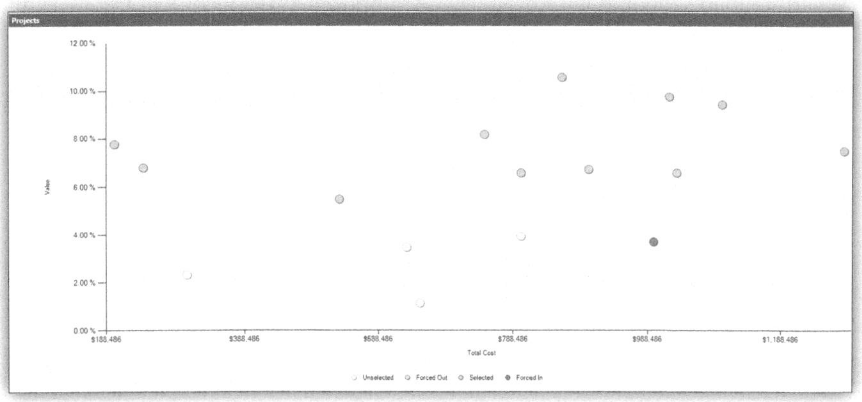

Fig. A.71 Scatter chart

Once an optimum portfolio is reached through comparison of several scenarios and executive consensus, the portfolio can now be committed.

Further to the cost constraint analysis, the selected projects could be passed through a Resource Constraint Analysis where it can be analyzed if the selected project portfolio could be achieved from a Resource perspective.

For the sake of brevity of this section, this discussion of Resource Constraint Analysis is omitted. Please refer to this Microsoft TechNet white paper on Portfolio Analysis, for further reference on this topic.

A.13 Earned Value Management

One of the key questions, that any project manager is commonly required to answer is how their project is performing against the planned schedule. The default percent complete in Microsoft Project provides the current progress but that does not really tell us whether the project is on schedule, or ahead or behind the schedule.

To be able to measure project performance against the Planned progress and costs, we will need to use **Earned Value Management (EVM)**.

MS Project provides several equivalent fields that are required to perform a EVM, out of the box, and also calculates those values for. This section will cover some of the key elements of this process and how they could be achieved using the tool.

A.13.1 Basic Elements Required to Conduct an EVM Analysis

Before we can perform Earned Value Management (EVM) on the project schedule, the schedule needs to have few basic elements in place.

- A well-defined schedule with properly decomposed WBS to reflect the work that is being done as part of the project.
- A baseline set at the correct stage for the project schedule.
- A time-phased budget for the project schedule. In a resource loaded schedule, since most of the cost comes from resource costs, it becomes imperative that the resources have an hourly standard rate associated to them.
- A regular and well-defined method for updating the project progress.

Once you have the above elements in place, you can use MS Project to calculate the earned value. The following table some of the fields required for EVM, and how they map to the fields in MS Project.

Field	MS Project field	
Planned Value (PV)	Budgeted Cost of Work Scheduled (BCWS)	The BCWS (budgeted cost of work scheduled) fields contain the cumulative time-phased baseline costs up to the status date or today's date. The time-phased versions of these fields show values distributed over time
Earned Value (EV)	Budgeted Cost of Work Performed (BCWP)	The budgeted cost of work performed (BCWP) fields contain the cumulative value of the task's, resource's, or assignments' percent complete multiplied by the time-phased baseline costs. BCWP is calculated up to the status date or today's date
Budget at Completion (BAC)	Baseline Cost	The Baseline Cost fields show the total planned cost for a task, a resource for all assigned tasks, or for work to be performed by a resource on a task. Baseline cost is also referred to as budget at completion (BAC), an earned value field.
Actual Cost (AC)	Actual Cost of Work Performed	The actual cost of work performed (ACWP) fields show costs incurred for work already done on a task, up to the project status date or today's date
Schedule Performance Index (SPI)	Schedule Performance Index (SPI)	The schedule performance index (SPI) field shows the ratio of the budgeted cost of work performed to the budgeted cost of work scheduled (BCWP/BCWS). SPI is often used to estimate the project completion date. The time-phased version of this field shows values distributed over time
Cost Performance Index (CPI)	Cost Performance Index (CPI)	The cost performance index (CPI) fields show the ratio of budgeted (or baseline) costs of work performed to actual costs of work performed, up to the project status date or today's date. The time-phased version of this field shows values distributed over time

For a more comprehensive list of fields that make up the Earned Value Management Calculations from MS Project, please refer to: https://support.office.com/en-us/article/Earned-value-fields-53ac5f70-6eec-4a09-8859-fc83e3509ed7

A.13.2 An Example

Consider the following schedule, which we will use an example for demonstrating the Earned Value Calculations in MS Project.

To set the method of Earned Value Calculation that MS Project should use,

- Navigate to ***File*** >> ***Options*** >> ***Advanced*** and set the following earned value options

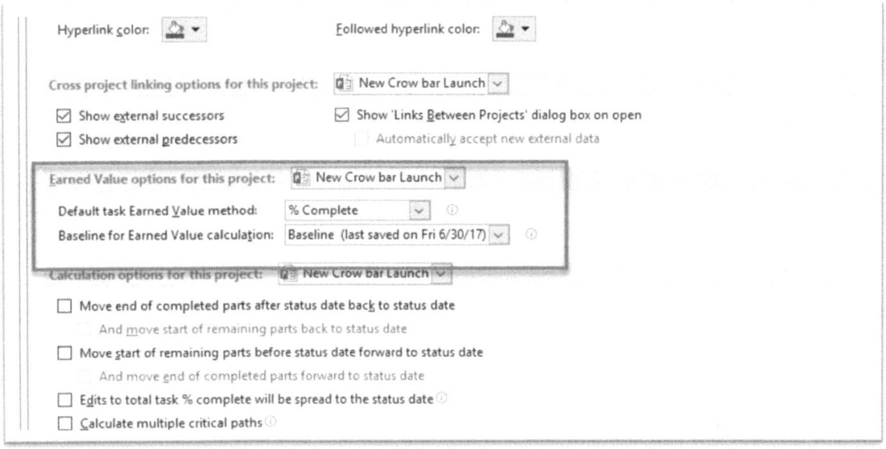

Fig. A.72 Setting earned value calculation options

- Make sure that the project schedule is built per instructions in previous chapters, baselined and resource loaded, with resource costs.

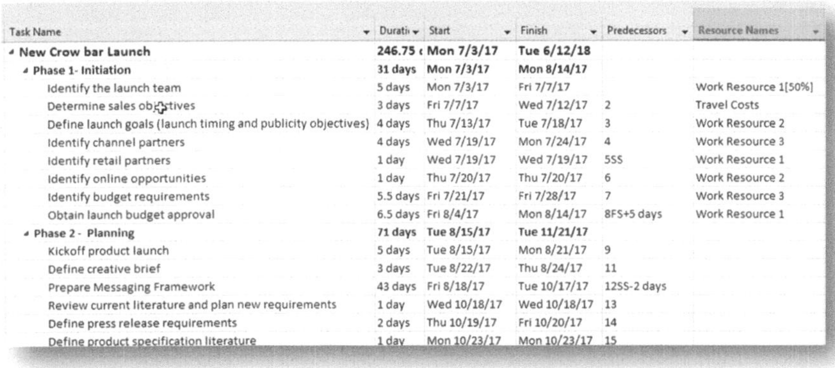

Task Name	Durati	Start	Finish	Predecessors	Resource Names
New Crow bar Launch	246.75 (	Mon 7/3/17	Tue 6/12/18		
Phase 1- Initiation	31 days	Mon 7/3/17	Mon 8/14/17		
Identify the launch team	5 days	Mon 7/3/17	Fri 7/7/17		Work Resource 1[50%]
Determine sales objectives	3 days	Fri 7/7/17	Wed 7/12/17	2	Travel Costs
Define launch goals (launch timing and publicity objectives)	4 days	Thu 7/13/17	Tue 7/18/17	3	Work Resource 2
Identify channel partners	4 days	Wed 7/19/17	Mon 7/24/17	4	Work Resource 3
Identify retail partners	1 day	Wed 7/19/17	Wed 7/19/17	5SS	Work Resource 1
Identify online opportunities	1 day	Thu 7/20/17	Thu 7/20/17	6	Work Resource 2
Identify budget requirements	5.5 days	Fri 7/21/17	Fri 7/28/17	7	Work Resource 3
Obtain launch budget approval	6.5 days	Fri 8/4/17	Mon 8/14/17	8FS+5 days	Work Resource 1
Phase 2 - Planning	71 days	Tue 8/15/17	Tue 11/21/17		
Kickoff product launch	5 days	Tue 8/15/17	Mon 8/21/17	9	
Define creative brief	3 days	Tue 8/22/17	Thu 8/24/17	11	
Prepare Messaging Framework	43 days	Fri 8/18/17	Tue 10/17/17	12SS-2 days	
Review current literature and plan new requirements	1 day	Wed 10/18/17	Wed 10/18/17	13	
Define press release requirements	2 days	Thu 10/19/17	Fri 10/20/17	14	
Define product specification literature	1 day	Mon 10/23/17	Mon 10/23/17	15	

Fig. A.73 Project schedule with WBS, baseline and resource loading

- Change the table in the view to ***Earned Value*** by navigating to ***View*** >> ***Tables*** >> ***Earned Value*** >> ***Apply***

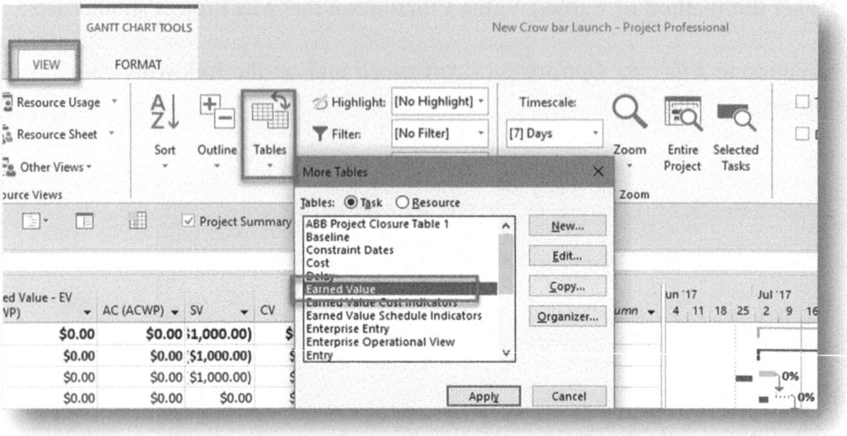

Fig. A.74 Project schedule with WBS, baseline and resource loading

- The table changes to Earned Value measures as shown below

	Task Name	Planned Value - PV (BCWS)	Earned Value - EV (BCWP)	AC (ACWP)	SV	CV	EAC	BAC	VAC
0	**New Crow bar Launch**	**$0.00**	**$0.00**	**$0.00**	**$0.00**	**$0.00**	**$9,800.00**	**$9,800.00**	**$0.00**
1	**Phase 1- Initiation**	**$0.00**	**$0.00**	**$0.00**	**$0.00**	**$0.00**	**$9,800.00**	**$9,800.00**	**$0.00**
2	Identify the launch team	$0.00	$0.00	$0.00	$0.00	$0.00	$1,000.00	$1,000.00	$0.00
3	Determine sales objectives	$0.00	$0.00	$0.00	$0.00	$0.00	$0.00	$0.00	$0.00
4	Define launch goals (launch timing and publicity objectives)	$0.00	$0.00	$0.00	$0.00	$0.00	$3,200.00	$3,200.00	$0.00
5	Identify channel partners	$0.00	$0.00	$0.00	$0.00	$0.00	$2,240.00	$2,240.00	$0.00
6	Identify retail partners	$0.00	$0.00	$0.00	$0.00	$0.00	$400.00	$400.00	$0.00
7	Identify online opportunities	$0.00	$0.00	$0.00	$0.00	$0.00	$800.00	$800.00	$0.00
8	Identify budget requirements	$0.00	$0.00	$0.00	$0.00	$0.00	$560.00	$560.00	$0.00
9	Obtain launch budget approval	$0.00	$0.00	$0.00	$0.00	$0.00	$1,600.00	$1,600.00	$0.00
10	**Phase 2 - Planning**	**$0.00**	**$0.00**	**$0.00**	**$0.00**	**$0.00**	**$0.00**	**$0.00**	**$0.00**
11	Kickoff product launch	$0.00	$0.00	$0.00	$0.00	$0.00	$0.00	$0.00	$0.00
12	Define creative brief	$0.00	$0.00	$0.00	$0.00	$0.00	$0.00	$0.00	$0.00
13	Prepare Messaging Framework	$0.00	$0.00	$0.00	$0.00	$0.00	$0.00	$0.00	$0.00
14	Review current literature and plan new requirements	$0.00	$0.00	$0.00	$0.00	$0.00	$0.00	$0.00	$0.00
15	Define press release requirements	$0.00	$0.00	$0.00	$0.00	$0.00	$0.00	$0.00	$0.00
16	Define product specification literature	$0.00	$0.00	$0.00	$0.00	$0.00	$0.00	$0.00	$0.00
17	Define sales presentations	$0.00	$0.00	$0.00	$0.00	$0.00	$0.00	$0.00	$0.00
18	Define internal communication needs	$0.00	$0.00	$0.00	$0.00	$0.00	$0.00	$0.00	$0.00
19	Align product release timing with marketing plan	$0.00	$0.00	$0.00	$0.00	$0.00	$0.00	$0.00	$0.00
20	Prepare for volume manufacturing according to sales goals	$0.00	$0.00	$0.00	$0.00	$0.00	$0.00	$0.00	$0.00
21	Plan sales force staffing and training to support sales goals	$0.00	$0.00	$0.00	$0.00	$0.00	$0.00	$0.00	$0.00

Fig. A.75 Project schedule with earned value fields

- Now, switch back to the entry table, and navigate to ***Project*** >> ***Status Date*** and set the status date

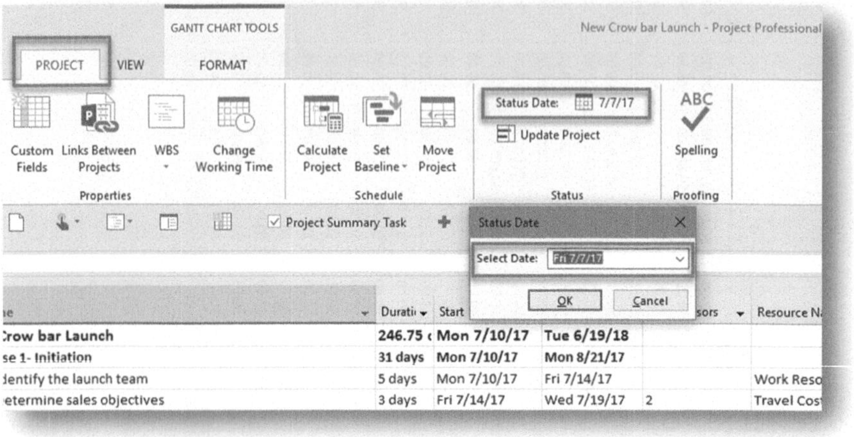

Fig. A.76 Setting the project status date

- Start marking progress for some of the tasks, and switch to the Earned Value table to see the calculations happening.

	Task Name	Planned Value - PV (BCWS)	Earned Value - EV (BCWP)	AC (ACWP)	SV	CV	EAC	BAC	VAC	SPI	CPI
0	**New Crow bar Launch**	**$3,160.00**	**$800.00**	**$5,800.00**	**($2,360.00)**	**($5,000.00)**	**$38,570.25**	**$5,320.00**	**($33,250.25)**	**0.25**	**0.14**
1	**Phase 1- Initiation**	**$3,160.00**	**$800.00**	**$5,800.00**	**($2,360.00)**	**($5,000.00)**	**$38,570.25**	**$5,320.00**	**($33,250.25)**	**0.25**	**0.14**
2	Identify the launch team	$200.00	$200.00	$5,200.00	$0.00	($5,000.00)	$5,200.00	$200.00	($5,000.00)	1	0.04
3	Determine sales objectives	$400.00	$200.00	$200.00	($200.00)	$0.00	$400.00	$400.00	$0.00	0.5	1
4	Define launch goals (launch timing and publicity objectives)	$800.00	$400.00	$400.00	($400.00)	$0.00	$800.00	$800.00	$0.00	0.5	1
5	Identify channel partners	$560.00	$0.00	$0.00	($560.00)	$0.00	$560.00	$560.00	$0.00	0	0
6	Identify retail partners	$400.00	$0.00	$0.00	($400.00)	$0.00	$400.00	$400.00	$0.00	0	0
7	Identify online opportunities	$800.00	$0.00	$0.00	($800.00)	$0.00	$800.00	$800.00	$0.00	0	0
8	Identify budget requirements	$0.00	$0.00	$0.00	$0.00	$0.00	$560.00	$560.00	$0.00	0	0
9	Obtain launch budget approval	$0.00	$0.00	$0.00	$0.00	$0.00	$1,600.00	$1,600.00	$0.00	0	0

Fig. A.77 Earned value fields with recalculated earned value indicators

As you can see:

- The Schedule Variance (SV) is negative for some tasks that the task is delayed per the plan.
- The Schedule Performance Indicator (SPI) is 1 for some tasks, meaning that the task completed as planned. However, for some tasks, it is less than 1, showing that the that the project is not doing well against the schedule.
- The Cost Performance Indicator (CPI) is 1 for some tasks, meaning that the task completed as planned. A value of 1 or above 1 means that the project is doing well against the budget. Conversely, a value below 1 means that the project/task is performing poorly and over budget.
- The Estimate at Completion (EAC) value shows an estimate of the project cost, considering the performance projection of the current progress. As you can see in the above example, the estimate at completion is much different than the current baseline cost.

This section provides a quick summary of Earned Value Analysis using MS Project. However, it is suggested to gain an understanding of the basic calculations of Earned Value to take the best advantage of the tool.

A.13.3 Earned Value S Curves

The Earned value S curves help analyze the progress and performance of the project. While an S Curve could be drawn against many parameters of the project schedule, the earned value S Curves take into account the following parameters

- Budget at Completion (BAC)
- Earned Value (EV)
- Planned Value (PV)
- Actual Costs (AC)

Since Microsoft Project provides all these values and the cumulative values as well, the plotting of S curves has become much simpler in MS Project.

With the rich reporting capabilities provided in MS Project, you can either use the out of the box report for Earned Value or generate your own variation using the fields available.

A.13.4 Accessing the S Curve Reports in MS Project

To access the earned value over time report,

- Navigate to Reports >> Costs >> Earned Value

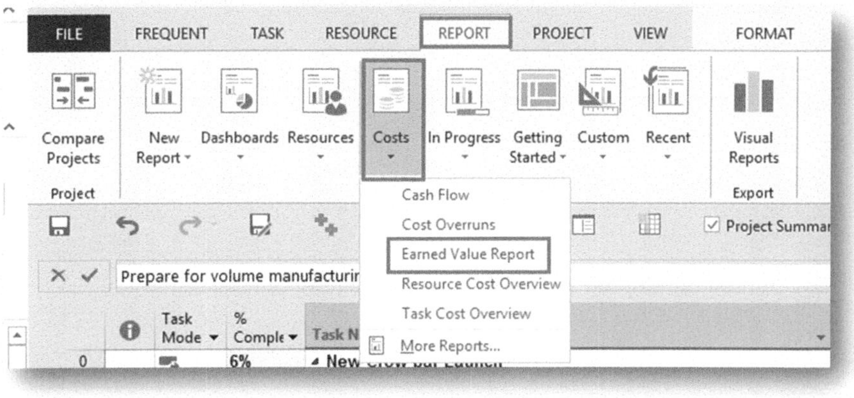

Fig. A.78 Accessing the earned value report in MS Project

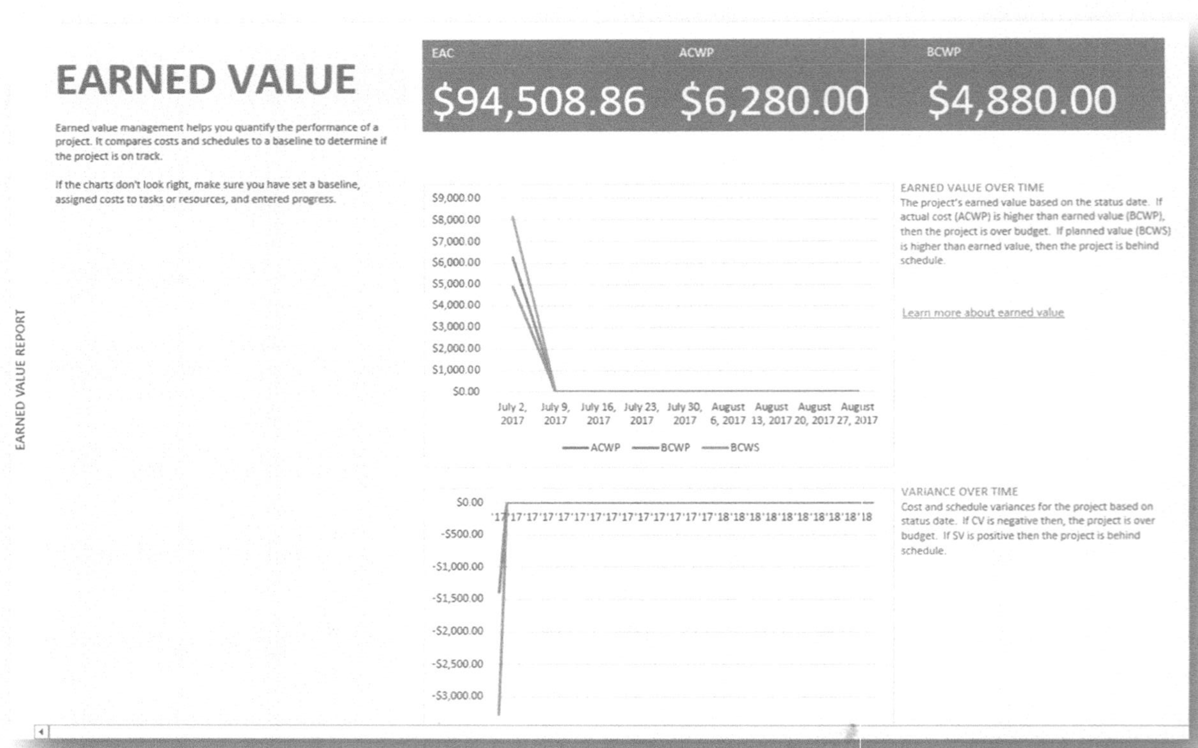

Fig. A.79 Earned value report in MS Project

A.14 Using Task Usage and Resource Usage Views

The 'Usage' views in MS Project display the assignment information in a time-phased manner. This is extremely useful when you are trying to view and analyze your project work or cost distribution across the timeline of your projects.

By default, the Work field displays in the time phased portion of the Resource Usage view and Task Usage view. If you don't see the information you need in the time phased portion of either usage view, you can display different information.

The Task Usage view shows the distribution of work, cost and related information over time for each assignment per **task**, whereas the resource usage view shows the distribution of work, cost and related information over time for each assignment per **resource**.

A.14.1 Accessing the Usage Views

- On the View menu, click Task Usage or Resource Usage.

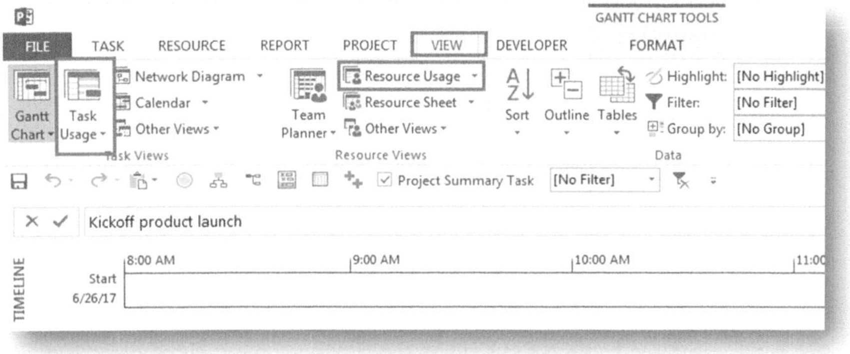

Fig. A.80 Accessing the usage views in MS Project

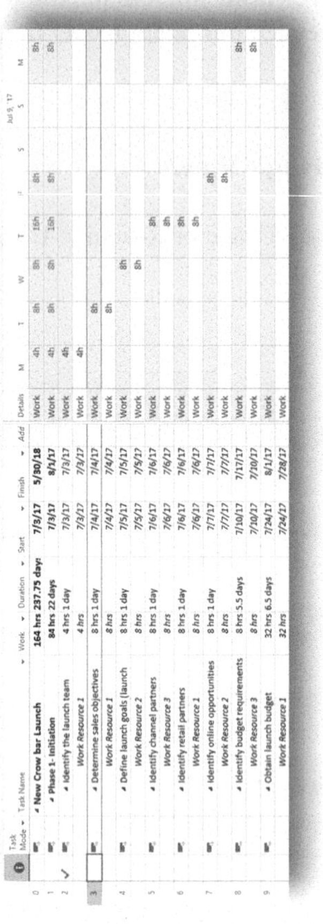

Fig. A.81 Task usage view

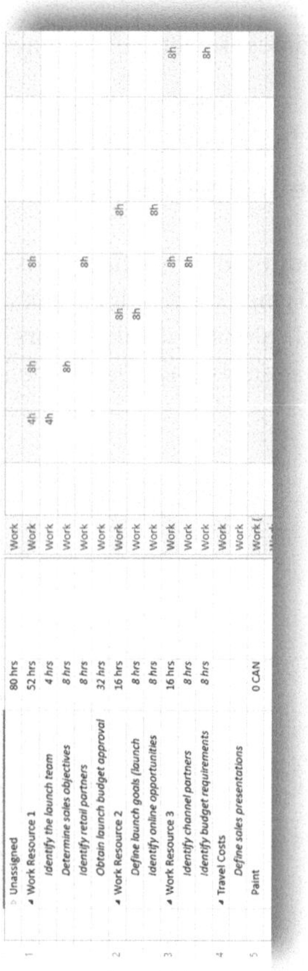

	▷ Unassigned	80 hrs	Work									
1	◢ Work Resource 1	52 hrs	Work	4h	8h		8h					
	Identify the launch team	*4 hrs*	Work	4h								
	Determine sales objectives	*8 hrs*	Work		8h							
	Identify retail partners	*8 hrs*	Work				8h					
	Obtain launch budget approval	*32 hrs*	Work									
2	◢ Work Resource 2	16 hrs	Work			8h		8h				
	Define launch goals (launch	*8 hrs*	Work			8h						
	Identify online opportunities	*8 hrs*	Work					8h				
3	◢ Work Resource 3	16 hrs	Work				8h				8h	
	Identify channel partners	*8 hrs*	Work				8h					
	Identify budget requirements	*8 hrs*	Work								8h	
4	◢ Travel Costs		Work									
	Define sales presentations		Work									
5	Paint	0 CAN	Work (									

Fig. A.82 Resource usage view

A.14.2 Adding Additional Fields to the Usage Views

- If you don't see the field that you want to add, you can additional fields.
- On the Format menu, click Detail Styles, and then click the Usage Details tab.
- In the Available fields list, click the field you want to show, and then click Show.
- Under Field settings, select a font, cell background, and pattern for the field.
- To view field names on the time phased portion of the view, click the Usage Properties tab, and then click ***Yes*** in the Display details header column box.
- To repeat field names on all assignment rows, select the Repeat details header on all assignment rows check box.
- To display abbreviated field names, select the Display short detail header names check box.

A.14.3 Advantages of the Usage Views

In addition to showing the distribution of work and costs across time, Usage Views are extremely useful in identifying over allocations of resources. The over allocations are highlighted in red, along with the hours that are causing the over allocation. This makes it easier to identify and correct the issues or reprioritize tasks.

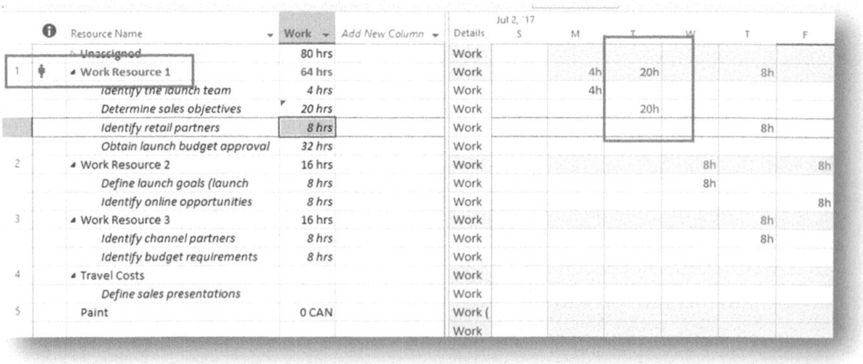

Fig. A.83 Using usage views to view resource overallocations

Web Sites

http://cmaanet.org (*Construction Management Association of America*)
http://gdi.ce.cmu.edu (*Carnegie Mellon Green Design Institute*)
http://rpmi.marc.gatech.edu (*Rapid Prototyping and Manufacturing Institute*)
www.agilealliance.com (*Agile Alliance*)
www.amsup.com (*PM and NPD Techniques*)
www.cadcamnet.com (*CAD and CAM*)
www.capterra.com/project-management-software (*PM Softwares*)
www.eiasm.be (*see: International Product Development Management Conference*)
www.elsevier.com/locate/ijproman (*International Journal of Project Management*)
www.epci.no (*European Institute of Advanced Project and Contract Management*)
www.fmeca.com (*Failure Mode & Effect Analysis Facilitator*)
www.iamot.org (*International Association for Management of Technology*)
www.inderscience.com/browse/index.php?journalID=96 (*International Journal of Project Organisation and Management*)
www.infogoal.com/pmc/pmcswr (*The Project Management Center*)
www.iol.ie/~mattewar/CIPPM (*Center for International Project & Program Management*)
www.ipma.ch (*International Project Management Association*)
www.microsoft.com/project (*The PM Software*)
www.pdma.org (*Product Development & Management Association*)
www.pdma.org/journal (*Journal of Product Innovation Management*)
www.plmic.com (*Product Lifecycle Management Information Center*)
www.pmforum.org (*The International Forum on Project Management*)
www.pmi.org (*Project Management Institute*)
www.pmi.org/Resources/Pages/Project-Management-Journal.aspx (*Project Management Journal*)
www.primavera.com (*The PM Software*)
www.prince-officialsite.com (*Prince2*)
www.project-management-software.org (*PM Softwares*)
www.qfdi.org (*Quality Function Deployment Institute*)
www.rims.org (*Risk and Insurance Management Society*)
www.tandf.co.uk/journals/titles/09544828.asp (*Journal of Engineering Design*)
www.thesoftwarenetwork.com/Project-Management-Software (*PM Softwares*)
www.tonchia.it (*author's web site*)
www.valuation-opinions.com (*Earned Value Website*)

S. Tonchia, *Industrial Project Management*, Management for Professionals,
https://doi.org/10.1007/978-3-662-56328-1

The manufacturer's authorised representative in the EU is Springer Nature Customer Service Centre GmbH, Europaplatz 3, 69115 Heidelberg, Germany. If you have any concerns regarding our products, please contact ProductSafety@springernature.com

Printed and bound by CPI Group (UK) Ltd, Croydon, CR0 4YY
01/10/2026
02225865-0012